T0324831

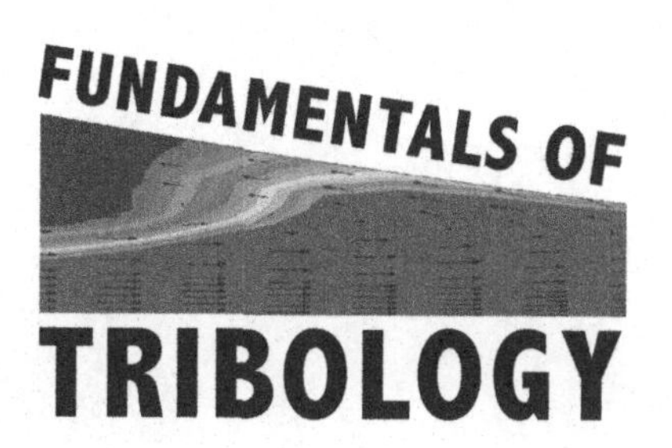
FUNDAMENTALS OF
TRIBOLOGY

R. Gohar
Imperial College London, UK

H. Rahnejat
Loughborough University, UK

FUNDAMENTALS OF TRIBOLOGY

ICP

Imperial College Press

Published by

Imperial College Press
57 Shelton Street
Covent Garden
London WC2H 9HE

Distributed by

World Scientific Publishing Co. Pte. Ltd.
5 Toh Tuck Link, Singapore 596224
USA office: 27 Warren Street, Suite 401-402, Hackensack, NJ 07601
UK office: 57 Shelton Street, Covent Garden, London WC2H 9HE

British Library Cataloguing-in-Publication Data
A catalogue record for this book is available from the British Library.

FUNDAMENTALS OF TRIBOLOGY

ISBN-13 978-1-84816-184-9
ISBN-10 1-84816-184-0

Editor: Tjan Kwang Wei

Typeset by Stallion Press
Email: enquiries@stallionpress.com

Printed in Singapore.

PREFACE

This textbook on Tribology, or Lubrication and Wear, as the subject was previously called, is the outcome of research and teaching by the authors over many years to undergraduate mechanical engineering students at Imperial College, London and Loughborough University. The book represents our ideas on how Tribology should be taught to modern engineering students who, unlike their predecessors, now generally have at their disposal the support of comprehensive computer systems. We hope the book will also be of use to practicing engineers who frequently encounter Tribology-centered problems, and who require quick, but adequate solutions.

Below is a summary of our approach to the teaching of Tribology:

- Because Tribology covers such a broad field, embracing engineering surfaces, through to their dry contact friction and finally fluid film lubrication, we have attempted to explain it in such a way that we demonstrate to the reader that there is often a close interaction between these distinct disciplines.
- The beginnings of some chapters introduce the reader to historical examples of Tribology and its economic impact on society as well as its engineering relevance.
- When discussing the properties of lubricants we had noted that in the past specially scaled graph paper was often needed. Our approach is to assume that none is available and only well-known empirical expressions describing lubricant behavior must be used instead. Some of these expressions can easily be employed in numerical solutions by using the widely available Mathcad or similar software.
- The book shows you how to develop simple mathematical models that can be used to find approximate solutions to Tribology related problems. For example, where fluid flow theory must first be employed, we show you how to derive an incomplete form of Reynolds equation using only the physics of lubricant flow theory. The relative significance of each of the variables involved in a complete solution then becomes immediately apparent.

- When we arrive at hydrodynamic bearing design, for a realistic solution, temperature effects should be included. In the case of a complete thrust bearing there is no point in studying an isolated wedge pair, as its performance is only of academic interest in the design process. Instead, the pair is studied in the context of the *whole* bearing, again helped by Mathcad, when seeking an approximate numerical solution.
- The worked examples and chapter questions set throughout the book are not always of the sterile examination type, where only substitution in a formula is needed for a solution. Instead, a computer program is occasionally necessary in order to solve a set of nonlinear governing equations. Again Mathcad does this rapidly in a few lines, using its vectorize and graphing facilities. Additionally, it shows you visually where there has been poor convergence, usually suggesting an unsatisfactory choice of some design variables.
- Where difficult Mathematics is encountered, we only summarize the procedure sufficiently for the reader to obtain the gist of the solution method before utilizing the resulting simplified equations. On other occasions, a regression formula resulting from a numerical solution derived elsewhere may be used.
- There are also chapters on Bio-Tribology (contributed by Professor Duncan Dowson) and an introduction to the recent science of Nano-Tribology.
- Finally, there is a chapter on fluid film lubrication applied to internal combustion engines. It demonstrates that in the real world, Tribology problems are often not as straightforward as those set in an engineering undergraduate course. Nevertheless, utilizing the equations supplied in the book, approximate numerical solutions may be obtained.
- There are solved examples in each chapter, supplanted by questions at the end of the book. A Book of Solutions is also provided at the end of the book, dealing with most of the set questions.

Acknowledgments

We are indebted to our university colleagues, some of whose course questions we have used, and also present and former postgraduates, undergraduate students and staff who have read and commented on the text. In particular we wish to acknowledge Dr. Patricia Margaret Rahnejat for proof-reading of the text and her comments, Dr. Philipa Cann for providing materials on interferrometry, Prof. Hugh Spikes for some questions and data, Prof. Stathis Ioannides for details of bearing fatigue life calculations for Chapter 11, Dr. Mircea Teodorescu for effort put in Chapters 12 and 13, and consultation on nano-tribology, Dr. Sashi Balakrishnan for some numerical results presented in Chapter 12 on pistons, Sebastian-Howell Smith for examples of laser etching on advanced cylinder liners and Dr. Manu Kushwaha for numerical results of transient cam-tappet contact, also for Chapter 12. Finally, we are extremely grateful to Professor Duncan Dowson for contributing a chapter on his speciality of Bio-Tribology.

NOTATION

Roman Symbols:

a	Line contact footprint half width and elliptical contact semi minor axis length, or circular footprint radius, or pocket radius aerostatic thrust bearings or acceleration (Chapter 12) or molecular diameter (Chapter 13)
A	Real contact area of a rough surface (Chapters 3 and 13)
A	Plan view area of a thrust bearing (Chapter 7)
$\bar{A}$	Hydrostatic thrust bearing load factor
A_i	Contact area of a single roughness feature summit
A_o	Nominal surface area of a rough surface
A_0	Contact area from direct load only
A_h	Hamaker constant
A_t	Throat area of a jet or orifice
b	Elliptical contact semi major axis half-width or Aerostatic thrust bearing outer radius (Chapter 9)
B	Hydrodynamic bearing width in flow direction
B_1, B_2	Taper land bearing wedge and land widths
Bo	Bond number
$\bar{B}$	Hydrostatic thrust bearing flow shape factor
c	Radial clearance
c_p	Specific heat at constant pressure
c_v	Specific heat at constant volume
Ca	Capillary number
C_d	Discharge coefficient
C_g	Eccentricity of the gudgeon pin
C_p	Eccentricity of the piston
d	Distance between smooth surface and reference plane of rough surface
d_c	capillary diameter
d_o	Orifice diameter
d_f	Surface deformation

d_s	Shell thickness
D_c	Capillary diameter in Chapter 13
D_d	Droplet diameter
$\bar{d}$	d/σ_s
e	Eccentricity or Weibull slope (Chapter 11) or electronic charge (Chapter 13)
e_b	Eccentricity at the piston skirt bottom
e_t	Eccentricity at the piston skirt top
E	Young's modulus of elasticity
E^*	Reduced Young's Modulus $\frac{1}{E^*} = \frac{1-v_1^2}{E_1} + \frac{1-v_2^2}{E_2}$
EHL	Elastohydrodynamic Lubrication
E	Fractional energy loss in a dry rolling contact (Chapter 9)
f_0	Natural frequency of a journal bearing (vibrations/s)
f_{con}	Connecting rod force
f	Friction force per unit area
f_g	Shape factor in aerostatic bearings or Gas force (Chapter 12)
f_{gg}	Gas force applied to the gudgeon pin
f_{gp}	Gas force applied to the piston
f_{ig}	Gudgeon pin inertial force
f_{ip}	Piston inertial force
f_{r1}	Piston skirt-cylinder contact force on the major thrust side
f_{r2}	Piston skirt-cylinder contact force on the minor thrust side
f_p	Flow due to Poiseuille pressure
F	Tangential traction or Friction force
F_a	Friction force to break an adhesive bond
F_d	Deformation (ploughing) friction
F_e	Elastic force (Chapter 12)
F_0	Friction force at thrust bearing runner
F_i	Inertial force (Chapter 12)
F_h	Friction force at thrust bearing pad
F_m	Meniscus force
F_n	Net normal force (Chapter 13)
F_p	Pressure factor in Aerostatic bearings (Chapter 9) or valve spring preload (Chapter 12)
F_r	Nonlinear viscous shear strain rate
G	Shear modulus (**suffix** *0*: oil)
G_l	Global factor in aerostatic bearings
g	Gravitational acceleration

h Film thickness (**suffix** c: at maximum pressure)
h_i Inlet film thickness
h_0 Minimum film thickness (thrust bearings, journal bearings, non-conforming rigid surfaces)
h_0 Conjunction film thickness (EHL)
h_g Undistorted gap height
h_s Elastic gap height above contacting rigid plane
h_1 Maximum film thickness (hydrodynamic bearings) or film thickness (**suffices:** 1: with major thrust side, 2: with minor thrust side (Chapter 13))
h_m Film thickness associated with maximum stiffness (Chapter 9),
h_m Minimum film thickness (Chapter 10)
$\bar{h}$ h/h_0
H Indentation hardness (contact mechanics)
H_{st} Viscosity Index constant
I Mass moment of inertia
j_θ Geometrical acceleration
k Maximum bulk shear stress at first yield or valve-spring stiffness (Chapter 12)
k_a Asperity geometry coefficient
k_f Coefficient describing failure to produce a wear particle
k_o Journal bearing film stiffness
k_r Rotor bending stiffness
k_t Thermal conductivity
K Inclination factor for plane thrust bearings or Boltzmann Constant (Chapter 13)
K_a Asperity geometry coefficient
K_{ad} Wear coefficient
K_c Viscous flow coefficient (hydrostatic bearings)
K_g Gauge pressure ratio
K_m Maximum stiffness of an aerostatic bearing
l Rolling element path length or or piston skirt length (Chapter 12)
L_c Capillary length
L Hydrodynamic bearing transverse length or Fatigue life in millions of revolutions of the inner race (Chapter 11) or connecting rod length (Chapter 12)
L_s Fatigue life in millions of revolutions for a probability of survival in a group of bearings of $(1 - S)\%$ (Chapter 11)

L_s	Rough surface sampling length (Chapter 2)
L_{st}	Viscosity index constant
L_{10}	Fatigue life in millions of revolutions corresponding to the probability that 90% of a group of bearings will survive
m	Mass flow in an aerostatic bearing (**suffix** m: maximum stiffness), equivalent sliding mass (Chapter 13)
m_{eq}	Equivalent mass (Chapter 12)
m_g	Mass of gudgeon pin
m_p	Mass of piston
m_v	Mass of valve
m_x	Mass flow per unit length x direction
M	Moment
M_{rot}	Rotor natural frequency vibration/min
M_{eff}	Rotor-bearing natural frequency vibration/min
M_{fr1}	Hydrodynamic moment at the major thrust side
M_{fr2}	Hydrodynamic moment at the minor thrust side
n	Number of asperity summits under load W_e or, refractive index (Chapter 10) or Number of bearings in a batch (Chapter 11)
n_s	Number of bearings in a batch that have successfully endured L_S revolutions in tests
N	Number of stress cycles endured (Chapter 11) or Total number of roughness summits over area A (Chapter 3)
N_A	Avogadro's number
N_c	Height count at discrete interval Δx over length L_s
p	Pressure (**suffices:** 1: with major thrust side, 2: with minor thrust side (Chapter 12))
p_a	Atmospheric pressure
p_d	Pocket absolute pressure (Aerostatic bearings)
p_M	Manifold absolute pressure (Aerostatic bearings)
p_r	Hydrostatic bearing recess pressure (Chapter 9)
p_r	Reduced pressure
p_s	Supply pressure, hydrostatic bearings
p_s	Supply absolute pressure, aerostatic bearings
p_{iso}	$1/\alpha$
p_m	Mean pressure
p_0	Maximum pressure (concentrated contacts)
p_c	Maximum pressure (thrust bearings)
$(p_0)_Y$	Maximum pressure at yield
$(p_m)_Y$	Mean pressure at yield

$p_{electro}$ Electrostatic pressure

p_s Solvation pressure

p_{vdw} Van der Waal's pressure

P Power

P' Force per unit length

Pe Peclet number

q Charge

q_x, q_y, q_z Volume flow per unit length in the x, y, z directions

Q Volumetric flow at inlet to a thrust bearing pair (long bearings)

Q Radial flow in hydrostatic thrust bearings (Chapter 9)

Q_{sl} Volume per unit sliding distance

Volumetric side flow (Poiseuille) from a finite length hydrodynamic bearing

Q_{se} Effective value of Q_s corresponding to effective viscosity η_e

Q_1 Total volumetric flow entering the film wedge in a finite length hydrodynamic bearing

Q_2 Volumetric flow leaving the film wedge end in a finite length hydrodynamic bearing

r, θ Polar coordinates

r Inter-molecular distance (Chapter 13), Crank-pin radius (Chapter 12)

r_p Average asperity tip radius

r_g Offset of disc mass center from its geometric center

r_r Inner race radius at groove bottom

R Reduced radius of curvature ($1/R = 1/R_1 + 1/R_2$) or Radius representing R_1 in journal bearings (Chapter 8) or Hydrostatic thrust bearing outer radius (Chapter 7) or Radius of hemispherical asperity (Chapters 4 and 13) or base circle radius (Chapter 12)

R_0 Hydrostatic thrust bearing recess radius

R_1, R_2 Respective radii of contacting cylinders or respective journal and bearing bush radii

R_x, R_y Respective reduced radii of curvature in xz and yz planes

R_{sk} Skewness

R_a	average height of roughness profile
R_p	Ball bearing ball pitch radius
R_{x1} and R_{x2}	Radii of curvature respectively in the xz and yz planes
R_x and R_y	Principle radii of curvature
$\bar{R}$	R/R_0
Re	Reynolds number
$\Re$	$c_p - c_v$
s	Half width of footprint stick region under traction, cam lift (Chapter 12) distance between the solids (Chapter 14)
$s_{\max}$	Maximum valve lift
S	Probability of survival (or reliability) expressed as a fraction or a percentage
$S(x)$	Undistorted gap in a touching contact
S	Sommerfeld Number
t	Time
t_p	Taper
T	Tractive reaction force or Temperature in deg Kelvin (Chapter 13)
u	Rotating disc offset (Chapter 8) or Number of stress cycles per revolution (Chapter 11)
U	Surface speed or, in Chapter 10, entrainment speed ($U = (U_1 + U_2)/2$)
$U_1,\ U_2$	Surface speeds in the x direction
U_t	Test oil viscosity (Chapter 5)
v	roller center velocity (Chapter 4), valve velocity (Chapter 12)
V	Volume or sliding velocity (Sec. 13.7)
V_o	Volume under stress (Chapter 11)
V_e	Magnitude of velocity vector $\overrightarrow{\mathrm{V_e}}$ (Chapter 10)
V_1, V_2	Surface speeds in the y direction
VI	Viscosity index
w	Deflection at x (**suffix** 0: centerline deflection in EHL) or interaction potential (Chapters 13 and 14)
W	Load (applied or contact)
W_a	Normal adhesion force (Sec. 13.8)
W_c	Dynamic load rating (Chapter 11)
W_i	Load on a single asperity
W_e	Total load on n asperities under an elastic contact

We	Weber number
W_m	Aerostatic bearing load at maximum stiffness
W_x, W_y	x and y components of W
x, y, z	Cartesian coordinates
x_i	Inlet distance
x_p	Pivot position, where $h = h_c$ (Chapter 7)
x'	Distance measured from start of an EHL conjunction
X	Pivot distance from thrust bearing pad leading edge
X^*	Dimensionless pivot distance
$[X_i]$	Concentration
Y	Tensile stress at yield
z	Height of a point on roughness profile from the mean height line (Chapter 2)
z_p	Valve-spring preload
z_0	Maximum wear depth of asperities
z_i	Valency
z^*	z/σ
z'	Height of a point on the roughness profile from the datum line (Chapter 2)
z_s	Height of a roughness summit above the reference plane
$\bar{z}_s$	z_s/σ_s
Z	Pressure-viscosity index
Z_0	depth below the load carrying track where τ_0 occurs (Chapter 11)

Greek Symbols:

α	Pressure-viscosity coefficient
β	Viscosity-temperature coefficient, Tilt angle (Chapter 12) or fraction of kinetic energy deforming asperities (Chapter 13)
δ	Centerline deflection in a concentrated contact
δ_i	Compliance of the ith deformed roughness summit E
$\bar{\delta}$	δ_i/δ_s
$\Delta_{1/2}u$	Sliding velocity with major and minor thrust sides
Δt	Time step size
Δx	Discrete distance interval
$\Delta\theta$	Temperature rise
ε	Eccentricity ratio
ε_0	Vacuum permittivity

ε_r	Relative permittivity
$\varepsilon_{x/y/z}$	Direct shear strains
$\phi(z)$	Probability Distribution function (PDF)
ϕ^*	Summit distribution with unit standard deviation
ϕ	Surface charge density (Chapter 13)
γ	Surface tension/surface energy
γ_{lv}	Surface tension at liquid-vapour interface
γ_{sl}	Surface tension at solid-liquid interface
γ_{sv}	Surface tension at solid-vapour interface
$\gamma_{xy/zx,yz}$	Shear strains
$\dot{\gamma}$	Shear rate (**suffices:** e: elastic shear rate, v: viscous shear rate, t: total shear rate)
η	viscosity
η_0	inlet or isoviscous viscosity
κ	Diffusivity
κ	Shielding length (Chapter 13)
κ^{-1}	Debeye length
λ	Wavelength
λ_a	Wavelength in air
λ_c	Capillary compensated hydrostatic bearing stiffness
λ_f	Constant flow hydrostatic bearing stiffness
λ_s	(Film thickness)/(roughness RMS height) (Chapter 1) or Feeding factor in aerostatic bearings (Chapter 9)
μ	Coefficient of friction
μ^*	$\mu(R/C)$
ν	Kinematic viscosity or coefficient of thermal expansion (Chapter 6)
θ	Temperature, Cam angle (Chapter 12), Contact angle (Chapter 13)
θ_e	Effective temperature
θ_s	Finite length bearing side flow temperature
θ_{se}	Finite length bearing effective side flow temperature at viscosity η_e
θ_0	Finite length bearing supply temperature
θ_1	Bearing wedge entry temperature
θ_2	Bearing wedge exit temperature
Θ	Temperature (°K)
Θ_s	Finite length bearing side flow temperature (°K)
Θ_M	Absolute temperature in manifold (Chapter 9)

ρ	Density (**suffices** *0*: base density, M: manifold density, ∞: bulk in Chapter 13) or charge density
σ	Root Mean Square of roughness heights
σ_1, σ_2	Principal stresses
σ_x, σ_y	Principal stresses (circular footprint contact)
τ	Shear stress
τ_c	Maximum tangential shear stress (Chapter 4) or limiting EHL shear stress (Chapter 10)
τ_s	Average shear stress at an asperity junction
τ_{xz}	x-direction shear stress in the plane normal to the z-axis
τ_{yz}	y-direction shear stress in the plane normal to the z-axis
$\tau_{\max}$	Maximum shear stress
$\tau_{xz}(\max)$	Maximum shear stress in an elliptical contact
τ_ℓ	Average shear strength of the fluid film (Chapter 13)
τ_0	Maximum orthogonal shear stress (Chapter 11) or Eyring reference stress (Chapter 10)
$\bar{\tau}$	Average shear stress
υ	Poisson's ratio
ω	Orbital angular velocity of rotor center or cam (Chapter 12)
ω_n	Natural frequency of rotor-disc
ω_0	Natural frequency of journal bearing
Ω	Angular velocity of rotor and its journal bearings
ξ	Dipole moment
Ψ	Plasticity Index
ψ	Attitude angle
$\Psi(z)$	CDF (running integral of PDF)
ξ	Creep Ratio

Dimensionless Groups

Roman Symbols:

$\bar{A} = \frac{1}{2}\left(\frac{1-(1/\bar{R})^2}{\ln(\bar{R})}\right)$ (Load factor, Chapter 9)

$\mathrm{Bo} = \frac{D_c^2 \rho g}{\gamma}$ (Bond number)

$\bar{B} = \frac{\pi}{6\ln(R/R_0)}$ (Flow shape factor, Chapter 9)

$Ca = \frac{\mathrm{We}}{\mathrm{Re}}$ (Capillary number)

$e_p^* \approx \left(\frac{R_x}{R_y}\right)^{2/3}$ (Ellipticity ratio)

$D = (\eta(U/2G_0a))$ (Deborah number, Chapter 10)
$F^* = \frac{F_0h_0}{LB\eta U_1}$ (Chapter 7)
$G^* = E^*\alpha$ (Materials' parameter, Chapters 10 and 12)
$h^* = h/R$ (Non-dimensional film thickness)
$H = \frac{\pi E^*h}{P'}$
$H_0 = \frac{\pi E^*h_0}{P'}$
$\bar{K} = \frac{p_d}{p_s}$ (Gauge pressure ratio, Chapter 9)
$Pe = \left(\frac{k_t}{\rho c_p}\right)/\frac{U_1h^2}{2B}$ (Peclet Number, Chapter 6)
$\bar{P} = \bar{B}/\bar{A}^2$ (Power coefficient, Chapter 9)
$\bar{p} = h_0^2p/6U_1\eta B$ (Chapter 6)
$\mathrm{Re} = \frac{\rho UL}{\eta}$ (Reynolds number)
$\mathrm{We} = \frac{\rho U^2L}{\gamma}$ (Weber number)
$\bar{R} = R/R_0$ (Chapter 9)
$S = \frac{W}{NDL\eta}\frac{c^2}{R^2}$ (Sommerfeld Number, Chapter 8)
$U^* = \eta_0(U/E^*R)$ (EHL speed or rolling-viscosity parameter)
$w_s^* = \frac{1}{U}\frac{\partial h}{\partial t}$ (Squeeze-roll ratio)
$W^* = \frac{Wh_0^2}{6U_1\eta LB^2}$ (Load factor, Chapter 7)
$W^* = P'/E^*RL$ (long line contacts: Load parameter Chapters 10 and 12)
$W^* = P'/E^*R^2$ (elliptical point contacts: Load parameter Chapters 10 and 12)
$\bar{x} = x/\sqrt{2Rh_0}$ (Chapter 6) or x/a (Chapter 10)

Greek Symbols:

β^* $\frac{Ua\rho\sigma}{2k}$ (Dimensionless speed, Chapter 4)
λ_s $h_{\min}/\sigma$ (Stribeck factor or film parameter, Chapter 1)
μ^* $\mu(R/c)$ (Chapter 8)
θ^* F^*/Q^* (Chapter 7)
Ψ $\frac{E^*}{H}\left[\frac{\sigma}{R}\right]^{1/2}$ (Plasticity Index, Chapter 3)

CONTENTS

CHAPTER 1

INTRODUCTION TO TRIBOLOGY

1.1. Introduction

The generic word *Tribology*, in the title of this book, is derived from ancient Greek. Its root is found in a verb of that period *trivein* which itself was based on the word *pedo*, meaning 'the character formation of privileged children by their home tutors' (*pedotriveis*). *Trivein* meant *rubbing* in the context of shaping the personalities of these children (cf '*rub it in*' to someone).

More recently, a 1964 Government appointed committee was set up to find ways of reducing the untoward effects of friction on the British industrial economy. The committee invented the word *Tribology* to emphasize the scientific nature of studying the interactions of solid contacting surfaces in relative motion, these being covered by the three disciplines of Friction, Lubrication and Wear, so making the prefix *Tri* appropriate. Friction can be considered as a part of physics or mechanical engineering, Lubrication is covered by mechanical engineering and chemistry, whilst Wear is a part of material science.

Tribology is, therefore, a multidisciplinary subject that draws upon a large section of the syllabus of a typical undergraduate science or engineering course.

In our everyday encounters with the solid environment, we often encounter the disciplines embraced by Tribology. For example, friction gives us the ability to walk and to grasp objects, whilst lubrication is essential for the proper functioning of the joints in our skeletons. Even though the skeleton's joints have the property of healing through re-growth, the untoward effects of wear can sometimes prevent them from functioning properly if their lubrication mechanism breaks down.

By means of effective lubrication, mechanical design seeks to minimize friction in devices such as engines, skis and computer hard disc drives. On the other hand, high friction is essential for the traction and braking of rubber car tyres against the road surface.

For the ancient Egyptians, control of friction was important when they were hauling slabs of stone up inclined slopes during the construction of the pyramids. It is equally important today when seeking ways to improve efficiency of automobile internal combustion engines. These involve mechanisms that have sliding contacts, such as the reciprocating motion of the engine pistons in their cylinders (see Chapter 12). Here, the solution resides in reduction of friction through lubrication, an example of which is given below.

One of the earliest examples of the use of lubricants was in the transportation of immense monuments in ancient Egypt. Circumstances often compelled the Egyptians to drag their loads, as depicted in Fig. 1.1. The bas-relief shows workers dragging a monumental statue, resting on a wooden sled, along a wooden track. A man on the front of the sled is pouring liquid from a flask onto the track ahead of it. This could well have been a vegetable oil or even plain water, either of which would reduce the sliding friction and wear of the contacting surfaces.

Historically, the invention of the wheel was one of the greatest achievements of man because of its impact on transport. The low rolling friction, associated with the wheel, enabled loads to be transported both rapidly and cheaply as compared with previously, when humans or animals had to carry or drag them. However, for its proper functioning the wheel always requires a suitable surface upon which to roll. This was not always the case in ancient times. For example, because their country's mountainside tracks were poor, the early Peruvian civilization did not appear to utilize the wheel for transport.

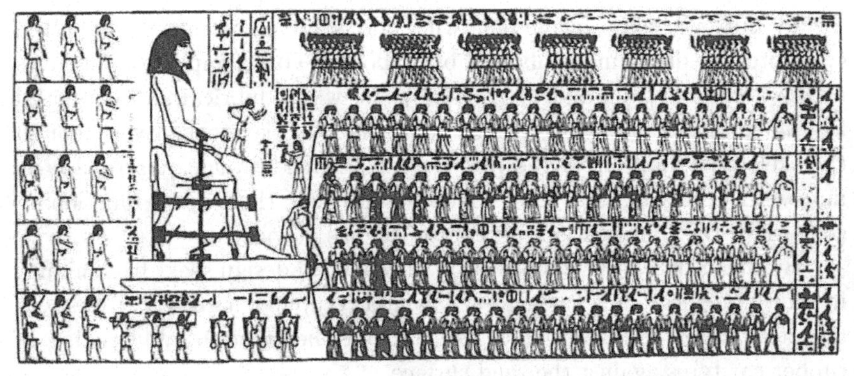

Fig. 1.1. Transporting an Egyptian statue. A wall drawing in the grotto of El Bersheh.

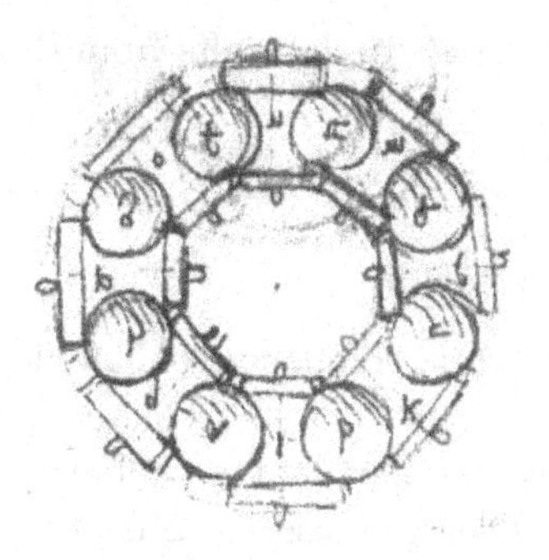

Fig. 1.2. Bobbin type separators designed by Leonardo (circa 1495).

An example of a device using rolling friction, was conceived by Leonardo Da Vinci[1] around 1495. His ingenious design was of a low friction thrust ball bearing using bobbin type separators shown in Fig. 1.2. Leonardo's full description of the device is given in Question 4.2 where we ask reader, with the aid of his own sketches, to elaborate further on the probable design of the bearing and its separators.

As we mentioned earlier, although sometimes high friction is needed in order to fulfil certain functions in our everyday lives, the need for reducing it is extremely important not the least because of the energy it consumes. Consequently, much scientific effort and ingenuity continue to be devoted to achieving this latter objective. Therefore, ways of reducing friction and wear between solid surfaces will be the main aim of our book. Apart from naturally occurring skeletal surfaces (when we discuss biomechanics), throughout the book, the solid surfaces will generally be of prepared steel, it being the material of choice in engineering and science.

Throughout the history of Tribology, the approach for finding ways of reducing friction has mainly been by trial and error or by experiments. These have produced some fundamental scientific laws, the basis of which we still use today. However, the complexity of modern engineering and the rivalry that pervades the market-place means that companies must continually develop existing products while conceiving new ones in a limited time period.

Often scientific research in Tribology is based on complex computer programs that are not always freely available to industry. A solution is, therefore, to devise adequate and simple design, diagnostic and remedial methods producing results as quickly as possible. To do this, the book sometimes utilises existing research results from the literature to obtain a solution.

1.2. Regimes of Lubrication

As lubrication is an important component of Tribology, here is a convenient point to introduce the reader to its structure in a very simple and approximate way. We will assume that two rubbing (or rolling) solid surfaces are like steel on steel. They are also **rough**, because a perfectly smooth surface cannot exist. The act of lubrication is to have between the surfaces a layer of a different material that reduces the friction force between them, either by being softer than the surfaces, or by being a coherent liquid lubricant or gas, entrained between the two surfaces by their relative movement. Now most engineering structures are in the atmosphere, so the primary friction reducing layer is a low shear strength oxide film formed naturally by the interaction of the steel surfaces with the atmosphere. (In a clean environment or a vacuum, the surfaces have no oxide film, resulting in a high friction force.)

Stribeck[2] originally devised a convenient way of relating roughness and film thickness by a parameter $\lambda_s = \frac{\text{film thickness}}{\text{roughness height}}$, the roughness height being some representative value of the undistorted roughness features, and the film thickness being measured from it (Chapter 2). He devised the **Stribeck curve**, shown in Fig. 1.3, where the coefficient of friction ($\mu = \frac{\text{friction force}}{\text{load}}$) is plotted against λ_s.

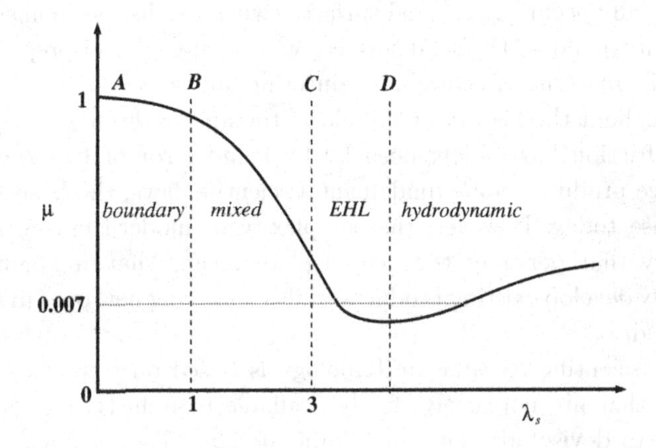

Fig. 1.3. Stribeck curve (not to scale).

The frictional contact of clean surfaces is represented by point A at the beginning of the **boundary lubrication** regime in Fig. 1.3. The subsequent formation of an oxide film, or additionally by the controlled deposition of a softer material, insertion of special boundary lubricants such as tallow, castor oil or additives to a mineral oil, all these will reduce the friction force. Regime $\boldsymbol{AB}$ represents this behavior.

At point B, where the surface roughness equals the film thickness, we enter the **mixed** or **partial lubrication** regime $\boldsymbol{BC}$. In this case the load is partly supported by a coherent flowing oil film and partly by a regime AB type contact or alternatively, by thin micro films formed between the distorted surface features themselves. Regime BC is one of the most common in practice. In some ways, it represents a design failure because a coherent lubricant film is not present everywhere. As in regime AB, in regime BC the surfaces wear, perhaps producing debris. Analysis of the forces at such contacts is discussed in Chapter 4.

The progressive reduction in the value of μ carries further into the regime $\boldsymbol{CD}$, where a coherent **elstohydrodynamic (EHL)** lubricant film is produced. Here, the values of λ are large enough for the surface features not to influence much the lubricant film thickness. The high pressures encountered in this zone cause the surfaces to distort elastically and the lubricant viscosity to increase. Rolling element bearings and gears fall into this category. The word elastohydrodynamic clearly defines the lubrication mechanism. It is characterized by a film thickness of one micron or less, requiring very smooth surfaces. Modern manufacturing methods and design improvements have increased this zone at the expense of zone BC. Remarkably, **Dowson**[3] points out that it has also been Nature's choice for the EHL regime of lubrication to occur in the synovial joints of creatures ranging from elephants to shrews!

The next regime, beyond D, covers the **hydrodynamic** and **externally pressurized** type of bearing. An example is a pair of hydrodynamic journal bearings used to support a steam turbine, where the film thickness is about 15–30 microns. In certain cases, increasing speeds and reducing loads in Zone CD can cause a reduction in elastic distortion of surfaces that were previously in EHL contact, making them also candidates for the zone beyond D. This situation sometimes occurs in ball bearings, where some of the rolling elements have become unloaded during their orbital path.

All these regimes are covered in the book.

1.3. Layout of the Book

The way this book is laid out is as follows: We start by discussing the characteristics of rough surfaces and their dry contact behavior under deformation forces. Then we determine the friction force and wear that occur when two rough surfaces in relative motion are loaded against each other.

In order to reduce the untoward effects of friction and wear, our attention then turns to separating two **smooth** loaded surfaces in relative motion by a viscous fluid. To do this, we introduce the concept of viscosity followed by a discussion of the principles of fluid film hydrodynamic lubrication. Then, application of these principles is made to fluid film thrust and journal bearings, taking into account both thermal and vibration effects. The aim of such procedures is to enable the reader to produce, from a given specification, designs of bearings that are quite close to results from the more complicated computer solutions currently available. Such procedures are also helpful for those of you who are practicing engineers and who need to get a rough idea of how the bearings to be employed will perform.

The next part of the book introduces the reader to the important, and still evolving, topic of elastohydrodynamic lubrication and its applications (Chapter 10), including approximate methods to forecast fatigue failure of machine elements.

In recent years we have witnessed miniaturization of many machines and mechanisms, some operating at very low loads such as the emerging fields of micro-engineering and nano-technology, an example being micro-electro-mechanical systems (MEMS, Chapter 13). It is clear that under a correspondingly low load and in a contact of typically one square micrometer area, the lubricant film thickness is of the order of a few to several molecular diameters. Chapter 13 is therefore devoted to **Nano-Tribology**.

Finally, there is a chapter on the important and topical subject of **Bio-Tribology**, (Chapter 14) contributed by Professor Duncan Dowson.

At the end of the book, under each chapter heading, there is a small selection of practical questions (with solutions where appropriate). These are designed to enhance the students' knowledge of the subject and assist them in the preparation for their examinations.

Throughout the book, results of experimental research and numerical solutions, derived elsewhere, are included where appropriate.

References

1. Da Vinci, L., Madrid, M. S. I (BNM), fol. 20 V, Ca. 1490–1495.
2. Stribeck, R., Die Wesentliechen ichen Eigenschaften Gleit und Rollen Lager or: Ball bearings for various loads. *Trans. ASME* **29** (1907) 420–463.
3. Dowson, D. and Neville, A. Bio-Tribology and Bio-Mimetics in the Operating Environment, *Journal of Engineering Tribology* **220**, 3 (2006) 333–339.

CHAPTER 2

THE NATURE OF ROUGH SURFACES

Big fleas have little fleas on their backs to bite 'em.
The little ones have lesser ones and so ad-infinitum.

2.1. Introduction

From a great height, a glance out of an aeroplane at the earth's surface might reveal that the large area we see appears to be quite smooth, because details of its features are unclear. However, as we lose altitude, the smaller though more detailed surface area below us appears to be rougher with gentle undulations perhaps mixed with hills having themselves smaller hillocks or shoulders on their slopes, all being covered with vegetation, snow or ice, depending on their height above the sea level. A walk on the same surface area would yield far more intimate detail of some of these features, such as soil or wind blown sand. Finally, if we were geologists digging deeply into part of that surface or studying nearby exposed rock faces, different types of strata would be encountered, our investigation still only being confined to only a small depth in relation to the thickness of the earth's crust.

Using appropriate measuring instruments, an investigation of the top layers of a machined steel specimen shows a similar topography, further magnification revealing more and more details of its features. In this case, above the bulk crystalline structure of the metal, there exists a deformed layer created by the manufacturing process with its lower part containing voids (inclusions) and its topmost part harder than the rest. It will be shown later that these voids can become the sites of fatigue cracks that may commence there. Finally, lying on this 'skin', there is generally an oxide layer caused by reaction with the environment above. This might be air, which itself is carrying impurities, such as dust particles, or it might be a liquid lubricant film separating this surface from another.

2.2. Surface Roughness

Surface texture, defines the random deviation of the peaks and valleys that make up the three-dimensional topography of a surface. As implied in the earth surface analogy made above, this texture is multi-layered, in that a close examination of an area reveals a complex pattern of many short wavelength features on top of longer wavelength ones, and so on. A qualitative representation of what one might find is shown in Fig. 2.1, which depicts a machined shaft (a), showing its surface features to increasing scales. The wavy features (b) are often present and can themselves have features of shorter wavelength on them. Such waviness may be the result of vibrations or work-piece deflections. Closer inspection of these wavy surfaces reveals small irregularities (c). If these irregularities are themselves further magnified, they show what are called the **surface roughness** features (d). Like the other textural properties, they are the result of the production process. The more refined the process the smaller these roughness features become.

2.3. Measurement of Surface Texture

In manufacturing engineering, we need to predict, measure and control surface texture of machine parts. This is either in order to specify the

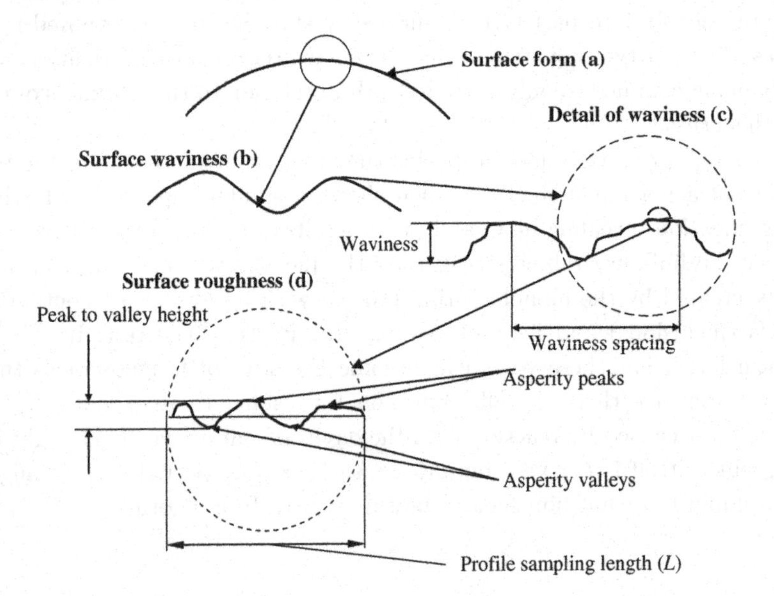

Fig. 2.1. Nature of surfaces.

surface quality at the design stage, to measure a given specimen surface, or to maintain an acceptable surface quality during manufacture.

The measurement of surface texture is normally carried out on a sample of the given surface. The size of this sample must be sufficient to represent statistically the whole area. The most commonly available instrument for this purpose is an instrument called a **Profilometer.** It usually employs a diamond tipped stylus, of radius less than 2 μm, which follows the profile over the length of the sample chosen. The vertical movement of the stylus, i.e. the ordinates defining the rough surface, are measured electronically from some chosen datum and amplified from $\times 200$ to $\times 100,000$ according to the required precision. There is also a horizontal magnification (typically $\times 100$), which is the ratio of the actual distance moved by the stylus to the distance moved by the recorder (that is, their speed ratio). Such a recorded profile does not show the true inclination of the **surface roughness** flanks because of the difference in vertical and horizontal scales.

Figure 2.2 illustrates the effects of the distorted and true scales (actual traces are found in Ref. 1). In (a) the true roughness to scale is shown. In (b) and (c), the length scale has been progressively reduced. The measuring instruments used usually depict the roughness by (c) because it shows far more detail. However, remember that what you see is *not* what it actually is.

We need also to register a datum for the profile. One method is to employ a flat shoe, upon which the stylus arm is pivoted. This shoe generates the general level of the profile, so enabling the roughness to be measured relative to this datum.

A characteristic of this type of profilometer is the finite radius of the stylus tip. The shortest measurable feature is of the order of the tip radius where it is more likely to miss tiny local roughness heights on the sample,

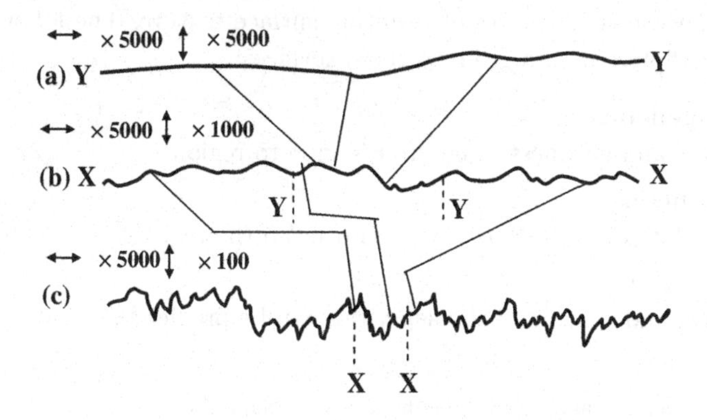

Fig. 2.2. Illustration of the effect of scale distortion.

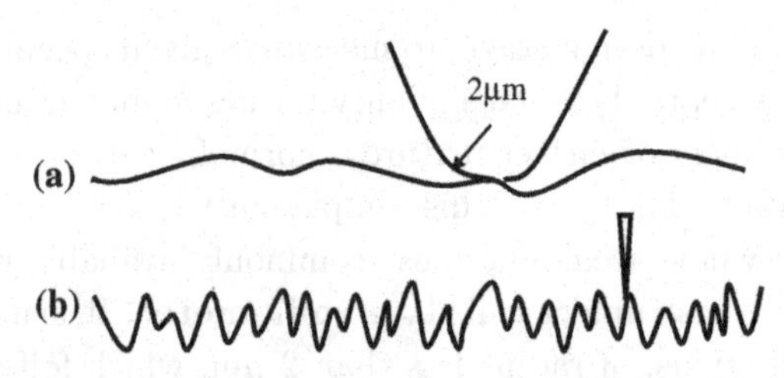

Fig. 2.3. Illustration of the profilometer stylus tip radius effect. (a) Relative sizes of stylus and surface. (b) Relative size with vertical magnification of 5000.

as well as to cross shoulders and narrow valleys without recording them. Figure 2.3(a) illustrates a stylus of the correct size and in Fig. 2.3(b), is its magnified relationship to these roughness features.

Moreover, the sampling length, L, in Fig. 2.1, can be electronically controlled so that, if it is sufficiently long, the recording will include obvious longer wave undulations. Thus, controlling L is a means of filtering out certain longer wave features on the surface when measuring the short wave features, which are deemed to represent the true roughness of the surface.

There are also more accurate profilometers available on the market. These may employ non-contact optical methods, such as **interferometry**, for measuring directly a limited area. Other highly accurate surface measuring instruments are the three dimensional non-contact Scanning Tunnelling Microscope and the low and constant contact force Atomic Force Microscope.[2] Both have a vertical resolution down to 0.02 Nm.

2.4. Types of Engineering Surface

The machining process employed governs the type of surface created. Thus, there are various categories of resulting surface texture. The following are some descriptive names given to these surfaces:

Inhomogeneous

Roughness and hardness* vary from region to region.

Homogeneous

Roughness and hardness are uniform over all the regions.

Deterministic

Both form and spacing of the asperities can be predicted.

*The significance of hardness will be discussed in Chapter 3

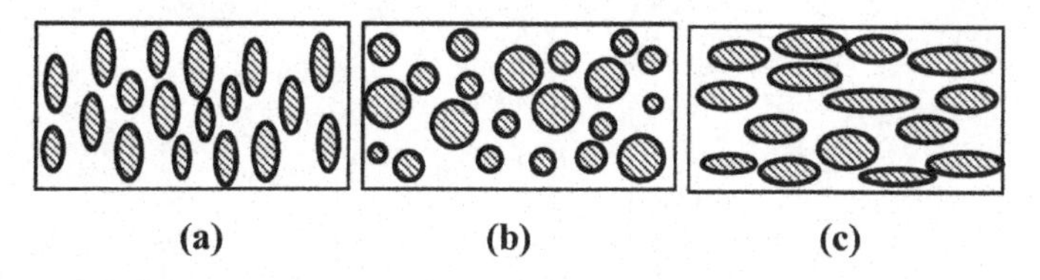

Fig. 2.4. Significance of asperity planforms when a fluid flows past them.

Random
Form and spacing of the asperities cannot be predicted

Isotropic
Whatever its orientation, a section through the surface has the same geometrical characteristic, for example: a surface composed of hemispherical asperities, illustrated in Fig. 2.4(b).

Anisotropic
Depending on its orientation, a section through the surface reveals a different geometrical characteristic, such as the ellipsoidal asperities illustrated in Figs. 2.4(a) and 2.4(c). Here, the orientation of surface features is clearly important. For example, a crankshaft is always ground against the direction of its rotation. Note also that, as demonstrated by Figs. 2.4(a) and (c), the measured roughness produced will depend on the stylus trace direction.

Table 2.1 shows some steel surfaces following different machining processes.

The top **unfiltered profile**, P, shown in Fig. 2.5, possesses both waviness and roughness characteristics.

The wavelengths present in a measured signal depend on the surface irregularities. For example, if a profile filter in the machine suppresses any wavelengths above 0.8 mm (a *cut-off* of 0.8 mm), then the irregularities with this spacing, or less, will appear on the profile trace within a sampling length (L_s) of 0.8 mm. The result is that smaller features have been removed leaving only waviness features on the trace.

2.5. Mathematical Representation of Surface Features

2.5.1. *Analogue solutions*

We will now discuss how to represent a surface mathematically. Consider a fictitious profilometer trace of sampling length L_s in Fig. 2.6. A **mean height line**, AA, through the profile and distance m from the datum line,

Table 2.1. Types of surface produced by different machining processes.

Process	Inhomogeneous	Homogeneous	Deterministic	Random	Anisotropic	Isotropic
Grinding	× (varying hardness)			×	×	
Shot blasting	× (uneven)	× (if even)				×
Turning		× (if no tool wear)	×			
Shaping			×			
Milling			×			
Honing				×	×	
Lapping				×		

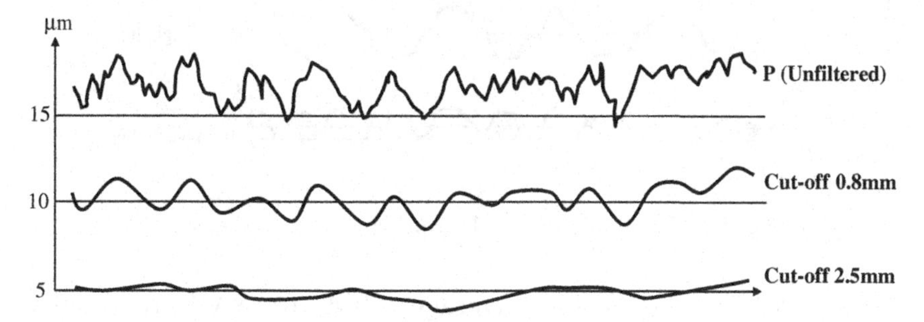

Fig. 2.5. Illustration of different sampling lengths effect. (Actual traces are found in Ref. 1.)

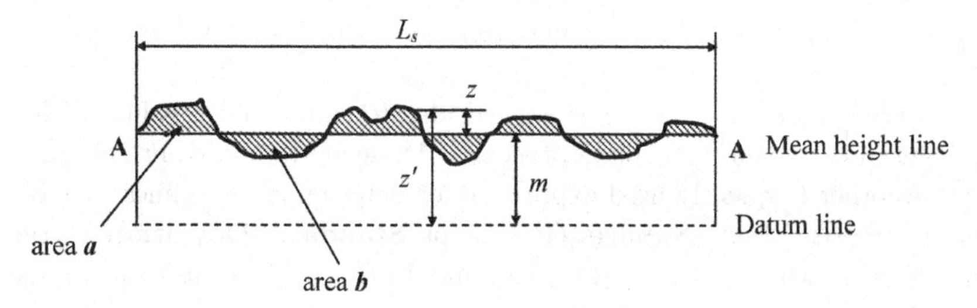

Fig. 2.6. Obtaining the mean height line.

can be found such that:

$$\sum \text{area } a = \sum \text{area } b.$$

Expressed mathematically in terms of a continuous profile, m can be found by the expression:

$$m = \frac{1}{L_s} \int_0^L z' dx, \tag{2.1}$$

where z' describes the roughness ordinates measured by the profilometer from the Datum line. Having obtained m from Eq. (2.1), if z describes the profile height measured from the mean height line, then:

$$z = z' - m. \tag{2.2}$$

The average height of the continuous profile (analogue solution) over length L is then given as:

$$R_a = \frac{1}{L} \int_0^L |z| dx. \tag{2.3}$$

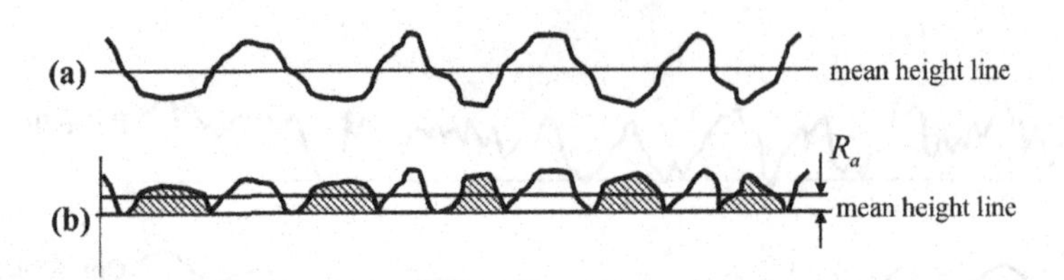

Fig. 2.7. Modified profile to determine the center line average.

R_a is called the **Center Line average (CLA, or AA** in the United States). Figure 2.7 demonstrates that the CLA is obtained by inverting all the profile portions below the mean height line in (a), now shown shaded in (b). Thus, R_a is the arithmetic mean height of the *absolute* departures of the roughness profile from the mean height line over sample length L.

Perhaps, at this stage, it is a good idea to show typical values of R_a, produced by various machining processes. These are tabulated in Table 2.2.

Another frequently used expression for determining roughness heights is the **Root Mean Square (RMS)** or **Standard Deviation** of the distribution with respect to the mean line. In terms of a continuous profile (*analogue* solution) it is:

$$\sigma = \left[\frac{1}{L_s} \int_0^L z^2 dx \right]^{1/2} . \tag{2.4}$$

Table 2.2. Roughness values from different machining processes.

Process	Roughness (R_a) μm									
	0.5 to 3nm	0.05	0.1	0.2	0.4	0.8	1.6	3.3	6.3	12.5
Cleaved Si	↔									
Superfinishing		←	—	→						
lapping		←	—	—	→					
polishing			←	—	→					
honing			←	—	—	→				
grinding			←	—	—	—	→			
boring					←	—	—	—	→	
turning					←	—	—	—	→	
drilling							←	—	→	
extruding						←	—	→		
drawing						←	—	—	→	
milling						←	—	—	→	
shaping						←	—	—	—	→

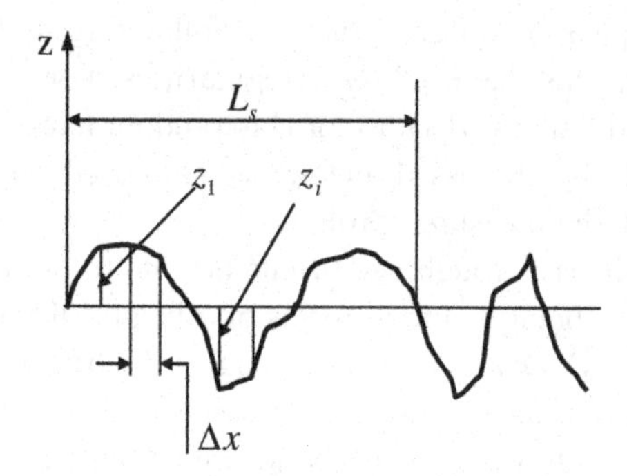

Fig. 2.8. How to determine the all ordinate digitized function.

2.5.2. *Discrete interval solutions of profile height*

If the profile is digitised over length L_s, heights $z_i, i = 1$ to N_c, are read at discrete intervals Δx, as in Fig. 2.8.

Then:

$$\Delta x = \frac{L_s}{N_c - 1}. \tag{2.5}$$

Equations (2.1), (2.3) and (2.4) follow that:

$$m = \frac{1}{N_c} \sum_{i=1}^{N} z_i, \tag{2.6}$$

$$R_a = \frac{1}{N_c} \sum_{i=1}^{N} |z_i|, \tag{2.7}$$

$$\sigma = \frac{1}{N_c} \left(\sum_{i=1}^{N} z_i^2 \right)^{1/2}. \tag{2.8}$$

In a typical trace along the mean line, there might be 4000 discrete height readings taken at intervals $\Delta x = 2$ microns over the sampling length L. The value chosen for Δx depends on the amount of surface detail required.

2.5.3. *Statistical representation of surface texture*

It can be shown that some measured profiles can have the same R_a and sometimes the same σ values, even though they have differing shapes, frequencies and peak-to-valley values. It follows that R_a can give no information about the shape of the irregularities, nor can it distinguish between peaks and valleys. Because of this random nature of real surfaces a statistical analysis, discussed below, is employed to give additional information about the measured profile.

We consider the true roughness profile heights to be random variables, their representation being made in terms of a height distribution curve, as shown by Fig. 2.9. Over an evaluation length, L_s, the profile heights are recorded by a series of narrow horizontal bands of height δz (typically 15). The amounts of profile within each band are added together and divided by L to obtain the relative fraction of the total profile sample that has the height z.

Thus:

$$\phi(z) = \frac{a + b + c + - - - -}{L_s}. \tag{2.9}$$

The resulting curve is called the **Height Distribution Curve** or **Probability Distribution Density Function (PDF)**. Put in terms of statistics, $\phi(z)$ represents the probability that the height of a particular point on the profile will lie between z and $z + \mathrm{d}z$

Furthermore, if we take the running integral of the PDF, as demonstrated graphically in Fig. 2.10, we obtain $\Psi(z)$, the **Material Curve or Cumulative Distribution Function (CDF)**. Here, a\, b\, etc., now represent the incremental *areas* under the PDF that produce ordinates a\, a\ + b\, etc. of the CDF. The readings of the PDF are generally normalized so that the total area enclosed by $\phi(z)$ and, therefore, the maximum ordinate of $\Psi(z)$, are both unity, ensuring that z must fall somewhere between the limits of $z_{\max}$ and $z_{\min}$. When suitably normalized,

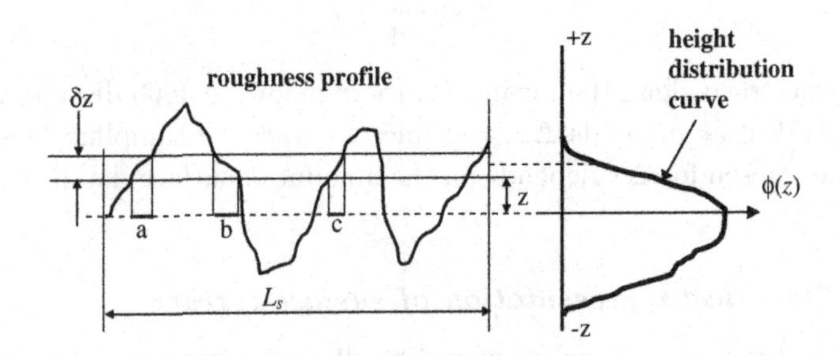

Fig. 2.9. Construction of a height distribution curve (PDF).

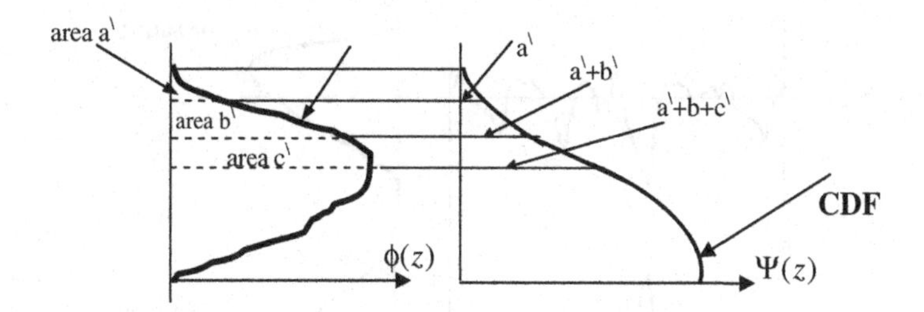

Fig. 2.10. Construction of the Cumulative Distribution Function (CDF).

the maximum ordinate of $\Psi(z)$, is unity. In terms of statistics, $\Psi(z)$ is the probability that a height on the profile exceeds z.

In many cases, such as for ground surfaces, the PDF of the roughness profile, shown in Fig. 2.9, is a normal Gaussian distribution that is typical of the top 10% of asperities.[3] We will see later that these are the significant asperities when two rough surfaces are loaded against each other. If $z^* = z/\sigma$, a Gaussian all ordinate distribution curve in its normalized form can be represented approximately by:

$$\phi(z^*) = \frac{1}{(2\pi)^{1/2}} \exp\left(\frac{-z^{*^2}}{2}\right). \tag{2.10}$$

Just as we did graphically above to obtain Ψ_z, if we integrate $\phi(z)$ with respect to z, the CDF for a continuous profile is obtained as:

$$\Phi(z) = \int_{-\infty}^{\infty} \phi(z)dz. \tag{2.11}$$

One example of a statistical function that can be obtained from the profile is to take the third moment of $\phi(z)$ about the mean line. We get the **skewness** or the **degree of symmetry** of $\phi(z)$ as:

$$R_{sk} = \frac{1}{\sigma^3} \int_{-\infty}^{\infty} z^3 \phi(z)dz. \tag{2.12}$$

Equation (12) gives more information about the symmetry of the height distribution of the profile. A high *negative* value of R_{sk}, suggests that there are a large number of minima below the mean height line over the sampling length. This is shown up by a negative skewness in the PDF curve as compared with a normal Gaussian distribution, which has zero skewness (a symmetrical profile). Examples of positive and negative skewness of PDF's are sketched in Fig. 2.11. Note that the negative skewness has the deeper valleys below the mean line, while the positive skewness has higher peaks.

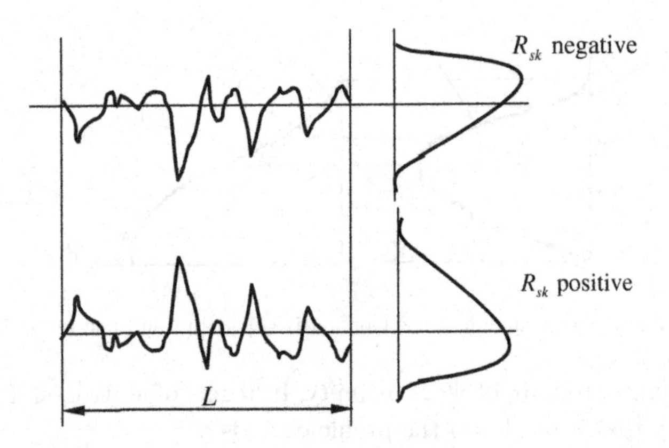

Fig. 2.11. Two asymmetrical profiles with different PDF's (not to scale).

We see that the shapes of both Gaussian and non-Gaussian height distributions can offer plenty of information on the topography of random surfaces. There are many other descriptive expressions available for defining the characteristics of rough surfaces, such as the RMS of the *summit* heights, the mean RMS profile *slope* and the **autocorrelation function**,[2] but those we have described above are sufficient for our elementary analysis.

2.6. Worked Example

A rough surface is represented by a sine wave of amplitude A and wavelength λ. With the origin at zero amplitude, height samples are taken at equal intervals over one complete wavelength of the profile.

(a) Calculate the CLA, found from the continuous profile (analogue) solution,
(b) Recalculate the CLA, for the profile described, by discrete sampling intervals: $\lambda/2$, $\lambda/4$, $\lambda/8$, $\lambda/16$,
(c) Sketch the height distribution curve of the profile.

Solution

(a) As the continuous profile is a sine wave, $z = A \sin \frac{2\pi}{\lambda} x$. Furthermore, all negative values are made *positive* for calculating the mean height R_a

which, from Eq. (2.3), can be written as:

$$R_a = \frac{2}{\lambda}\int_0^{\lambda/2} A\sin\left(\frac{2\pi}{\lambda}\right) x dx.$$

Integration produces $R_a = 0.636A$ (**Answer**)
The discrete solution is found from Eq. (2.7): $R_a = \frac{1}{N}\sum_1^N |z_i|$

For intervals $\Delta x = \lambda/2$, $R_a = 0$ because the z_i heights are only at π and 2π of the sine wave. This is an example of intervals that are too large to define the profile at all!

For interval $\Delta x = \lambda/4$ $R_a = \frac{A}{4}\sum_1^4 \sin\pi/2 + \sin\pi + \sin 3\pi/2 + \sin 2\pi = 0.5A$ (**Answer**)

For interval $\Delta x = \lambda/8$ $R_a = \frac{A}{8}\sum_1^8 \sin\pi/4 + \sin 2\pi/4 + - - + \sin 2\pi = 0.6035A$ (**answer**)

For interval $\Delta x = 16$ $R_a = \frac{A}{16}\sum_1^{16} \sin\pi/8 + \sin 2\pi/8 + - - + \sin 2\pi = 0.629A$ (**Answer**)

(For this regular profile, doubling the number of intervals, brings R_a closer to the analogue solution).

2.7. Closure

In this chapter, we introduced you to the nature and measurement of undistorted surface texture together with some of the terminology employed. Forces that arise from the loaded contact of two deformable bodies cause distortion, and with relative tangential motion, there are frictional forces between their surfaces that can result in wear. We, therefore, need to study their **static contact** behavior before any relative motion is introduced. This is the purpose of the next chapter.

References

1. Taylor Hobson Precision Ltd. *Exploring Surface Texture* (1997).
2. Bhushsan, B. *Handbook of Tribology, Materials and Coatings.* McGraw Hill (1999).
3. Arnell, R. D., Davies, P. B., Halling, J. and Whomes, T. W. *Tribology Principles and Design.* Macmillan (1991).

CHAPTER 3

ELASTIC SOLIDS IN NORMAL CONTACT

3.1. Introduction

In Chapter 2 we discussed the nature of surface texture. Later in the book we will need to study the performance of some solid bodies that are under elastic contact or are contacting through the medium of a lubricant film. Some examples are engineering components, such as ball bearings and gear teeth. To do this we usually treat them as having ideally *smooth* contacting surfaces. On the other hand, occasionally we must consider their surfaces as *real*, implying that they are rough. Therefore, in this chapter we will cover both conditions. Initially, we will study the contact behavior of some individual basic geometrical shapes that can simulate either roughness features on a small scale, or finite engineering parts generally assumed to have smooth contacting surfaces.

3.2. Deformation Characteristics

Let us start by recalling ideal and real stress–strain behavior of some engineering materials. Referring to Fig. 3.1, a cylindrical body, such as a testing laboratory tension specimen, deforms under increasing axial load. If the specimen is considered to be perfectly rigid-plastic, as in (a), then its deformation starts at some critical stress, H, where it is immediately assumed to become fully plastic, deforming with no further increase in stress. In case (b), which is typical of most metals, the deformation is initially elastic, with the slope giving Young's Modulus, E. At the elastic limit, Y, it begins to yield plastically and **work hardens** during the process before eventually fracturing. If before fracture, the load is reduced at some point A, the stress-strain curve is AB, parallel to the elastic line, leaving a residual strain OB. If the load is reapplied, line BA is followed approximately. The specimen has become **strain (or work) hardened**, with a higher elastic limit stress at A. We will need to comment on such stress/strain behavior occasionally in this chapter.

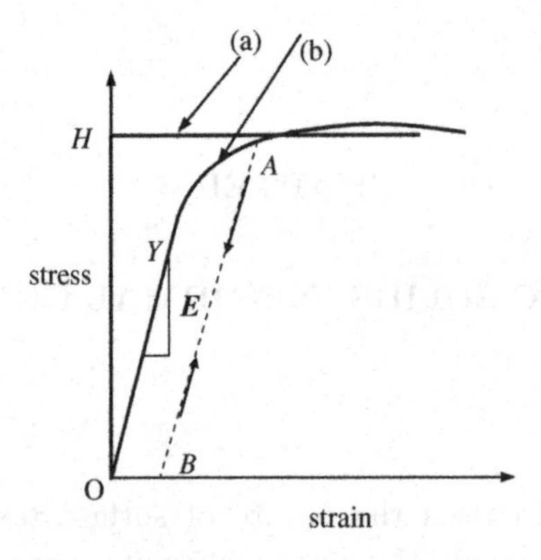

Fig. 3.1. Stress–Strain curve of a tensile specimen. (a) Rigid-fully plastic (b) elastic-plastic with strain hardening.

Now consider two elastic bodies having non-conforming surfaces that touch either at a point or along a line. Any loading will create a finite contact area. These non-conforming geometries may be used to simulate contacting surfaces either on a roughness scale or, if the surfaces are assumed to be perfectly smooth, they can simulate finite engineering shapes, such as ball bearing components or gear tooth flanks. The elasticity theory we will employ was derived by Heinrich Hertz[1] in 1881. It has been considerably simplified here, but is ample for this elementary textbook.

Hertz assumed that:

(1) The region of contact is small compared with the other dimensions of the bodies.
(2) The contact region created as a result of deformation, is much smaller than the radii of curvature and dimensions of the bodies, thus allowing for a small strain analysis. This effectively makes the bodies elastic half-spaces and the contact area plane with pressures applied normal to it.
(3) Any resulting deflections are much less than the dimensions of the contact area.
(4) The surfaces are frictionless. (This condition is relaxed when dealing with friction.)

Consider firstly a touching **convex contact** between two non-conformal elastic body surfaces of different radii of curvature and Young's modulus, as in Fig. 3.2. (By convex contact we mean that the centers of curvature of the respective surfaces are on *opposite* sides of the contact area. A **concave**

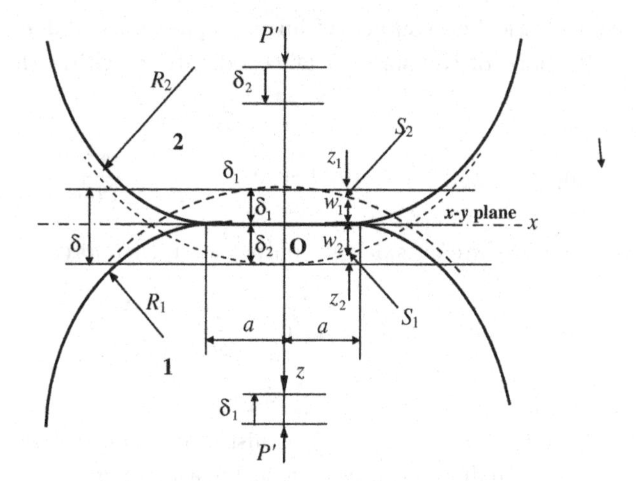

Fig. 3.2. Elastic surfaces in concentrated contact.

(or conformal) contact means that both surfaces have their centers of curvature on the *same* side of the contact area.) At O, through the x-y plane, there would initially be a touching point, if the bodies were spherical, or a line into the paper if they were cylindrical. A cylindrical shape, in general terms, is defined by $x^2 = (2R - z)z$, where z is the distance above the x-y plane at x, and R is the cylinder radius. Because of the assumed small contact region, $R \gg z$, we can replace the true cylindrical shape by a parabola. The gap is now defined by: $z = x^2/2R$.

Let there now be equal and opposite forces, P', applied at distant points* in the bodies, thus creating a finite contact area (called a **footprint**). It would be circular in plan-form passing through O for spherical surfaces, or would be a long rectangular band for cylindrical surfaces. In the latter case, it is called an **elastic line contact**. Relative to O, body (1) will deflect δ_1 upwards at point T_1 along the line of centers, and body (2), δ_2 downwards at point T_2. Had the surfaces not deformed, then the movements δ_1 and δ_2 would have caused the original undistorted profiles to overlap (dotted lines). If movements of any two points, S_1 and S_2, on these relative initial positions of the undeformed surfaces, are sufficient for them just to coincide under the load, then that common point is distance x from the line of centers.

*By distant points, we mean where the contact area deformation has no effect.

Taking an elastic line contact as an example, if w_1 and w_2 are the respective deflections of the surfaces at coordinate x, within the contact area, thus:

$$w_1(x) + w_2(x) = w_1(0) + w_2(0) - \frac{x^2}{2R_1} - \frac{x^2}{2R_2}. \tag{3.1}$$

However, if S_1 and S_2 are outside the contact area, therefore:

$$w_1 + w_2 < w_1(0) + w_2(0) - \frac{x^2}{2R_1} - \frac{x^2}{2R_2}, \tag{3.2}$$

indicating that they do not touch. Also, at $x = 0$, let $w_1(0) = \delta_1$ and $w_2(0) = \delta_2$. Therefore, from Eq. (3.1), if distance $x = a$ defines the edge of the contact area (half the contact width for a rectangular band or the radius, r, for a circle), dividing Eq. (3.1) throughout by a we get:

$$\left(\frac{\delta_1}{a} - \frac{w_1(a)}{a}\right) + \left(\frac{\delta_2}{a} - \frac{w_2(a)}{a}\right) = \frac{a^2}{2a}\left(\frac{1}{R_1} + \frac{1}{R_2}\right). \tag{3.3}$$

Now let:

$$\delta_1 - w_1(a) = d_{f1}, \quad \delta_2 - w_2(a) = d_{f2}.$$

Therefore, Eq. (3.3) becomes:

$$\frac{d_{f1}}{a} + \frac{d_{f2}}{a} = \frac{a}{2}\left(\frac{1}{R_1} + \frac{1}{R_2}\right), \tag{3.4}$$

d_{f1} and d_{f2} being a measure of the true deformation of the surfaces at the boundary $x = a$, that is , relative to their respective distant points on the bodies.

Equation (3.4) allows us to obtain an insight into how the deformations and pressures will vary under the applied forces. Johnson[2] explains this in the following way:

If we say that the strain in each body is characterized by the ratio d/a, where: $d_f = d_{f1} + d_{f2}$, by elementary elasticity theory:

$$\frac{d_f}{a} \propto \frac{p_m}{E}, \tag{3.5}$$

where p_m is the mean contact pressure and E is Young's Modulus.

Substituting Eq. (3.4) into Eq. (3.5):

$$\frac{p_m}{E_1} + \frac{p_m}{E_2} \propto a\left(\frac{1}{R_1} + \frac{1}{R_2}\right). \tag{3.6}$$

More generally, if υ is the Poisson's Ratio, we can also replace the Young's Modulus and radius of curvature of each contacting body by their *'reduced'* values as:

$$\frac{1}{E^*} = \frac{1-\upsilon_1^2}{E_1} + \frac{1-\upsilon_2^2}{E_2}, \tag{3.7}$$

$$\frac{1}{R} = \frac{1}{R_1} \pm \frac{1}{R_2}. \tag{3.8}$$

It is *plus* if the radii of curvature are on opposite sides of the contact footprint, producing a **convex** or **counterformal** or **external contact**, and *minus* if the radii are on the same side, producing a **concave** or **conformal** or **internal contact**. These reduced values are effectively replacing two curved surfaces by one surface, with radius of curvature, R and Young's modulus E^*, contacting a rigid plane.

Using these substitutions, Eq. (3.6) becomes:

$$p_m \propto \frac{aE^*}{R}. \tag{3.9}$$

As an example, take two long cylinders initially in convex line contact (Fig. 3.3). If a uniform load per unit length, P', is applied, the original touching line will become a thin band of width $2a$ and length L, except

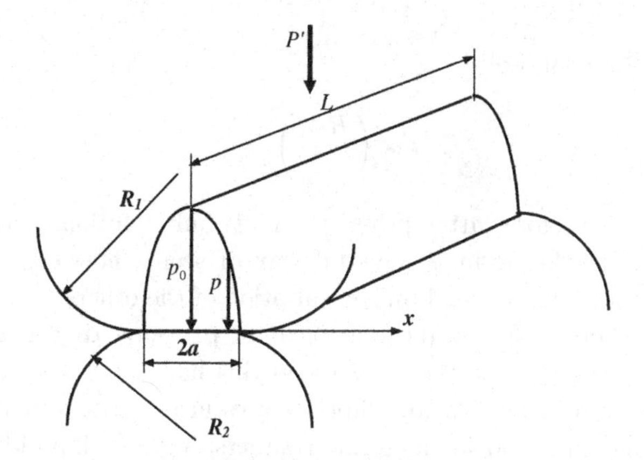

Fig. 3.3. Elastic line contact of long cylinders.

close to its ends. If $L \gg a$, we can assume that the bodies are under plane strain in the region under consideration, being far from their ends.

As $P' = 2ap_m$, from Eq. (3.9), it follows that:

$$\frac{P'}{a} \propto E^* \frac{a}{R},$$

or:

$$a^2 \propto \frac{P'R}{E^*}. \tag{3.10}$$

Alternatively, from purely dimensional considerations, if k and n are constants. Let:

$$a = k\left(\frac{P'R}{E^*}\right)^n. \tag{3.11}$$

Letting F $\equiv$ force, L $\equiv$ length we can say in terms of dimensions that:

$$\left(\frac{P'R}{E}\right) \equiv \frac{(\mathrm{F/L})\mathrm{L}}{\mathrm{F/L}^2} = \mathrm{L}^2.$$

Hence, in order to make the right hand side of Eq. (3.11) dimensionally compatible with its left hand side, $n = 1/2$. Let us take another example of the simple stress analysis approach, applied to the elastic contact of two spherical surfaces under a normal force W. If p_m is the mean contact pressure, then:

$$W = \pi a^2 p_m.$$

Therefore, from Eq. (3.9):

$$a \propto \left(\frac{RW}{E}\right)^{1/3}. \tag{3.12}$$

We see that, before attempting to find full solutions, having the relationships in this form is a useful way of seeing how one dependent external variable will depend on the variation of the others.

We have not yet discussed the shape of the **pressure distribution** over the contact area because the need to do this had not yet arisen. Hertz[1] showed that dimensionally, an elliptical pressure distribution is required over the footprint band for long line contacts, and an ellipsoid is needed over the circular footprint for contacting spherical surfaces. In the general

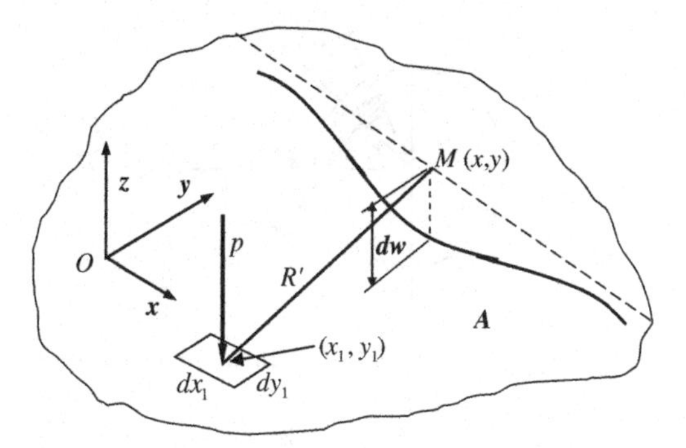

Fig. 3.4. Deflection of an elastic half space due to a pressure element.

case of a point contact with differing **principal radii of curvature** along orthogonal axes, such as a ball in an annular groove (a ball/race contact is an example), we require an ellipsoidal pressure distribution over an elliptical footprint. If p_0 is the maximum pressure at a contact footprint center, the resulting pressure distributions for various geometries are given in Appendix Table 3.1 at the end of the chapter.

3.3. Surface Deformation in a Spherical Contact

There is insufficient space for the derivations of all the various expressions obtained from Hertz's contact theory. Full solutions can be found from Johnson[2] and Gohar.[3] They are based on two governing equations. The first is the geometry, Eq. (3.1) and the second, (3.13), because of its similarity to electrical potential, is called the **Potential Equation**.. It relates an arbitrary pressure distribution, covering a footprint of area A on the surface of an **elastic half-space**, with the deflection at a point, M, anywhere on its surface, caused by a pressure at point distance R' from M (Fig. 3.4). In its integral form it is:

$$w(x, y) = \frac{1}{\pi E} \iint_A \frac{p dx_1 dy_1}{R'}, \tag{3.13}$$

where:

$$R' = [(x - x_1)^2 + (y - y_1)^2]^{1/2}.$$

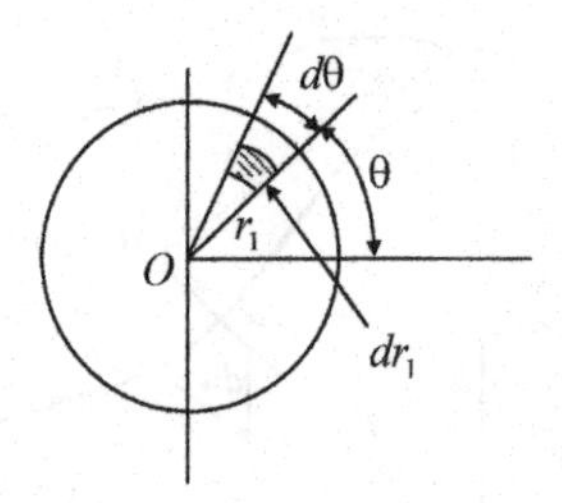

Fig. 3.5. Circular footprint: Polar coordinates.

It is quite straightforward to derive complete solution to Eq. (3.13) for the particular case of the deflection at the center of a circular footprint with an axisymmetric ellipsoidal pressure distribution (Table Appendix 3.1). Referring to Fig. 3.5 and expressing Eq. (3.13) in polar coordinates the center deflection is:

$$w(0,0) = \frac{1}{\pi E}\int_0^{2\pi}\int_0^a p dr_1 d\theta, \tag{3.14}$$

where $R' = r_1$ is now the distance between the footprint center, and $rdrd\theta$ has replaced $dxdy$. As the pressure distribution is symmetrical and $p = p_0(1-(r_1^2/a^2))^{1/2}$, integration with respect to θ yields:

$$w(0) = \frac{2p_0}{E}\int_0^a \left(1-\frac{r_1^2}{a^2}\right)^{1/2} dr_1. \tag{3.15}$$

A further integration with respect to r_1 yields:

$$w(0) = \frac{\pi p_0 a}{2E}. \tag{3.16}$$

If this pressure distribution is caused by the contact of two bodies, E in Eq. (3.15) can be replaced by a **reduced Young's modulus** E^* given by Eq. (3.7).

We can also use Eq. (3.1), reproduced below at $r = 0$:

$$w_1(0) + w_2(0) = \delta. \tag{3.1}$$

The center deflection, δ, is known as the *compliance* in concentrated contacts.

Thus, the mutual approach of distant points on contacting elastic spherical bodies is, from Eqs. (3.1) and (3.16):

$$\delta = \frac{\pi p_0 a}{2E^*}. \tag{3.17}$$

Furthermore, using Eq. (3.1) with the substitutions:

$$w_1(0) = \delta_1, \quad w_2(0) = \delta_2, \quad \delta = \delta_1 + \delta_2 \quad \text{at } r = 0,$$

$$w_1(r) + w_2(r) = \delta - \frac{r^2}{2R}. \tag{3.18}$$

From Ref. 3, at any radius $r \leq a$, the full solution to Eq. (3.14) is given by:

$$w(r) = \frac{\pi p_0}{4aE^*}(2a^2 - r^2). \tag{3.19}$$

Therefore, for the contact of two spherical bodies at $r = 0$, from Eqs. (3.18) and (3.19) we will again obtain Eq. (3.17) above. Furthermore, using Eq. (3.18) this time at $r = a$ gives:

$$w(a) = \delta - \left(\frac{a^2}{2R}\right). \tag{3.20}$$

Hence, from Eqs. (3.17) and (3.19) at $r = a$, and using Eq. (3.20):

$$\frac{\pi p_0}{4E^*} = \frac{a}{2R}. \tag{3.21}$$

The footprint radius a, in Eq. (3.21), can be found in terms of the other external variables if we note that $W = \pi a^2 p_m$ and, for an ellipsoidal pressure distribution on a circular footprint, $p_o = (3/2)p_m$ giving:

$$p_0 = \frac{3W}{2\pi a^2}. \tag{3.22}$$

Thus, from Eqs. (3.21) and (3.22), the complete solution to Eq. (3.12) is:

$$a = \left(\frac{3WR}{4E^*}\right)^{1/3}. \tag{3.23}$$

Moreover, from Eqs. (3.22) and (3.23) we get:

$$p_0 = \left(\frac{6WE^{*^2}}{\pi^3 R^2} \right)^{1/3} . \tag{3.24}$$

This completes our analysis of circular and line elastic contact expressions, but for completeness we should consider firstly different types of elastic conforming contacts before venturing right up to the elastic limit to see what lies beyond.

3.4. Various Contact Geometries

3.4.1. *Line or circular footprint contacts*

We have so far dealt only with the simple examples of two elastic spherical surfaces in point contact, or two rollers in line contact. Equation (3.8) allowed us to distinguish between *external* and *internal* contact, using the plus sign when the bodies' centers of curvature are on opposite sides of the contact tangent plane (**counterformal contact**) and minus when they are on the same side (**conformal contact**). For such situations Fig. 3.6 shows some practical examples of contact geometry, where the values of the radii of curvature have been altered to produce cases (a) through (d).

Application of a normal load leads to a circular footprint for balls and a narrow band footprint for rollers. For long rollers, as shown in Fig. 3.3, only one section plane far from the ends is needed to define such

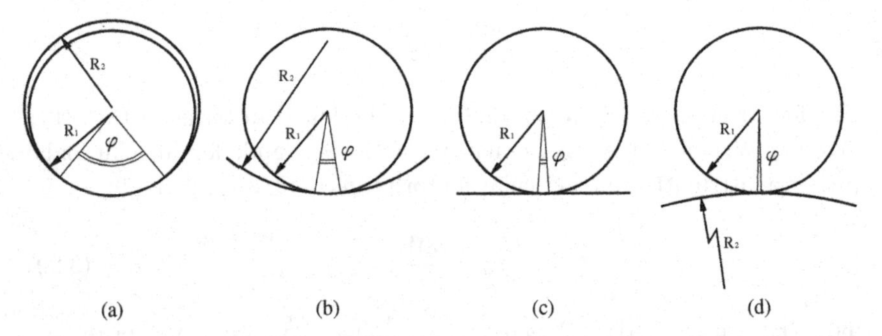

Fig. 3.6. Degree of contact conformity. (a) Closely conforming (journal bearings), (b) conforming (piston-to-cylinder bore, ball to race) (c) cylinder or ball on a flat, (d) counterformal (roller or ball to race).

elastic contact behavior. The same situation applies to contacting spherical surfaces because of axial symmetry.

In Fig. 3.6(a), the surfaces are clearly closely conformal. An example of this is an artificial hip joint assembly where the degree of closeness of the surfaces there will determine what sort of mathematical procedure is needed for a solution. If a sufficient load is applied, both the footprint and pressure dimensions will vary radially, with the footprint width being very large (defined by angle φ in the figure) and becoming comparable to the radii of curvature. In this case the Hertz's assumptions (1) and (2) above may no longer apply, so we must be cautious in considering them as the equivalent of the external contact between a cylinder or ball and a plane! In Fig. 3.3(b), the conformity is much less. An example is the internal contact between two involute gear teeth or a piston skirt in its cylinder. We have already covered examples 3.6(b) to 3.6(d) in the Hertzian contact analysis above.

3.4.2. *Elliptical footprint contacts*

Elliptical footprint contacts are quite common in engineering. Consider again the example in Fig. 3.5(b). Let it now be an end view of a ball contacting the annular groove in the inner race of a radial ball bearing. We will explain the procedure for a solution by means of a numerical example. The left hand figure below (in the zx plane) shows this end view of the ball in its annular inner race groove with the outer race groove at the top. However, the side view (in the zy plane) is different. The bottom contact surface, of radius R_{y1}, is that of the race at the bottom of its groove, so a different geometry results in that plane. When a load is applied, this differing geometry, in the principal planes, creates an **elliptical footprint** with the major axis being in the zx plane, because of the conformity there. A solution procedure using a worked example is given below. This includes a simplified theory for elliptical contacts.

3.4.3. *Worked Example (1)*

The deep groove ball bearing, which is designed primarily to accommodate high radial loads, has the balls guided between concave section annular grooves machined into the inner and outer races. The groove radius of curvature, being larger than that of the balls yields an elliptical shaped footprint with its major axis transverse to the rolling direction.

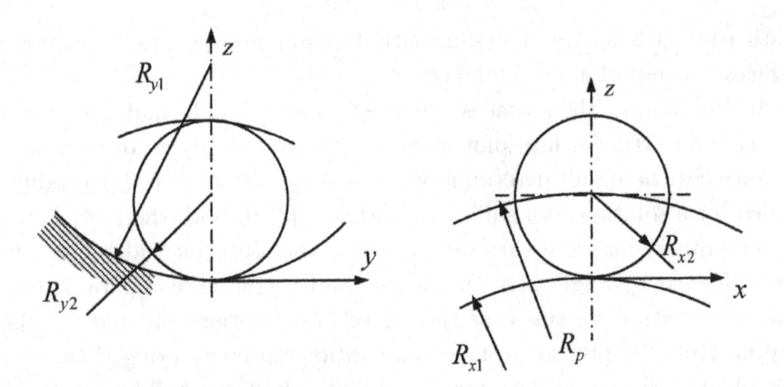

Given: $R_{x2} = R_{y2} = 12\,\text{mm}$, Groove radius $R_{y1} = 14.5\,\text{mm}$ and pitch radius of ball set, $R_p = 30\,\text{mm}$, find the contact footprint dimensions.

Solution:

Considering each plane separately we can determine the principal radii of curvature of the contacting surfaces:

$$R_{x1} = (R_p - R_{x2}) = (30 - 12) = +18,$$
$$R_{y1} = -14.5, R_{x2} = R_{y2} = 12\,\text{mm}$$
$$R_y = \left(\frac{1}{R_{y1}} + \frac{1}{R_{y2}}\right)^{-1} = \left(-\frac{1}{14.5} + \frac{1}{12}\right)^{-1} = 69.9\,\text{mm}$$
$$R_x = \left(\frac{1}{R_{x1}} + \frac{1}{R_{x2}}\right)^{-1} = \left(\frac{1}{18} + \frac{1}{12}\right)^{-1} = 7.2\,\text{mm},$$
$$\frac{R_y}{R_x} = 9.712.$$

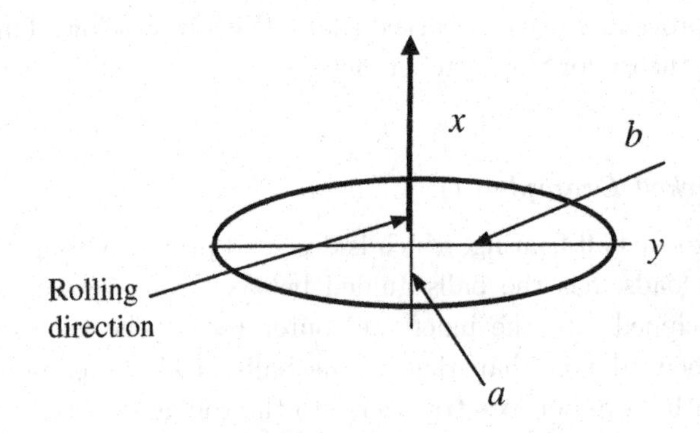

The general solution for a contact between surfaces that produces an elliptical footprint is complicated by the need to employ complete elliptical integrals in a solution. Johnson[2] describes the relevant theory, while Gohar[3] gives a solution procedure. However, in cases where the principal radii of curvature are orthogonal to each other, as they are in this example, and the footprint has an aspect ratio $\frac{b}{a} \leq 5$, an approximate expression is given by

$$\frac{b}{a} \approx \left(\frac{R_y}{R_x}\right)^{2/3}.$$

Substituting the calculated values of

$$R_x \text{ and } R_y \quad \frac{b}{a} = 4.545 \text{ (\textbf{Answer})}.$$

If maximum pressure and footprint dimensions are needed, Table Appendix 3.1 supplies the necessary expressions.

3.5. Onset of Yield

3.5.1. *Cylindrical surfaces*

Figure 3.1(b) has shown that, for a tensile test specimen, prior to hardening, an increasing load will eventually cause the onset of plastic deformation when a yield stress (Y) is reached. A similar behavior occurs for the contact of bodies with non-conformal surfaces. The one with the softer material will start yielding when the maximum shear stress within it reaches a critical value there. For the plane strain elastic line contact of cylinders, the **maximum shear stress**, $\tau_{\max}$, is given by[4]:

$$\tau_{\max} = \left\{\frac{(\sigma_x - \sigma_z)^2}{4} + \tau_{xz}^2\right\}^{1/2}, \tag{3.25}$$

where σ_x and σ_z are the direct stresses within the body at point x, z and τ_{xz} is the shear stress in the x direction in a plane normal to the z-axis. Because of plane strain conditions, $\tau_{\max}$ can also be expressed in terms of the principal stresses as:

$$\tau_{\max} = \frac{1}{2}(|\sigma_1 - \sigma_2|). \tag{3.26}$$

We see that, $\tau_{\max}$ is half the difference between the maximum and minimum principal stresses. In Fig. 3.7, contours of $\tau_{\max}/p_0$ are plotted for an elastic

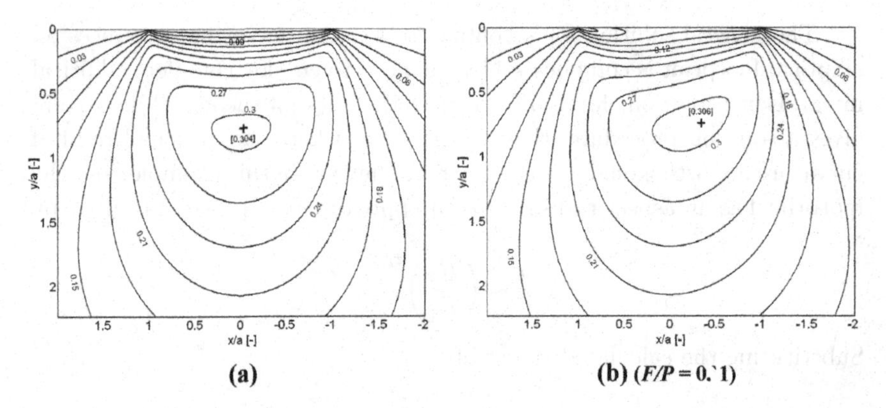

Fig. 3.7. Lines of constant $\tau_{\max}/p_0$ for a two-dimensional elastic line contact.

line contact under direct load P per unit length. The critical value of $\tau_{\max}$ at yield is defined by k. In this case, $k = 0.3p_0$, occurring on the z-axis at $0.78a$ below the footprint.

Now the **Tresca yield criterion** (see for example Case and Chilver[5]), for ductile materials in plane strain under normal load only, states that:

$$\max|\sigma_1 - \sigma_2| = 2k = Y = 0.6p_0, \tag{3.27}$$

the softer material starting to yield on the z-axis at a depth of $0.78a$. The corresponding maximum pressure at yield is therefore:

$$(p_0)_Y = \frac{4}{\pi}(p_m)_Y = 3.3\,k = 1.67Y \tag{3.28}$$

If the normal load continues to increase, the plastic zone will enlarge from its nucleus below the footprint until eventually it reaches its surface. This process is quite slow, because the surface elements below the footprint are under orthogonal compressive elastic stresses, creating a constraining hydrostatic effect on the spread of the plastic zone. Eventually, the plastic zone will reach the footprint when p_m has grown approximately to $6k[= 2.3(p_m)_Y]$. The mean pressure for this final fully plastic condition is defined as the material **indentation hardness** value, H.

Thus:

$$H \approx 6k \approx 3Y, \tag{3.29}$$

$$(p_0)_Y \approx 0.6H. \tag{3.30}$$

The above theory applies to a normal load only. If, in addition, there is an applied tangential traction F per unit length, in the x direction, the lines of constant maximum shear stress alter their orientation. The effect of these combined loads is shown in Fig. 3.7(b). They cause the maximum value to approach the contact surface, thus facilitating the more rapid spread of plastic deformation below the footprint.

3.5.2. *Spherical surfaces*

In the case of contacting spherical surfaces, the maximum shear stress occurs beneath the footprint on its polar axis of symmetry.

On this axis, the principal direct stresses are now σ_z, σ_r and $\sigma_\theta (= \sigma_r)$.

For steel ($\nu = 0.3$) the maximum value of the principal shear stress there is[6]

$$k = \frac{1}{2}|\sigma_z - \sigma_y| = 0.31 p_0 \tag{3.31}$$

at a depth of $0.48a$ below the surface on the footprint polar axis of symmetry.

Therefore, the value of p_0 at yield by the Tresca yield criterion is:

$$(p_0)_Y = \frac{3}{2}(p_m)_Y = 3.22k = 1.60Y \approx 0.6H. \tag{3.32}$$

We see that the value of the maximum pressure at yield is almost the same for both elastic line and circular contacts. One difference is the position of the yield points for the two geometries ($0.78a$ and $0.48a$ respectively). Another concerns the distribution of tensile stresses. For an elastic line contact they are zero everywhere. On the other hand, for circular contact footprints, there exists a maximum radial tensile stress round the footprint edge. Johnson[2] points out that this stress can cause **ring cracks** when the materials are brittle. There is a similar situation when the contact footprint is elliptical, as in a ball bearing. Present are radial tensile stresses at the ends of the contact footprint ellipse major and minor axes.

Moreover, as for line contacts, a tangential traction, in addition to the normal load, will allow any plastic deformation to occur more readily in the region below the footprint. This situation becomes significant when we deal with friction forces on roughness features in Chapter 4.

3.6. Nominally Flat Rough Surfaces in Contact

3.6.1. *Idealized rough surfaces*

The aim of this section is to give the reader an idea of the behavior of contacting rough surfaces, firstly employing idealized models based on the expressions derived above for circular contact footprints. This simplification assumes that both surfaces are nominally flat, but one of them has on it isotropic roughness features (see Fig. 2.4(b).† We initially assume that these comprise identical separate spherically shaped asperities with a reduced modulus, E^*, all of reduced radius, R, and the same initial summit height, z_s. The assumption appears reasonable if we refer to Fig. 2.1 and remember that the true roughness shape is composed of low slope bumps. The other contacting surface is now assumed to be a rigid smooth plane. Also, any vertical displacement due to load each feature suffers is considered not to influence the deformation of its surrounding neighbors. The assumed rough surface is depicted in Fig. 3.8 with the smooth rigid surface penetrating equally the asperity tips.

Let z_s define the position of the undistorted asperity peaks depicted in Fig. 3.8. As the two surfaces are loaded together, the center deflection (compliance) of the asperities, δ, corresponds to the current position of the top surface, d, with respect to the rough surface reference plane (the separation).

The compliance is:

$$\delta = z_s - d.$$

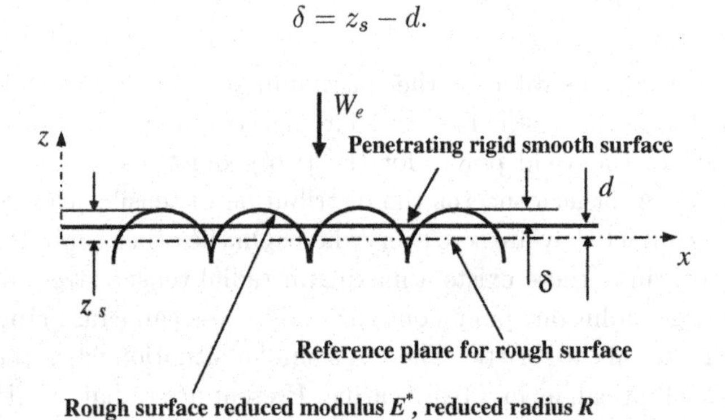

Fig. 3.8. Regular patterned ideal rough surface.

---

†Greenwood and Tripp[7] have shown that this assumption gives similar results when *both* surfaces are considered rough and each has a Gaussian distribution.

If W_i is the normal load on each asperity, and there are n asperities per unit area, then the total load is:

$$W_e = nW_i. \tag{3.33}$$

We can now apply our derived expressions for circular contacts to obtain an expression for the load on a single elastic spherical asperity in terms of its deflection.

From Eqs. (3.17), (3.22) and (3.23):

$$W_i = \frac{4}{3}(E^* R^{1/2} \delta^{3/2}). \tag{3.34}$$

We will also need an expression for a circular footprint area. The area is:

$$A_i = \pi a^2. \tag{3.35}$$

Combining Eqs. (3.23) (for W_i), (3.34) and (3.35), we get:

$$A_i = \pi R \delta. \tag{3.36}$$

Equation (3.36) shows that, for the elastic contact between a spherical surface and a rigid plane, the resulting circular footprint area is half the area obtained by assuming the sphere is fully plastic. (In that case the circular footprint, of a roughness feature, would have a radius equal to the chord radius at distance δ above the touching contact plane.) Also, the total elastic footprint area, obtained from all the equal spherical features, is:

$$A_e = nA_i. \tag{3.37}$$

Hence, the total load, W_e, in terms of A_e, is found from Eqs. (3.33), (3.34), (3.36) and (3.37) to be:

$$W_e = \left(\frac{4}{3}\right) E^* R^{1/2} n^{-1/2} \left(\frac{A_e}{\pi R}\right)^{3/2}. \tag{3.38}$$

That is:

$$A_e \propto W_e^{2/3}. \tag{3.39}$$

Other terms involve material and geometrical properties only. Equation (3.39) is important. It shows that for rough surfaces with idealized and identical features in elastic contact the true contact area is *not* proportional to the normal load, something we always assume when dealing with smooth surfaces. We should, therefore, investigate the problem further after assuming that there are *real* rough surfaces.

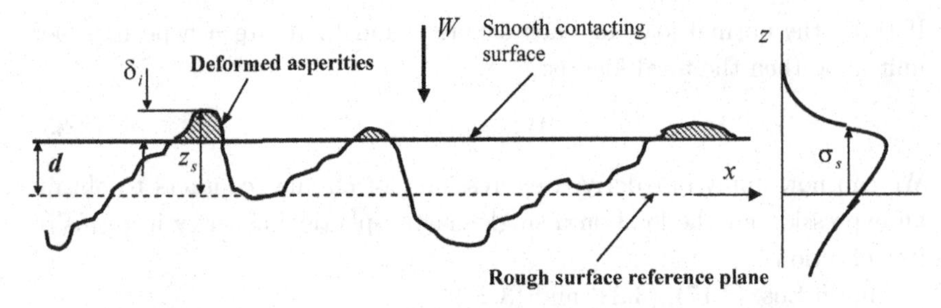

Fig. 3.9. Real surface penetrated by a rigid smooth surface.

3.6.2. *Contact between real rough surfaces*

Let us now assume that the contacting rough surfaces are real, that is their asperities are of varying height and radii, having an equivalent standard deviation (RMS) of their summits of (see Chapter 2):

$$\sigma_s = (\sigma_1^2 + \sigma_2^2)^{1/2}. \tag{3.40}$$

Thus, σ_s is the RMS roughness of an equivalent rough surface contacting a smooth plane.

The roughness features are distributed randomly following some probability distribution, as in Fig. 3.9. The analysis below follows Greenwood and Williamson.[8]

In general terms, we have for each feature:

$$A_i = \pi a_i^2 = f(\delta_i), \tag{3.41}$$

$$W_i = g(\delta_i), \tag{3.42}$$

where $f(\delta_i)$ and $g(\delta_i)$ are functions that depend on the material and geometrical properties of the surfaces.

We can carry out our modification by employing the methods described in Chapter 2 for random rough surfaces. Let the summit height distribution curve be $\phi(z_s)$. If the compliance is: $\delta_i = z_s - d$, then it means that those asperities with heights exceeding the separation, d, will have been penetrated (shaded area in Fig. 3.9).

Letting ϕ be some summit distribution function, the probability that any asperity of height z_s within a nominal surface area A_0, has been penetrated, is described by:

$$\text{prob } (z_s > d) = \int_d^{\infty} \phi(z_s) dz_s. \tag{3.43}$$

If there are N summits within A_0, and the number of them that have been penetrated is n, from Eq. (3.43):

$$n = N \int_d^\infty \phi(z_s) dz_s. \tag{3.44}$$

For the real area of contact and load, we can still employ the idealized spherical surface geometry of Sec. 3.4(a) as a model, if we assume that the equivalent surface is represented by its *mean* asperity radius and *mean* contact area, as defined in Sec. 3.4(a).

Thus, in general terms, over a nominal surface area A_0, Eq. (3.41) and (3.42) become respectively for a real contact area, A:

$$A = N \int_d^\infty f(\delta_i)\phi(z_s) dz_s, \tag{3.45}$$

$$W = N \int_d^\infty g(\delta_i)\phi(z_s) dz_s. \tag{3.46}$$

As a demonstration of the importance of the various parameters involved in the contact of rough surfaces, let us assume for simplicity that there is an exponential height distribution. This assumption approximates to a Gaussian distribution for the top 10% of asperity summits.[2] The exponential distribution can be written non-dimensionally if we scale z_s with the standard deviation of the peak height. Therefore let $\bar{z}_s = z_s/\sigma_s$. Likewise, let $\bar{d} = d/\sigma_s$ and $\bar{\delta} = \delta_i/\sigma_s$. Thus:

$$\phi^*(\bar{z}_s) = \exp(-\bar{z}_s). \tag{3.47}$$

In Eq. (3.47), ϕ^* has been scaled to make its standard deviation unity. To expand Eqs. (3.44) to (3.46), let us change the lower integration limit using the substitution $\bar{\delta}_i = \bar{z} - \bar{d}$ so that at $\bar{z} = \bar{d}, \bar{\delta}_i = 0$. With this substitution, Eq. (3.44) has become a definite integral as:

$$\begin{aligned} n &= Ne^{-\bar{d}} \left(\int_0^\infty e^{-\bar{\delta}} d\bar{\delta} \right), \\ \therefore n &= Ne^{-\bar{d}}. \end{aligned} \tag{3.48}$$

Likewise, using Eq. (3.48), Eqs. (3.45) and (3.46) respectively become:

$$A = Ne^{-\bar{d}} \int_0^\infty f(\delta_i) e^{-\bar{\delta}} d\bar{\delta} = nI_f, \tag{3.49}$$

$$W = Ne^{-\bar{d}} \int_0^\infty g(\delta_i) e^{-\bar{\delta}} d\bar{\delta} = nI_g. \tag{3.50}$$

The definite integrals in Eqs. (3.49) and (3.50), I_f and I_g, are always constants, independent of separation, $\bar{d}$, and of whatever distortion regime the deforming asperities follow. Also, because mean pressure $p_m = W/A$, both A and p_m are always proportional to the number of asperity summits and to each other. As confirmation of this, supposing the mode of deformation is below the elastic limit (signified by suffix e). In this case, from Eq. (3.36):

$$f(\delta_i) = \pi R \delta_i. \tag{3.51}$$

For the real contact area, substituting Eq. (3.51) into (3.49), thus:

$$A_e = \pi R \sigma N e^{-\bar{d}} \int_0^\infty e^{-\bar{\delta}} d\bar{\delta}.$$

Integrating and substituting $n = Ne^{-\bar{d}}$ from Eq. (3.48):

$$A_e = N\pi R\sigma e^{-\bar{d}} = n\pi R\sigma. \tag{3.52}$$

For the load, from Eqs. (3.34) and (3.42), with δ_i replaced by $\bar{\delta}$, thus:

$$g(\delta_i) = \frac{4}{3}\sigma^{3/2}(E^* R^{1/2} \bar{\delta}^{3/2}). \tag{3.53}$$

Substituting Eq. (3.53) into Eq. (3.50), integrating‡ and substituting $n = Ne^{-\bar{d}}$ from Eq. (3.46), it follows:

$$W_e = Ne^{-\bar{d}}(\pi R)^{1/2}\sigma^{3/2}E^* = n(\pi R)^{1/2}\sigma^{3/2}E^*. \tag{3.54}$$

Hence, $W_e \propto A_e \propto n$. It shows that, for an exponential height distribution and elastic deflection, load and real area are both directly proportional to the number of contact spots occurring, as we have normally assumed.

‡ $\int_0^\infty \bar{\delta}^{3/2} e^{-\bar{\delta}} d\bar{\delta} = \frac{3\pi}{4}$.

If we now take Eqs. (3.52) and (3.54), the mean pressure for elastic contact is:

$$\frac{W_e}{A_e} = p_m = \sqrt{\frac{\sigma}{\pi R}}E^*. \tag{3.55}$$

We see that at least for exponential surfaces deforming elastically, the mean contact pressure is constant, depending only on surface and material properties. What occurs is that, in order to maintain a constant mean pressure, more and more contact spots share the normal load as it increases. If some of the higher asperities reach their maximum shear stress, k, during the loading process, instead of going fully plastic, their condition remains close to the elastic limit because more of the lower height asperities are sharing the total load. Approximately, a similar conclusion is reached for a Gaussian distribution over a limited range of loading. This result, for an exponential height distribution, is unlike that of Eq. (3.39) for the idealized condition of identical asperities all deflecting equally. There, W_e is proportional to $(A_e^{2/3})$. The above results for rough elastic contacting surfaces become significant when, in Chapter 4, a tangential force is applied in addition to a normal force.

3.6.3. *Plasticity index*

It is useful to have some criterion (like the Reynolds number for fluids) to gauge, from their material properties, the extent of plastic deformation between nominally flat rough contacting surfaces. It can be obtained in the following way:

One consequence of the real area of contact being proportional to the load for an exponential probability height distribution is that the real mean pressure, p_m, must be constant. As $p_m = W_e/A_e$, from Eq. (3.55) we can write it again as:

$$(p_m)_Y = 0.564E^* \left[\frac{\sigma}{R}\right]^{1/2}. \tag{3.56}$$

Recall that the Yield Point for the spherical asperity model we are using, is given by Eq. (3.32) as $(p_0)_Y = 0.6H$ or, in terms of mean pressure, if $(p_m)_Y \geq 0.39H$, plastic flow will commence below the contact area. Inserting this condition on the left hand side of Eq. (3.56), we get Eq. (3.57). Greenwood and Williamson,[8] called the resulting dimensionless group, the **Plasticity Index** (Ψ). For this particular condition with exponential surfaces:

$$\Psi = \frac{E^*}{H}\left[\frac{\sigma}{R}\right]^{1/2} = 0.69. \tag{3.57}$$

The Plasticity Index describes the topographical and material properties of the contacting surfaces and is *independent* of load. In particular, $\frac{\sigma}{R}$ is a measure of the asperity slope, while $\frac{E^*}{H}$ describes the material elastic and work hardened properties. Thus, the pivotal value of ψ, for an *exponential* height distribution, is close to 0.69 at the yield point of the softer of the contacting materials. We can now say that, for an exponential height distribution, if $\Psi < 0.69$, the contact will be mainly elastic.

For the more realistic Gaussian height distribution, Greenwood and Williamson showed that if $\psi < 0.6$, the conditions at the contact are mainly elastic with considerable pressures necessary to cause plastic flow. For $\psi > 1$, there is some plastic flow in the contact region even at trivial loads. They pointed out that most engineering surfaces have a plasticity index exceeding 1.0. Also note that a low modulus material with a high hardness (rare) will delay the onset of plastic flow.

One conclusion we can reach for this approach is that we have used elasticity theory to describe plastic behavior. This approach is accurate provided only a small proportion of the contact region is plastic and the deflections are confined to elastic magnitudes.

Another more obvious conclusion, is that the rougher the surface, the *higher* the value of ψ because, topographically, rough surfaces have high RMS heights and relatively small radii of curvature, the converse being true for smooth surfaces.

3.6.4. *Fully plastic surface contacts*

As we mentioned above, when dealing with a single contact, plastic flow of some asperities commences at the point of maximum shear stress, its spread being restricted by the surrounding elastic material. Asperities that will initially experience plastic flow will be the ones with the highest summits that make first contact with the descending equivalent smooth surface in our model of Fig. 3.5. They will eventually have their individual loads reduced as additional asperities become deflected elastically to support the load. It would be interesting to find the relationship between the area of plastic contact and the loads over those particular asperities to see if they follow the same behavior as the elastically deflecting ones. Just as for elastic contacts, we can estimate the probability of a fully plastic contact as:

$$\text{prob}[z > (d + w_p)] = \int_{d+w_p}^{\infty} \phi(z)dz,$$

where w_p is the asperity deflection to cause fully plastic distortion.

Johnson[2] has given approximate expressions for the fully plastic state of a single spherical contact after work hardening. These are:

$$A_{iP} = 2\pi r\delta, \quad P_{ip} = 6\pi YR.$$

Therefore, the expected ingredients of a fully plastic contact between rough surfaces will be:

$$n_p = N \int_{\bar{d}+\bar{w}_{\bar{p}}}^{\infty} e^{-\bar{\delta}} d\bar{\delta} = Ne^{-(\bar{d}+\bar{w}_p)}, \tag{3.58}$$

$$A_p = 2\pi NR\sigma \int_{\bar{d}+\bar{w}_p}^{\infty} \bar{\delta}e^{-\bar{\delta}} d\bar{\delta} = 2\pi NR\sigma e^{-(\bar{d}+\bar{w}_p)} = 2\pi nR\sigma, \tag{3.59}$$

$$W_p = 6\pi YRN\sigma \int_{\bar{d}+\bar{w}_p}^{\infty} \bar{\delta}e^{-\delta} d\vec{\delta} = 6\pi\, YRN\sigma e^{-(\bar{d}+w_p)} = 6\pi YRn\sigma, \tag{3.60}$$

$$\therefore \quad \bar{p}_m = \frac{W_p}{A_p} = 3Y = H. \tag{3.61}$$

The above equations show us that, just as for elastic contacts, the fully plastic condition has the contact area and load *only* proportional to the number of contact spots but, unlike in Eq. (3.55), *not* proportional to the height distribution. Moreover, as would be expected, if the surfaces are rigid-plastic the mean pressure is constant at the hardness of the softer surface. Any fully plastic asperity contacts are highly significant when we later consider adhesive friction in Chapters 4 and 13, because the mean pressures occurring under plastic flow are usually associated with **cold welding** at their junctions.

3.6.5. *Worked Example (2)*

Two nominally flat ground steel contacting bodies with Gaussian surfaces, each has an RMS roughness of 1.13μm and a mean asperity summit radius of 7.62μm.

(a) By finding the plasticity index, investigate whether the asperity deformation is predominantly plastic or elastic.
(b) If the surfaces are now lapped until each has an RMS roughness of 0.046μm and summit radius of 480μm, find the modified plasticity index and comment on the alteration of roughness shape between (a) and (b).

Take their hardness to be 8 GPa and $E^* = 110$ GPa.

Solution

(a) From Eq. (3.40), $\sigma = \sqrt{(\sigma_1^2 + \sigma_2^2)} = \sqrt{(1.13^2 + 1.13^2)} \times 10^{-6} = 1.6\mu$m. The summit radius of each surface is 300μm, so the reduced radius for a convex contact is $R = 150\mu$m. Equation (3.56) will give us the plasticity index.

$$\Psi = \frac{E^*}{H}\left[\frac{\sigma}{R}\right]^{1/2} = \frac{110 \times 10^9}{8 \times 10^9}\left[\frac{1.559 \times 10^{-6}}{3.81 \times 10^{-6}}\right]^{1/2} = 1.559\ (\textbf{Answer}).$$

Therefore, the contact will be predominantly plastic for these surfaces
(b) When the surfaces are lapped, $R = 240\mu$m, $\sigma = 0.065 \times 10^{-6}$ m giving $\Psi = 0.226$ (**Answer**).

The excellent finish has considerably reduced the waviness height and slope making the contact now predominantly elastic.

3.7. Contact Between Curved Rough Surfaces

The theory of nominally curved contacting rough surfaces is far more complex than when they are nominally flat, as they were in Sec. 3.6. The difference is that for contacting curved surfaces, the nominal contact area is now the footprint, which may be much smaller than the arbitrary nominal areas chosen for flat surfaces. Greenwood and Tripp[7] have shown that the same behavior roughly applies in the curved contact case. Again, the average real contact mean pressure remains constant as the overall load is increased. In both cases there are only a small proportion of the roughness summits that are in contact, the difference being that in the curved case, for the same load, these are more clustered together within the confines of the footprint.

3.8. Hertzian Impact

So far we have considered various contact conditions. However, there are many circumstances that an impact occurs between a pair of elastic solids of revolution, such as between snooker or billiard balls. A subset of such impacts is described by the impact theory developed by Hertz in extending his contact theory in 1881.[1] These impacts are considered as localized (i.e. obey the Hertzian assumptions, described in Sec. 3.2). Referring to Table Appendix 3.1 for contact center deflection, δ in the case of a circular point contact: $\delta = \left(\frac{9W^2}{16E^{*2}R}\right)^{1/3}$, which can be re-written in the form:

$$W = K\delta^{3/2} \quad \text{where} \quad K = \frac{4E^*\sqrt{R}}{3}. \tag{3.62}$$

K is a constant of proportionality and is known as the contact spring non-linearity and W is the contact load. The non-linearity indicates that the actual contact stiffness changes with the extent of deflection, that is: $k = \frac{\partial W}{\partial \delta} = \frac{3}{2}K\sqrt{\delta} = 2E^*\sqrt{R\delta}$. As the deflection increases, so does the stiffness, k. When a pair of snooker balls impact (or a rigid sphere of equivalent radius R impacts a semi-infinite elastic solid of modulus E^*), the impact kinetic energy $\frac{1}{2}mv^2$ (m is the mass of the equivalent sphere) is converted into strain energy of deformation. At maximum deflection, this stored strain energy is released to rebound the sphere. Note that the Hertzian impact assumes no loss of energy.

We can find the stored energy as: $E = -\int_0^{\delta_{\max}} W d\delta$ (this is Euler's equation, and the negative sign indicates stored energy). Replacing for W from Eq. (3.62), we get: $E = -\frac{2}{5}K\delta_{\max}^{5/2}$. The kinetic energy of the impacting solid is arrested gradually as: $\frac{1}{2}m(\dot{\delta}^2 - v^2)$. Thus, at any instant during penetration:

$$-\frac{2}{5}K\delta_{\max}^{5/2} = \frac{1}{2}m(\dot{\delta}^2 - v^2). \tag{3.63}$$

Hence, we can find the maximum deflection (when $\dot{\delta} = 0$, i.e. moment of rebound) due to an impact velocity v as:

$$\delta_{\max} = \left(\frac{5mv^2}{4K}\right)^{2/5} = \left(\frac{15mv^2}{16E^*\sqrt{R}}\right)^{2/5}. \tag{3.64}$$

The impact time is very short indeed, usually of the order of few tenths of millisecond. We can re-write Eq. (3.63) as: $\dot{\delta}^2 = v^2 - \frac{4}{5}\frac{K}{m}\delta^{5/2}$, thus:

$$dt = \left(v^2 - \frac{4}{5}\frac{K}{m}\delta^{5/2}\right)^{-1/2}$$

$$d\delta = \frac{1}{v}\left(1 - \frac{4}{5}\frac{K}{mv^2}\delta^{5/2}\right)^{-1/2} \tag{3.65}$$

$$d\delta = \frac{1}{v}\left(1 - \left(\frac{\delta}{\delta_{\max}}\right)^{5/2}\right)^{-1/2}$$

Now let: $x = \frac{\delta}{\delta_{\max}}$, $d\delta = \delta_{\max}dx$, then:

$dt = \frac{\delta_{\max}}{v}(1 - x^{5/2})^{-1/2}dx$, and $x = 0$ at $t = 0$ and $x = 1$ at $t = t_{\max}$ (impact time), then:

$$t_{\max} = \frac{\delta_{\max}}{v}\int_0^1 (1 - x^{5/2})^{-1/2}dx \approx 2.94\frac{\delta_{\max}}{v} \tag{3.66}$$

For a ball falling freely (under influence of gravity) from a height h onto a flat plane, $v = \sqrt{2gh}$, and thus knowing the physical and geometrical properties of the ball we can obtain the impact time. We can follow the same procedure for a roller, using the expressions given in Table Appendix 3.1. You can do this as an exercise, and should obtain:

$$t_{\max} = \frac{2\delta_{\max}}{v}\int_0^1 \frac{1}{\sqrt{1 - x^2}}dx = \frac{\pi\delta_{\max}}{v}. \tag{3.67}$$

Note that $\delta_{\max}$ here is different to that for circular point contact in Eq. (3.64). Hertzian impact theory applies to the dynamic behavior of solids of revolution below their modal behavior (no global deformation). Therefore, the theory does not apply to solids (such as hollow balls', cylinders, tubes and church bells), where the nature of the solid under impact conditions leads to modal behavior. Some of these structural modes coincide with their acoustic modes and result in sound propagation, such as in church bells.

3.9. Closure

Chapter 3 has covered most of the theory of deforming solids in normal contact that will be needed in the subsequent chapters. In Chapter 4, we will give a brief coverage of friction forces and wear that occur between contacting rough surfaces when there is impending or relative motion.

Table Appendix 3.1. Relationships between variables in elastic contacts.

Variable	Elastic line contact	Circular contact	Elliptical contact
Contact pressure distribution	$p = p_0\left(1 - \frac{x^2}{a^2}\right)^{1/2}$	$p = p_0\left(1 - \frac{r^2}{a^2}\right)^{1/2}$	$p = p_0\left(1 - \frac{x^2}{a^2} - \frac{y^2}{b^2}\right)^{1/2}$
Contact half width or radius	$a = \left(\frac{4P'R}{\pi E^*}\right)^{1/2}$	$a = \left(\frac{3WR}{4E^*}\right)^{1/3}$	$\sqrt{ab} = \left(\frac{3W\sqrt{R_x R_y}}{4E^*}\right)^{1/3}$
Maximum and mean contact pressures	$p_o = \frac{4}{\pi}p_m = \left(\frac{P'E^*}{\pi R}\right)^{1/3}$	$p_0 = \frac{3}{2}p_m = \left(\frac{6WE^{*2}}{\pi^3 R^2}\right)^{1/3}$	$p_0 = \frac{3p_m}{2} = \left(\frac{6WE^{*2}}{\pi^3 R_x R_y}\right)^{1/3}$
Load or load/unit length	$P' = 2ap_m$	$W = \pi a^2 p_m$	$W = \pi ab p_m$
Contact center deflection	$\delta = \frac{P'}{\pi E^*}\left[\ln\left(\frac{L^2 \pi E^*}{2RP'}\right) + 1\right]$	$\delta = \frac{\pi p_0 a}{2E^*} = \left(\frac{9W^2}{16E^{*2}R}\right)^{1/3}$	$\delta = \frac{1}{2}\left(\frac{9W^2}{2E^{*2}\sqrt{R_x R_y}}\right)^{1/3}$
Maximum shear stress	$0.3p_0, 0.78a$ below surface on contact center line	$0.31p_0, 0.48a$ below surface on contact center line	
Elastic deformation limit	$(p_0)_Y = 3.3k = 1.67Y = 0.6H$	$(p_0)_Y = 3.2k = 1.6Y = 0.6H$	

Appendices

Table Appendix 3.1 displays the complete expressions for the elastic contact problem relationships we have discussed in this chapter.

For elliptical contacts[§],

$$\frac{1}{R_x} = \left(\frac{1}{R_{x1}} + \frac{1}{R_{x2}}\right), \frac{1}{R_y} = \left(\frac{1}{R_{y1}} + \frac{1}{R_{y2}}\right), b/a \approx (R_y/R_x)^{2/3},$$

R_x and R_y are respectively the reduced relative radii of curvature in the xz and yz planes of the contact. Note that for elliptical contacts in Table Appendix 3.1, the expressions for a, b, p_0 and δ, are only accurate if $b/a \leq 5$. For larger values see Ref. 3.

References

1. Hertz, H. *Miscellaneous Papers by H. Hertz.* Macmillan, London (1896).
2. Johnson, K. L. *Contact Mechanics.* Cambridge University Press (1985).
3. Gohar, R. 'Elastohydrodynamics' *Imperial College Press* (2001).
4. Sackfield, A. and Hills, D. Some useful results in the classical Hertz contact problem. *Journal of Strain Analysis* **18** (1983) 101.
5. Case, J. and Chilver, A. H. *Strength of Materials* Edward Arnold Ltd. (1959).
6. Arnell, R. D., Davies, P. B., Halling, J. and Whomes, T. W. *Tribology Principles and Design.* Macmillan (1991).
7. Greenwood, T. A. and Tripp, J. H. *The Elastic Contact of Rough Surfaces. Trans ASME Series E* (1967) 417.
8. Greenwood, J. A. and Williamson, J. B. P. Contact between nominally flat surfaces. *Proc. R. Soc. A* **24** (1996) 300.

[§]All the radii of curvature are considered positive here.

CHAPTER 4

DRY FRICTION AND WEAR

4.1. Introduction

Friction is the tangential resistance to motion, or impending motion, between two contacting solid bodies. The direction of the friction force is parallel to the tangential component of the external force vector applied to one of the bodies. When there is relative motion, the resultant friction force is always along the relative tangential velocity vector between the two bodies. The occurrence of solid body friction is a part of everyday life. For instance, all mammals need it in order to walk, for without it they would slip. On the other hand, as discussed in Chapter 1, in many applications there is a need to reduce friction to a minimum.

The type of friction we will discuss in this chapter is called **dry friction**, meaning that there is no coherent liquid or gas lubricant film between the two solid body surfaces. These can be two rough plain surfaces or, for example, the contact between an elastic ball loaded against a plane. In 1699 Amontons discovered empirically, two important laws of friction.

The laws are:

(1) The friction force is independent of the nominal (apparent) area of contact between two solid bodies, such as a rectangular block sliding on a plane.

Thus, whatever face of different nominal area of one of the bodies is chosen to be in contact with the other body, the same friction force would result.

(2) The friction force is directly proportional to the normal component of the load, so that if it is doubled the force of friction is also doubled. Amontons expressed his second law mathematically as:

$$F = \mu W, \tag{4.1}$$

where W is the normal component of the load and μ, the constant of proportionality, is called the **coefficient of friction**. Let us see if we can justify these laws from what we have derived in Chapters 2 and 3 by assuming that:

(a) During sliding, if the friction force per unit area of contact, f, is constant and the *real contact* area is A, then clearly:

$$F = fA. \tag{4.2}$$

(b) The real contact area is proportional to the load, that is:

$$A = qW. \tag{4.3}$$

Eliminating A from Eqs. (4.1) and (4.2):

$$F = (fq)W. \tag{4.4}$$

Assumption (a) is true, because we have shown in Chapter 3 that A represents statistically the real contact area.

Again, from Chapter 3, we have shown that assumption (b) is justified for plastically deforming surfaces and elastic ones that are exponential in form, or a combination of both these states. Greenwood and Williamson[1] have shown that this assumption is also reasonably accurate for elastically deforming Gaussian surfaces.

A simple experiment of a block on an inclined plane can confirm law 2 (see Fig. 4.1).

If the plane is gradually tilted until at some measurable angle, θ, the block just starts to move, then:

$$\mu = \frac{F}{W} = \tan\theta. \tag{4.5}$$

Other experimental arrangements that allow measurement of F when there is a continuous relative speed between the surfaces, will give us

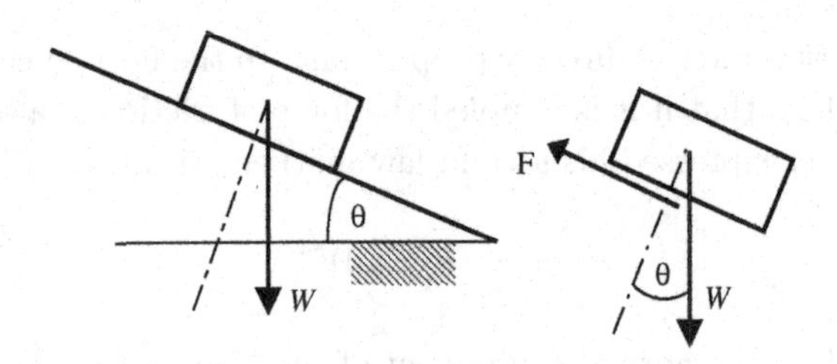

Fig. 4.1. Block and inclined plane experiment.

the *kinetic* coefficient of friction that generally is less than the static coefficient. Experiments carried out in the 18th century by Coulomb confirmed that the same relationship as Eq. (4.1) may be assumed for sliding bodies under relative motion.

(3) This brings us to a third law of friction first proposed by Coulomb in 1785. He stated that friction force is independent of sliding speed. Law (3) applies only approximately to dry surfaces for a reasonable range of moderate sliding speeds. Bhushsan[2] shows some experimental results that confirm this. The law has its uses in certain applications, one of which is suggested below as well as in one of this chapter's questions.

(4) We may also state that if we have two contacting bodies in relative sliding motion, the total friction force vector is in the same direction as their relative velocity vector.

Assuming that laws (3) and (4) are true, can we use them to design a low friction suspension?

The shaft shown in Fig. 4.2 is supported by two sleeves that can be independently driven in opposite senses at speed ω. If the shaft itself rotates and has its maximum speed, $\omega_s \ll \omega$, then the shaft is on a very low friction suspension with what are effectively equal and opposite friction torques applied at each end. This principle is employed in some gyrososcope support (gimbal) bearings.

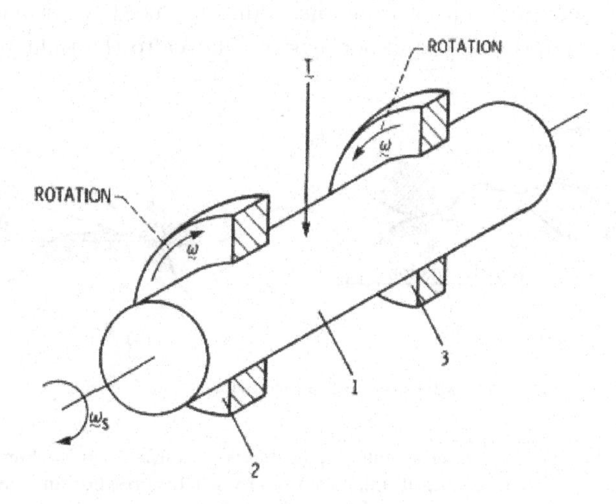

Fig. 4.2. Low friction machine.

4.2. The Basic Mechanisms of Dry Friction

4.2.1. *Adhesion and deformation*

There are two main factors that can contribute to friction between dry contacting surfaces in impending or sliding motion. They are **adhesion** and **deformation.**

Consider as a demonstration of adhesion, Fig. 4.3(a). It shows two rough surfaces with some local asperity contacts. Under normal load, W, there will be contact footprints created, with the tips of some contacting asperities becoming plastic that may cold weld to each other.* If the upper surface then moves to the right, or attempts to relative to the lower surface, there will be a tangential force F_a imposed on that surface.

If this force is sufficient to break these junctions and cause sliding, it becomes the source of the *adhesive* component of friction.

To demonstrate the *deformation* component of friction, Fig. 4.3(b) shows two opposing asperities in *oblique* contact near their summits at a small angle, θ. Let the lower asperity be much softer than the upper one. With load W only applied to it, there will be a contact footprint in the tangent plane at angle θ to the horizontal. With sufficient load, the lower asperity will become plastic and cold weld to the upper one under the intense contact pressure. Again, breaking this weld in the tangent plane by shearing it with force F_a, a component of the total friction force, F(not shown), is the cause of the adhesive friction contribution. But now there is also a deformation component, F_d, of F, which is caused by the **ploughing action** resulting from the additional need to displace a wall of the softer material as the upper asperity moves to the right relative to

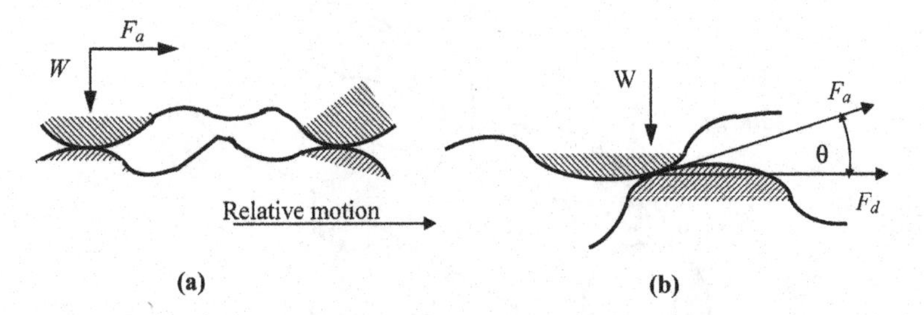

Fig. 4.3. Adhesion and deformation affect friction.

*Adhesion influences the inter-atomic forces between surfaces in contact. There is attraction between them on an atomic scale, increased by pressure and decreased by surface roughness and contaminant films.

the lower asperity. An often less significant form of component F_d, is when the asperities both deform sufficiently to cause one to ride over or past the other by further distortion and no ploughing action.

Assuming, as a simplification, that F_a and F_d do not interact and that angle θ is small, the total friction force is, therefore:

$$F = F_a + F_d. \tag{4.6}$$

For the case of a real surface with many such contacts, the adhesion component is the force needed to shear the welded junctions. Let the average shear strength of these junctions be τ_s over the real contact area, A. Then, the adhesive friction force is:

$$F_a = A\tau_s. \tag{4.7}$$

Furthermore, as in Fig. 4.3, if one of the surfaces is harder than the other, its roughness features will plough a groove through the softer material rather than simply deform the features further, so that we can more accurately state that:

$$F = A\tau_s + F_p. \tag{4.8}$$

We will now deal separately with these two components of friction.

4.2.2. *Adhesive friction*

Bowden and Tabor[3] first developed a comprehensive theory of adhesive friction. A simplified version is given below.

- As the load is being applied, contact is initially only made with some of the higher asperity tips, causing those on the softer material to suffer fully plastic yield.
- Further application of the load causes the total contact area to increase, both from growth of the plastically deforming asperities and from new, lower ones, which are beginning to make contact and are starting to deform elastically.
- Finally, when the loading is completed, the plastic contacting asperities will have cold-welded together, forming strong adhesive bonds. We will consider these, because they are responsible for the adhesive friction.

If the real contact area is A, and p_m is the mean contact pressure, then the load supported is:

$$W = Ap_m. \tag{4.9}$$

The conditions at the cold welded asperities must be close to the fully plastic condition, so assume that p_m equals the hardness of the softer material, H. Thus:

$$W \approx AH. \tag{4.10}$$

Considering Eq. (4.7), if F_a is the force needed to break these bonds, then it has been shown elsewhere[2] that τ_s is very close to the bulk maximum shear stress, k of the softer material, which cannot be exceeded. Hence, in this condition:

$$F_a = Ak. \tag{4.11}$$

The reason for τ_s approaching k in magnitude is because plastic yielding in the junction occurs as a result of a combination of both stresses. We see this in Fig. 3.7(b), where the maximum bulk shear stress has combined with the surface shear stress to become closer to the contact surface.

It follows, from Eqs. (4.10) and (4.11) that:

$$\mu_a = \frac{F_a}{W} \approx \frac{k}{H}. \tag{4.12}$$

Equation (4.12), like many of the previous equations, we have derived, is in terms of material constants. It shows that the **coefficient of adhesive friction** between two rubbing surfaces depends only on their material properties of maximum bulk shear stress and hardness, but not on the load or sliding speed. It, therefore, confirms the Amontons's empirical Eq. (4.1). However, Eq. (4.12) has certain inadequacies:

- As the coefficient of friction we have derived depends on material properties only, we would expect it to be always that of the softer of the two materials. This suggests that whatever the properties of the other material of the pair, the coefficient of friction would not change. However, such behavior does not occur in practice.
- For steel on steel, for example, Eq. (4.12) gives $\mu = 0.16$, which is far lower than what is found in practice, even in a normal atmosphere. This brings us to the next point.

- There is a large difference in the measured coefficient of friction, depending on whether the environment has the contaminants that are normally found in the atmosphere (like oxide films) or is free from them, such as in a vacuum chamber or as occurs in space applications.

On account of these shortcomings found in Eq. (4.12), Bowden and Tabor modified their theory of adhesive friction in the following way: Because they found excessive coefficients of friction between metals in a clean environment, they surmised that the true area of contact determined by the normal load, might in fact become much enlarged under the additional shear force. To explain this, Halling *et al.*[4] assumed that the material pair was composed of a soft, infinitely long asperity with a fully plastic tip, contacting a rigid plane.

Figures 4.4(a) and (b) represents the resultant two dimensional stress system, initially under load W and mean stress $p_m = W/A_0$. In Fig. 4.4(c) a tangential force, F has been applied, causing a shear stress τ_s on the slice *bcde*, anti-clockwise round it to prevent rotation. This situation was explained by Halling using the Mohr's Circle construction for a two dimensional system at its yield point, k. From the geometry of the construction, we can see that:

$$p_m^2 + 4\tau_s^2 = 4k^2. \tag{4.13}$$

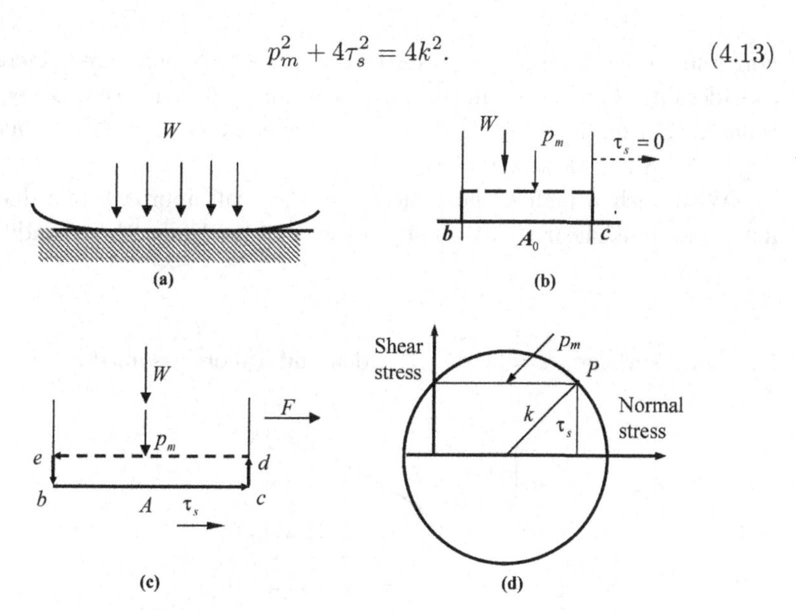

Fig. 4.4. Mean normal and shear stresses. (a) Asperity tip pressed onto a hard surface. (b) Forces on asperity tip. (c) Additional τ_s applied, causing bc to increase to area A. (d) Mohr's Circle for two dimensional stress.

We have shown above that, for asperity junctions under normal load W with full yield at their tips, $A_0 = W/H$, and the maximum bulk shear stress approximately equals k.

Upon application of an additional incremental tangential force, there will be further plastic flow at constant shear stress k, resulting in an incremental increase in the contact area of A. Bowden and Tabor[5] called this increase the **junction growth**. The effect is to reduce the mean pressure to $p_m = W/A$ so that by itself it is insufficient to cause plastic flow at the interface. However, the tangential stress, τ_s on area A, must increase to such a value that the combined effect with p_m maintains the plastic flow.

The increases in A and τ_s are illustrated in Fig. 4.5.

Equation (4.13) has a form based on the **Von Mises Yield Criterion**[4] of:

$$p_m^2 + \alpha\tau_s^2 = \beta^2. \tag{4.14}$$

The constants, α and β, in Eq. (4.14), can be determined by appropriate boundary conditions. These are, with a normal load only, $\tau_s = 0$ and $p_m = H$, making $\beta = H$.

Equation (4.14) then becomes:

$$p_m^2 + \alpha\tau_s^2 = H^2. \tag{4.15}$$

For the other boundary condition, if the junction area increases considerably, because of increasing shear force, it may well exceed the constant normal force. As both act over the same area, $\tau_s \gg p_m$, making $\alpha\tau_s^2 = H^2$ under these conditions.

With such a high surface shear force, τ_s will approach in value the maximum bulk shear stress, k, within the asperity. Under these conditions:

$$\alpha \approx \frac{H^2}{K^2}. \tag{4.16}$$

For metal surfaces, $\alpha = 36$, but Bowden and Tabor[5] assumed a value of 9.

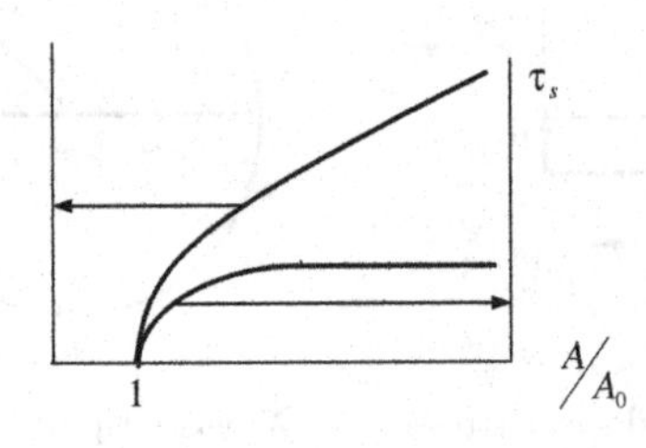

Fig. 4.5. Increase in contact area from a tangential force at the interface.

4.2.2.1. *Influence of contaminant films*

The above analysis applies only to clean surfaces with the junction summits initially deforming plastically in order to create the friction welds. The effect is to cause excessive friction forces that can even exceed the normal forces. In practice, friction forces are generally generated under a normal atmospheric environment that weakens the asperity junctions, because of the contaminant films on the surfaces. These films cause the maximum shear stress achieved there, τ_c, to be less than k, such that $\tau_c = ck$. Thus, when: $F/A = \tau_c$, sliding will occur, signalling the end of any further junction growth. With sliding, from Eqs. (4.15) and (4.16) it follows that:

$$p_m^2 + \tau_c \alpha^2 = \alpha k^2. \tag{4.17}$$

Letting $k = \tau_c/c$ in Eq. (4.17):

$$p_m^2 + \alpha\tau_c^2 = \frac{\alpha\tau_c^2}{c^2}. \tag{4.18}$$

The adhesive friction coefficient can now be written in terms of c as:

$$\mu_a = \frac{\tau_c}{p_m} = \frac{c}{[\alpha(1-c^2)]^{1/2}}. \tag{4.19}$$

If, from Eq. (4.19), μ is plotted against c for different values of $\alpha = (H/K)^2$, we obtain Fig. 4.6.

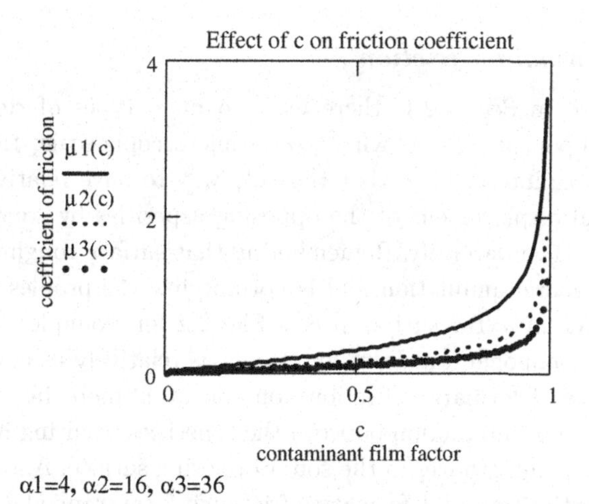

Fig. 4.6. Adhesive friction, effect of c on friction coefficient.

Figure 4.6 shows that for high values of c, such as with clean surfaces, excessive friction coefficients between metals can be expected. However, in normal atmospheres the coefficients are much less. In particular, if c is small, say less than 0.2, $\mu_a \approx ck/H$.

Because $ck = \tau_c$, we can state that in this case:

$$\mu_a = \frac{\text{critical shear stress of the interface}}{\text{plastic yield pressure of the substrate}}.$$

When $c \to 1$, we get Eq. (4.12).

The above relationship is explained by noting that for low values of c, the interfacial shear stress is never large enough for the junction to grow appreciably. Therefore, the true area of the contact is determined by load, W, with the plastic yield pressure here defined by H.

The value of c is also the basis of **boundary lubrication** (see also Chapter 1), where chemical additives are added to lubricants. These are useful as inhibitors of wear, especially in motor vehicles during a cold start process. Moreover, if we deliberately deposit a low shear strength film to separate the surfaces (again defined by c), such as white metal (such as Babbitt, see Chapter 12) on steel in journal bearings, we can create a low friction coefficient between the journal and the bearing bush at start up.

The above approximate analysis has discussed the adhesion component of dry friction. We will now investigate the deformation component.

4.2.3. *Deformation friction*

As mentioned in Sec. 4.2.1, there are two main types of **deformation friction** component. One is where, on a microscopic scale, the asperities on the surfaces interlock, so that the only way to have relative motion is through local displacement of the opposing asperities by squashing them or displacing them laterally. Remembering that surface roughness is really only a very gentle undulation and is nothing like the profiles we see on a distorted scale in a Talysurf scan (see Fig. 2.2 for example), this type of deformation component, at least for metals, is relatively insignificant.

The other deformation friction component is more important. It is from macroscopic interaction between the asperities resulting in the harder material ploughing grooves in the softer opposing surface. Another obvious candidate for this type of deformation friction is from trapped wear particles damaging the softer of the two surfaces. As explained by Fig. 4.2, a ploughing action is the main component of deformation friction with the harder of the two surfaces penetrating the softer material to the scale of the asperities. The resulting resistance to motion is a measure of the ploughing friction, all this occurring with the softer material yielding plastically at their contact.

The principle of the ploughing component is explained mathematically below.

4.2.3.1. *The ploughing component of deformation friction*

Consider the usual spherical asperity ploughing a groove of depth, h, through the softer material. The action is depicted in Fig. 4.7.

At height h: $A_1 = \pi d^2/8A$. In order to find the projected are A_2, of the groove, it is best to use Cartesian coordinates and remember that $h \ll R$, yielding to a first approximation:

$$A_2 = \frac{2hd}{3}. \quad \text{Alternatively } A_2 = \frac{d^3}{12R}.$$

Assuming that plastic yielding of the softer body takes place at hardness H, then: $W = HA_1$ and $F_p = HA_2$, from which the ploughing friction coefficient is found as:

$$\mu_p = \frac{F_p}{W} = \frac{A_2}{A_1} = \frac{16}{3\pi\sqrt{8}}\sqrt{\frac{h}{R}}.$$

Thus:

$$\mu_p = 0.6\sqrt{\frac{h}{R}}. \tag{4.20}$$

As an example, let Eq. (4.20) be applied to a typical spherical surface roughness feature having $R = 100\,\mu\text{m}$ near its tip. If $h = 1\,\mu\text{m}$ deep

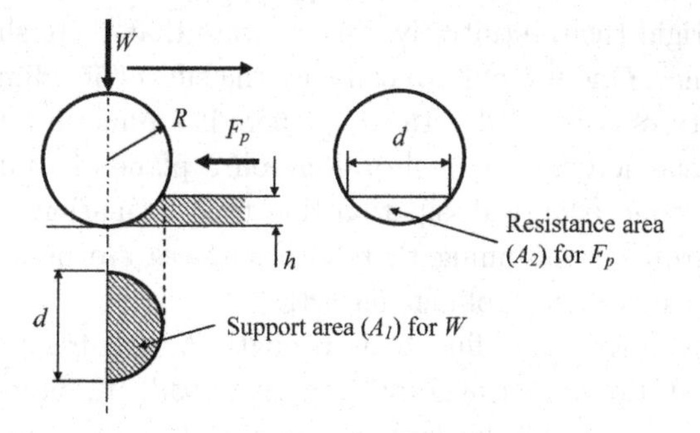

Fig. 4.7. Friction from Ploughing motion by a spherical asperity.

due to the ploughing action, then: $\mu_p = 0.06$. The corresponding adhesive component can be found approximately from Fig. 4.4. If $\alpha = 4$ and $c = 0.5$, then $\mu_a = 0.4$. Thus, μ_a is greater than μ_p. This is generally the case on a surface roughness scale, especially for hard contacting metals. The relative values may, however, be different if the roughness slope is high.

We see that compared with the adhesive component of friction, the ploughing component is relatively small, but can still be significant. Moreover, this approximate theory has neglected the additional resistance resulting from the piling up of the softer material ahead of the hard asperity. More sophisticated theories, discussed by Halling *et al.*,[4] include the interaction between the adhesion and ploughing components. Clearly, the ploughing component also results in wear of the softer surface, a topic we will discuss later.

All the above analysis of friction forces has covered the obvious effects of an incipient or complete sliding action between contacting rough surfaces. Below we will investigate a more occult type of friction, which is when the surfaces nominally have a mutual rolling action between them.

4.2.4. *Elastic rolling friction*

Because of its low value compared with that for sliding, **rolling friction** is a desirable feature in all sorts of machinery, the wheel being the most obvious design (see discussion of sliding coefficient of friction versus rolling resistance in Chapter 1). One of the most widely used low friction mechanisms is the rolling element bearing. We will have reason to discuss fluid film lubricated rolling contacts in Chapter 10.

In Chapter 3 we analyzed the Hertzian elastic contact behavior of spheres and rollers. Below, we will start by studying the tangential frictional forces when cylinders are in free rolling motion. If the contacting bodies are perfectly rigid then, intuitively, we can state that there should be zero rolling resistance. One has only to compare the efforts of riding a bike with fully inflated tyres compared with when there is a puncture. On the other hand, for an elastic cylinder freely rolling on a plane of similar material, there will be some energy dissipation due to deformation losses. Let us analyze the problem, assuming that both surfaces are perfectly smooth. Figure 4.8 explains such a rolling contact.

Let the cylinder, of radius R and length L, roll freely from left to right over the stationary surface with a normal load, W, on it. Considering only the forward part of the Hertzian footprint, that is to the right of

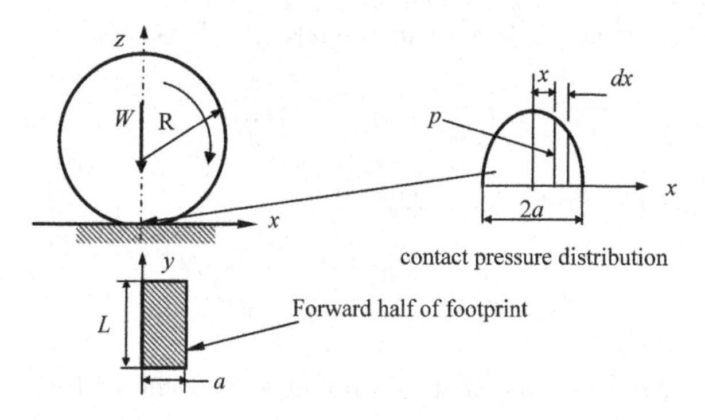

Fig. 4.8. Elastic rolling friction.

the y-axis, the resisting moment exerted from this part due to its forward compression is:

$$M = L \int_0^a pxdx. \tag{4.21}$$

Equation (4.21) can be expanded by using the elastic expressions below from Table 3.1 for a line contact:

$$p = p_0 \left(1 - \frac{x^2}{a^2}\right)^{1/2}, \quad p_o = \left(\frac{P'E^*}{\pi R}\right)^{1/2}, \quad a = \left(\frac{4P'R}{\pi E^*}\right)^{1/2}.$$

With these substitutions and remembering that $P' = \frac{W}{L}$, and integrating Eq. (4.21):

$$M = \frac{4}{3\pi^{3/2}} \left(\frac{W^3 R}{E^* L}\right)^{1/2}. \tag{4.22}$$

We can now investigate the work done, when the cylinder rolls freely a distance S, while turning through an angle θ. The elastic work done by the compression is idealy $M\theta$ and, for no slip, $S = R\theta$. A characteristic of elastic bodies being compressed and relaxed is that there is a nett loss of energy by internal friction, called **hysteresis,** discusssed by Halling in Ref. 6. This means that for the trailing part of the contact zone, not all the work done in the forward part of the contact is recovered. Therefore, let the actual work done against hysteresis be E$M\theta$, where E represents this fractional loss of energy. Furthemore, if the resisting force on the cylinder is F, then the actual work done also equals FS. Thus:

$$FS = \text{E}M\theta = \text{E}M\frac{S}{R}.$$

Substituting for M from Eq. (4.22):

$$\mu = \frac{F}{W} = \frac{4\text{E}}{3\pi^{3/2}}\left(\frac{W}{E^*RL}\right)^{1/2}. \tag{4.23}$$

Equation (4.23) shows us that the rolling coefficient of friction, caused by the cylinder deforming elastically, increases with load and decreasing radius. Having small wheels is one of the disadvantages of folding bikes or of adults riding child scooters. The coefficient of friction also increases with soft materials or a short roller length. For hard steel spherical surfaces in elastic contact, E is about 0.02. If we assume a similar E value for rollers, we can estimate a value for μ. With lubricated line contacts, to be discussed later in the book, the dimensionless load group, in the brackets of Eq. (4.23), is typically 10^{-6}. Using E $=$ 0.02, μ $=$ 5.65E-6, which is negligible. Much higher values of rolling coefficient of friction occur when W is sufficient to cause sub-surface plastic deformation. In that case there is substantially more energy dissipated than is caused by elastic hysteresis only. If such rolling contacts are repetitive, work hardening results, with the plastic deformation becoming elastic again. Such behavior is associated with the *shakedown process*.[7,8]

Let us now assume the roller has additionally a tractive force applied to it.

4.2.5. *Tractive rolling of an elastic cylinder*

Figure 4.9 shows a loaded cylindrical elastic wheel rolling on an elastic plane, of the same material, at speed V. In order to drive the wheel, a tractive reaction force, T, is exerted by the plane on the wheel at the contact footprint. The way it acts is that, if there is slip over the whole of the footprint, T equals the limiting friction μW, causing any rolling motion to cease. Such a situation sometimes occurs if the driving wheels of a car are failing to move it off an icy surface. At the other extreme, if $T = 0$, there is no slip anywhere over the footprint so that only free (pure) rolling can occur, for example when a car is rolling steadily with its gears in neutral. This condition has been discussed above in Sec. 4.3.1. The

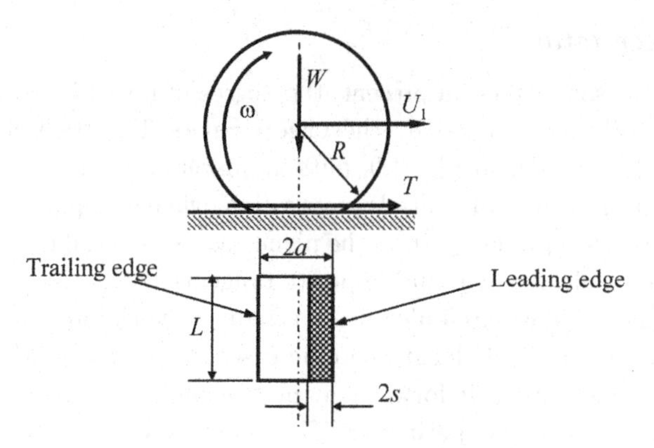

Fig. 4.9. Tractive rolling of a cylinder.

general tractive condition is between these two extremes, that is, in a part of the contact footprint, there is some slight slipping between the surfaces (called **microslip** or **creep**) and in the other part (called the locked or **stick** region) there is pure rolling.

Johnson[8] suggests that such behavior is similar to the creep of an elastic belt between a driving and a driven pulley. There, the velocity on the belt's tight-side exceeds that on its slack-side, with the frictional traction pulling the belt forward on the driving pulley, while dragging it back on the driven pulley. In order to convert the tight-side speed to the slack-side speed, there is a stick arc on the driving pulley, followed by a slip arc that reduces the belt speed to that of the slack-side. A similar situation occurs on the driven pulley to convert the belt speed back up to that of the tight-side.

As indicated in Fig. 4.9, it can be shown that the *stick* region starts at the leading edge of the contact footprint, with the slip region commencing at distance, $2s$, from there and continuing up to the trailing edge at distance $2a$. How the boundary at $2s$ varies with T, is given by Johnson as:

$$\frac{T}{\mu W} = 1 - \left(\frac{2s}{2a}\right)^2. \tag{4.24}$$

Thus, when $2s = 2a$, that is there is sticking over the whole footprint, $W = 0$ and no traction is being transmitted. When $2s = 0$, that is there is slipping over the whole footprint, $T = \mu W$, gross sliding occurring all over from the leading to the trailing edges.

4.2.6. *Creep ratio*

When there is some traction present, the degree of microslip over a rolling contact footprint, is defined by the **creep ratio**. The traction force, T, applied to the cylinder in Fig. 4.9, puts its material in the no slip zone of the footprint, into circumferential compression, while an equal and opposite force on the corresponding part on the plane has its material in tension. The effect of the resulting equal and opposite tangential strains is to make the cylinder behave as if its circumference was reduced, with the contacting part of the plane, as though its length had increased. These geometrical changes cause the cylinder to roll forward in one revolution a distance slightly less than its undeformed perimeter. The creep ratio, ξ, is defined as the fractional difference between the periferal velocity of the cylinder, caused by its reduction in length, and its forward speed. The relation between T and ξ, for contacting cylinders of the same material, is given by Johnson[8] as:

$$\frac{\xi R}{\mu a} = 1 - \left(1 - \frac{T}{\mu W}\right)^{1/2}. \tag{4.25}$$

Figure 4.10 plots Eq. (4.25) and shows how the regions of stick and slip vary with T.

4.2.7. *Other examples of rolling motion*

There is insufficient space in this elementary text book to cover other examples of rolling motion behavior. These include situations where the contacting bodies have different Young's moduli such as a rubber tyre

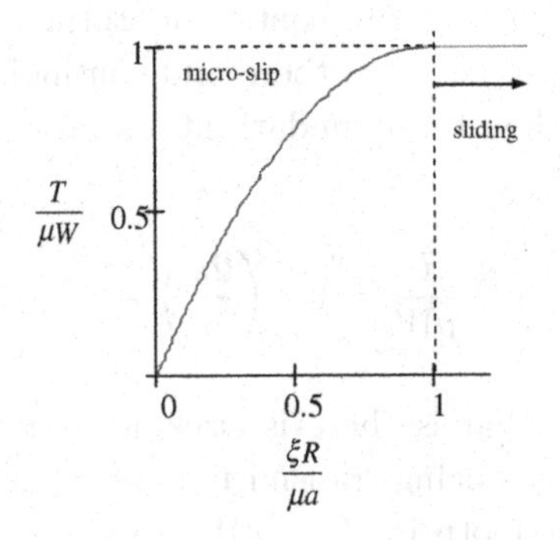

Fig. 4.10. Regions of stick and slip.

rolling and sliding on hard surfaced road (see Mavros *et al.*[9]), or when a ball is rolling in a conforming groove. In both examples there are quite complicated zones of stick and slip within the contact footprint that will affect the performance. For a discussion of these examples and further references, see Johnson.[10]

4.3. Thermal Effects of a Frictional Temperature Rise in Concentrated Contacts

A frictional temperature rise can be predicted for large smooth surfaced bodies in sliding contact or, on a small scale, between opposing surface features. Figure 4.11 shows an elastically distorted spherical surface sliding on smooth plane moving at speed, U. Energy is dissipated due to friction in the form of heat, with the highest temperatures occurring at the actual contact area.

In the case of surface features coming in and out of contact, the rise in temperature is usually limited to the asperity tips, with the local values well above the bulk temperatures of the bodies, but nevertheless of short duration. For this reason the phenomenon is called **temperature flash**, the faster of the bodies being the heat source for the slower body and vice versa. In the example below, we require the mean contact temperature θ_m.

Archard[11] has obtained an approximate solution for the mean temperature rise in a spherical contact, above the temperature of a point far from the contact area, as:

$$\theta_m = \left(\frac{0.31\mu WU}{ka}\right)\left(\sqrt{\frac{k}{\rho c_p U a}}\right). \tag{4.31}$$

In Eq. (4.31) k = thermal conductivity, ρ = density, c_p = specific heat, μ = coefficient of friction and W = load.

If the dimensionless speed is defined as: $\beta^* = \frac{Ua\rho c_p}{2k}$, then Eq. (4.31) only applies when $\beta^* > 5$. For example, let the following data apply

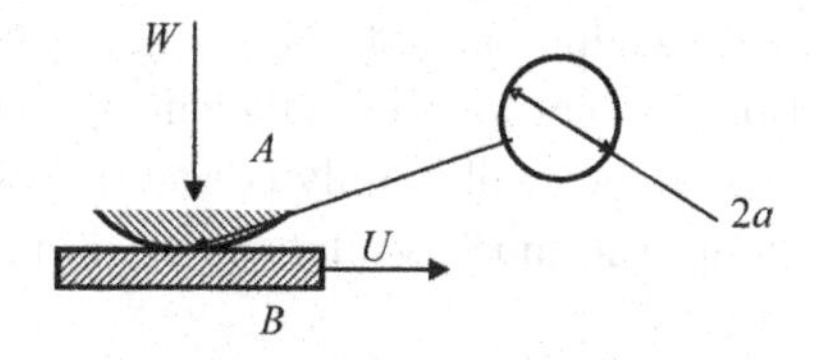

Fig. 4.11. Temperature rise between contacting bodies.

to body A:

$$R = 0.0127\,\text{m}, \quad \rho = 7850\,\text{kgm}^{-3}, \quad k = 52\,\text{Jm}^{-1}\text{s}^{-1\circ}\text{K}^{-1},$$

$$c_p = 460\,\text{m}^2\text{Kg}^{-1\circ}\text{K}^{-1}, \quad U = 1\,\text{ms}^{-1}, \quad W = 500\,\text{N},$$

$$\mu = 0.1, \quad E^* = 113\,\text{GNm}^{-2}$$

From this data, we find that the Hertzian contact radius is $a = (\frac{6WE^{*2}}{\pi^3 R^2})^{1/3} = 3.47 \times 10^{-4}\,\text{m}$.

Hence, the dimensionless speed is $\beta^* = 12.05$ so that, from Eq. (4.31), the mean temperature rise above ambient is $\theta_m = 175°\text{C}$, a very high value but nevertheless confined to the contact region and just beneath it.

4.4. Wear of Surfaces

In this section we will consider the visible consequences of friction on surfaces, which is called **Wear**. Wear in its widest sense is a part of the natural world. For example, wind erosion can alter parts of our landscape over the long term. Or when considering the contact of solids, one must sometimes buy a new pair of shoes, because the soles of the old ones have worn out, unless fashion only has dictated the need for a new purchase. Thus, there can be considerable economic involvement in the wear process.

The type of wear we will consider is mainly caused by the progressive loss of surface material from relative tangential motion at the contact between two solid bodies. It arises from the friction forces created by the resultant interaction of these bodies' surface roughness features. As engineers or scientists we are often striving to minimize wear, for example, with respect to the moving parts in a car engine. We do this by employing a liquid lubricant to separate the surfaces and hence reduce friction, even though some early wear is acceptable as a part of the engine running in process† (also see the Stribeck curve in Chapter 1). On the other hand, in the case of the car brakes, we expect their shoes to wear and be replaced more frequently than the engine parts, because the brake shoes depend on the high contact frictional forces for their efficient operation.

As we have done so far, we will mainly concentrate on the mechanical wear of metals, this being the most commonly used engineering material.

†In modern automobiles, the need to run in the engine is far less nowadays.

Because wear is a complex subject that depends heavily on experimental results, we will confine ourselves to simple theories that explain the physics of the process. The three main categories of wear are: **adhesive wear, abrasive wear** and **fatigue wear**, all these being directly related to friction. Other categories and types of wear will be listed with adequate references.

4.4.1. *Adhesive wear*

The basic theory of adhesive wear of dry surfaces was originally presented by Bowden and Tabor[3] and developed further by Archard.[11] It follows logically from the adhesive theory of friction that was covered in Sec. 4.2.1. Let us assume that the contact area between the bodies is composed of hemispherical asperities, with a normal load sufficient for some of the contacting tips on the softer material to have become plastic and hence to have cold welded onto those on the harder surface. Consider a single pair of these asperities, with a footprint of radius a, as shown in Fig. 4.12(a).

For sliding to occur under a tangential force, the asperity junction must be broken, either at the contact interface, or within the softer asperity. If the weld is sufficiently weak, shearing will take place at that interface, with the resulting wear being sensibly zero. This fact immediately suggests that the presence of an oxidized surface layer at the interface reduces friction, so weakening the weld (defined by the factor c in Sec. 4.2.2.1) and thereby guaranteeing that the break would occur there. On the other hand, if failure occurs within the bulk of the softer asperity, because of the high weld interface strength, a fragment might be plucked out and either carried away attached to the harder material or, by subsequent rubbing, become detached as debris. Because the region of full plasticity under the footprint in the softer material is approximately a hemisphere of radius a, shown shaded in Fig. 4.12(a), this is chosen to define the detached shape. It is

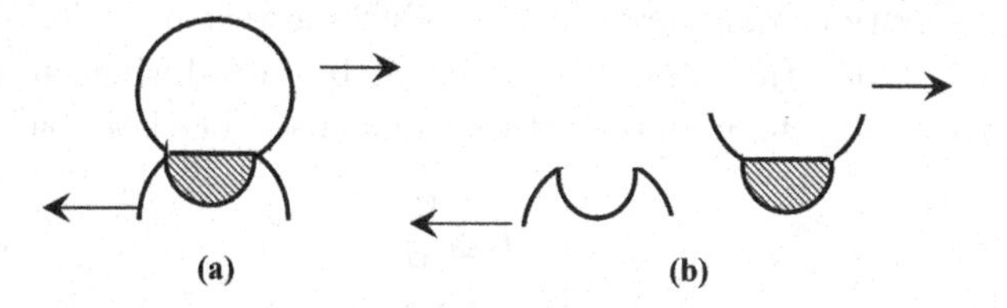

Fig. 4.12. Principle of adhesive wear.

the accumulation of such particles, taken from the worn surface, which is a measure of adhesive wear. Let us describe this process by a simple mathematical model.

The contact area at a single asperity is πa^2 and, assuming ideal elastic-plastic conditions with full plastic deformation, each asperity contact will support a load of $\delta W = \pi a^2 H$. The wear volume taken from an asperity is $\delta V = 2\pi a^3/3$, so that the volume produced by that asperity per unit sliding distance is:

$$\delta Q_{sl} = \frac{\delta V}{2a} = \frac{(2\pi a^3/3)}{2a} = \frac{\pi a^2}{3}.$$

Therefore, the total volume produced per unit sliding distance, for n asperities, is $Q_{sl} = n\pi a^2/3$, where Q_{sl} has units $1/(\text{length})^2$.

If total load, $W = n\delta W$, the total volume produced from these n asperities per unit sliding distance is:

$$Q_{sl} = \frac{W}{3H}. \tag{4.26}$$

Equation (4.26) enables us to define the laws of adhesive wear as:

The volume of worn material is proportional to:

1. The sliding distance.
2. The load.
3. Inversely proportional to the softer material hardness.

These laws have shortcomings:

The first and third laws have been confirmed experimentally over a wide range of conditions. The second law only applies at very low loads, after which a dramatic increase in wear rate is observed The transition occurs approximately when the mean pressure reaches $H/3$, this equalling the initiation of plastic yield within the softer asperity under normal stress only (see Sec. 3.4.2).

Another important point is that, if the interface is able to shear, not every asperity contact necessarily produces a wear particle. Archard[11] therefore reconciled theory with experiment by introducing an empirical wear coefficient K_{ad}, to account for these anomalies. The law now becomes:

$$Q_{sl} = K_{ad}\frac{W}{H}, \tag{4.27}$$

where: $K_{ad} \ll 1$.

Table 4.1. Wear coefficients.

Lubricant	Coefficient of friction (μ)	K_{ad}
Dry argon	0.5	10^{-2}
Dry air	0.4	10^{-3}
Gasoline	0.3	10^{-5}
Mineral oil (no additives)	0.12	10^{-7}
Engine oil	0.08	10^{-10}

Archard obtained typical values of K_{ad}, for iron on iron. These are given in Table 4.1, taken from lecture notes by Professor H. A. Spikes (Imperial College, London).

Note that in cases where oil is present μ has reduced, making K_{ad} extremely low. We will see later in the book that the lubricants can effectively separate the roughness features with a coherent liquid viscous film, thus reducing the tangential stress and hence the adhesive wear rate to nearly zero.

4.4.2. *Abrasive wear*

Abrasive wear, the second type in our classification, can be far more severe than adhesive wear if the penetration is sufficient. It is caused by a ploughing action on a softer surface by harder surface features, or by third bodies. The mechanism of ploughing friction for a spherical asperity has already been discussed in Sec. 4.2.3.1, but the ploughing wear produced by a spherical asperity is relatively low. A far more significant effect is obtained from a hard sharp feature ploughing through a softer surface. This action is called **two-body abrasion**. If the wear is instead caused by rough hard particles (debris), trapped between the two sliding surfaces, it is called **three-body abrasion**. A simple model to explain the principle of abrasive wear is a single conical asperity (like a pin tip used in some wear experiments). One is shown cutting its triangular sectioned groove in Fig. 4.13.

From the geometry of the figure, the volume of material cut out by the cone over length L is:

$$V = \frac{d^2 L}{4} \cot \theta. \tag{4.28}$$

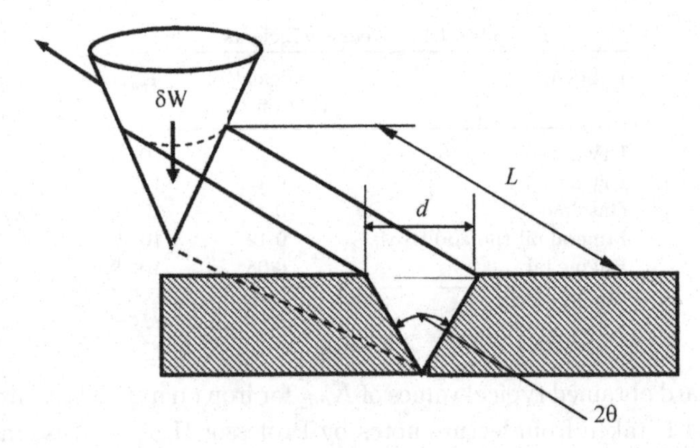

Fig. 4.13. Ploughing wear.

As the material is fully plastic and remembering that only half the cone's horizontal cross-sectional area is exposed, the mean pressure under the cone due to load δW is:

$$H = \frac{8(\delta W)}{\pi d^2}. \tag{4.29}$$

If $W = n(\delta W)$, k_f accounts for the events that fail to produce a wear particle, and $K_a = \frac{2k_f \cot\theta}{\pi}$ additionally defines the asperity geometry, from Eqs. (4.28) and (4.29), the volume per unit length cut out by n of these asperities is:

$$Q = \frac{2k_f n(\delta W)\cot\theta}{\pi H} = \frac{K_a W}{H}. \tag{4.30}$$

Like the coefficient in Eq. (4.27), K_a must be determined empirically. Note also that both equations take the same form. Other types of wear, that can be of low level but potentially serious, occur where there are repeated cycles of relative movement between the contacting bodies. These are discussed in the next section.

4.4.3. *Macroscopic fatigue wear*

There can be indications of wear when points on the surfaces of bodies, in rolling contact, are being loaded and unloaded cyclically. The ball bearing is a prime example, because all points on its contacting surfaces are repeatedly being loaded and unloaded as the elements rotate. In Sec. 4.2.4 the rolling contact of an elastically homogeneous and smooth cylinder on a smooth plane was studied. There, the elements, below the leading part of the footprint, are compressed elastically followed by their elastic recovery in the trailing part. In this ideal situation, no wear of the bodies should result, but in practice there may be wear after a large number of cycles, because no material is entirely homogeneous. This type of wear is called **macroscopic fatigue wear**. Even if the surfaces are exceedingly smooth, and the contacting bodies are operating below their yield strengths, signs of wear may eventually appear. The causes of this type of wear are limited to the hinterland immediately on and below the contact footprint, reaching its semi-minor axis in the case of a ball bearing and eventually showing signs of visible wear on the races. This wear is often caused by pre-existing inclusions and micro-cracks located near the point of maximum shear stress originating from the production process. Due to the cyclic loading, these faults grow, eventually breaking out onto the contact surfaces after a large number of cycles, where they appear as cracks and debris. Fatigue failure of the bearing is then deemed to have occurred and its life is pronounced over. The statistical method of forecasting this important event will be covered in Chapter 11.

4.4.4. *Microscopic sliding fatigue wear*

The other type of fatigue wear is called **microscopic sliding fatigue**. It occurs whenever there is some relative sliding between the rough surfaces of the two contacting bodies, causing the individual asperities to suffer repeated sub-surface shear stresses, even though the total load is apparently low enough to make their contact conditions nominally below their yield stress. This type of fatigue wear is limited to the region of the asperities themselves, starting with *pre-existing tiny cracks* within them. A practical example, where there might be microscopic fatigue wear, is a pair of meshing gear teeth that repeatedly come in and out of sliding/rolling contact during their rotation. One visible sign of this wear are the appearance of tiny depressions on their flanks called **micro-pits**.

The difference with rolling contact fatigue wear, discussed in Sec. 4.4.3, and in Chapter 11, are that with sliding fatigue wear, the asperities themselves are in fully plastic contact, with the welds joining them being broken and reformed under continuous sliding. This repeated action is a cause of crack formation and eventual failure. The cyclic nature of the process is demonstrated in Fig. 4.14.

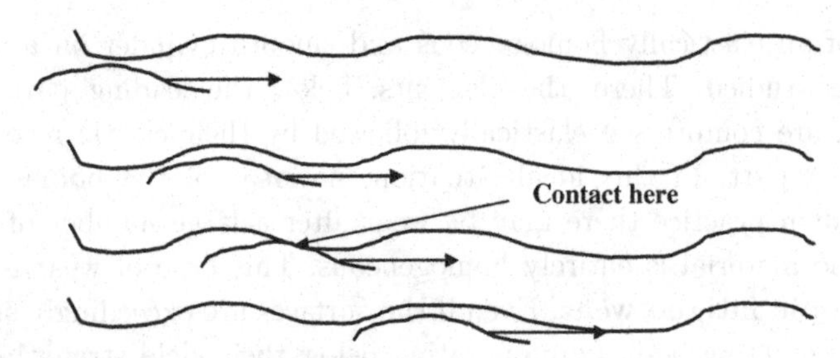

Fig. 4.14. Cyclic nature of fatigue wear (The faster lower surface is repeatedly contacting the upper surface asperities).

4.4.5. *Corrosive wear*

Chemically reacted surface layers can be friction reducing films of material, put down deliberately, or they can be formed as a result of contact with oxygen from the air (see also Chapter 1). Air is the most common corrosive medium, causing a chemically reactive surface layer to be formed over an initially clean metal surface (rust). Corrosive wear is the removal of this layer by the friction process. A corrosive environment can be dangerous for steel structures, such as off-shore platforms. They may weaken and form cracks under the combined action of wave impact and the intrusive effects of corrosion in the structure near the water line. In the case of sliding surfaces, an oxide film can be beneficial, because it weakens the welds being created between the plastically deforming asperity contacts. However, the process is complicated by the worn surfaces not having time to re-oxidize before the following asperity impacts occur, causing the wear rate to rise. Furthermore, the presence of oxidation at crack tips, just below the surfaces, tends to increase the rate of fatigue wear. The best solution is to try to avoid corrosive wear by applying the correct type and balance of additives that will maintain protection of the surfaces.

4.4.6. *Fretting corrosion*

Fretting corrosion is fatigue wear caused by cyclic relative tangential motion of tiny amplitude, between two nominally static metal surfaces in a medium such as air. It can occur, for example, at sites where there are insufficiently tightened nuts and bolts holding down components in a vibrating environment. Fretting corrosion creates trapped wear debris between the contact faces and if there are ferrous components present, the debris consists of various unstable oxides of iron, colored red or black. Such observations form a part of the forensic science of diagnosing the causes of industrial accidents.

4.5. Closure

This chapter has discussed the various forms of friction and wear between nominally dry contacting surfaces. It has shown that using simplified models the principles of friction and wear behavior can be explained. In the following chapters of this book, we will consider various aspects of fluid film lubrication.

References

1. Greenwood, J. A. and Williamson, J. B. P. Contact of nominally flat surfaces. *Proc. R. Soc. A* **24** (1996) 300.
2. Bhushsan, B. *Principles and Applications of Tribology.* John Wiley and Sons (1999).
3. Bowden, F. P. and Tabor, D. *The Friction and Lubrication of Solids Pt 1.* Clarendon Press (1954).
4. Arnell, R. D., Davies, P. B., Halling, J. and Whomes, T. W. *Tribology Principles and Design.* Macmillan (1991).
5. Bowden, F. P. and Tabor, D. *The Friction and Lubrication of Solids Pt 2.* Oxford University Press (1964).
6. Halling, J. *Introduction to Tribology.* Wykham Publications (London) Ltd (1976).
7. Harris, T. A. *Rolling Bearing Analysis.* John Wiley and Sons (1991).
8. Johnson, K. L. *Contact Mechanics.* Cambridge University Press (1985).
9. Mavros, G., Rahnejat, H. and King, P. Analysis of handling properties of a tyre based on the coupling of a flexible carcass-belt model with a separate tread incorporating transient-viscoelastic frictional properties. *Vehicle System Dynamics* **43** (2005) 199–208.
10. Johnson, K. L. 100 years of Hertz contact. *Proc. I. Mech. E.* **196** (1982) 363–378.
11. Archard, J. F. Contact and rubbing of flat surfaces. *J. Appl. Phys.* **24** (1953) 981–988.

CHAPTER 5

LUBRICANT PROPERTIES

5.1. Introduction

We have seen so far that very high friction forces and wear result when contacting clean solid surfaces are made to slide relative to each other. On the other hand, both friction and wear can be considerably reduced in a normal atmosphere because of the thin contaminant low shear stress film that forms between the surfaces.

Are there more effective ways of further reducing friction and wear? For most of the remaining part of the book we will deal with the mechanics of **viscous film lubrication**, which enables the surfaces to be separated by a thin fluid film, thus reducing the untoward effects of friction and wear.

Relative motion between two elements of a fluid is retarded by intermolecular interactions caused by **viscosity**, a real fluid property. This property is the most essential feature of lubricants, as it enables them, given the right conditions, to separate two solid bodies in relative motion.

5.2. Dynamic Viscosity

Consider an elastic element of solid of unit face-width and thickness dz, with a shear stress, τ, put on its opposing faces, as in Fig. 5.1(a).

The resulting movement of the top face relative to the base yields the relationship $G = \frac{\text{shear stress}}{\text{shear strain}} = \frac{\tau}{dx/dz}$, where G is the shear modulus of the material.

Now, let us consider a thin layer of oil in one-dimensional flow. From Fig. 5.1(b), if du = velocity difference between its top and bottom surfaces *of the layer*, then du/dz = velocity gradient across it. This is called **shear strain rate** or strain rate. To clarify strain rate further, as $u = dx/dt$:

$$du/dz = \frac{d}{dz}\left(\frac{dx}{dt}\right) = \frac{d}{dt}\left(\frac{dx}{dz}\right) = \frac{d\,(\text{shear strain})}{dt} = \text{strain rate.}$$

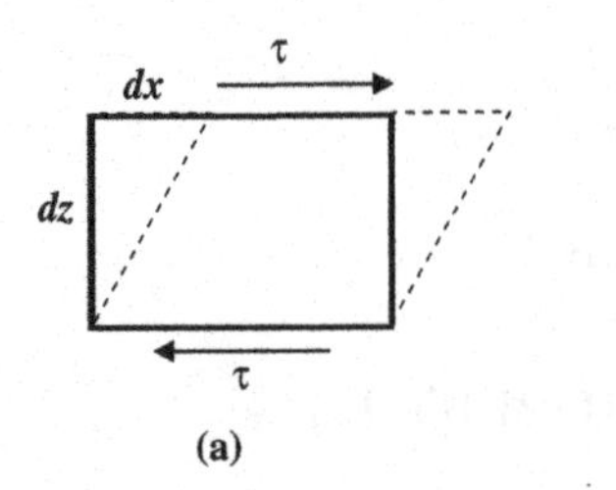

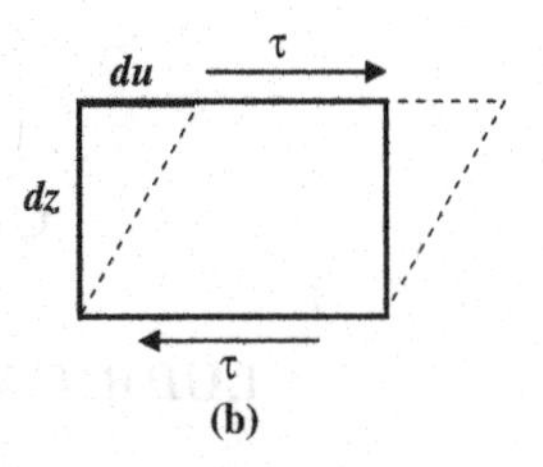

Fig. 5.1.

Thus, for fluids, shear strain is replaced by shear strain rate, du/dz, while G is replaced by the **dynamic viscosity**, η, of the fluid giving:

$$\eta = \frac{\text{shear stress}}{\text{shear strain rate}},$$

or

$$\eta = \frac{\tau}{(du/dz)}. \tag{5.1}$$

Equation (5.1) defines dynamic viscosity of a **Newtonian fluid**. An important property of a Newtonian fluid is that it must take the same velocity as any solid boundary it contacts. The units of dynamic viscosity in terms of (force, length, time) are:

$$\frac{(F/L^2)L}{(L/T)} = \frac{FT}{L^2}.$$

In the SI units system, dynamic viscosity is in: Newtons per square meter seconds ($\frac{N}{\text{m}^2}s$), or Pascal seconds (Pas).

In SI units in terms of mass, length and time, dynamic viscosity is $\frac{M}{LT} = \frac{\text{Kg}}{\text{m. sec}}$. This linear relationship between stress and shear rate, given by Eq. (5.1), is analogous to the Hook's law in solid body mechanics. Thus, Newtonian fluids are called as such after Isaac Newton who defined them.

Other frequently quoted units of dynamic viscosity, follow the cgs system. To define the unit of viscosity in this system, we can state that:

- If the shear stress is 1dyne/cm^2, and the velocity gradient, du/dz, is $1\,\text{sec}^{-1}$, then the absolute unit of viscosity is 1 *poise* (P).

The conversion from CGS to SI units of dynamic viscosity is obtained from:

- $1\ P = 0.1\text{Pas}$

Sometimes the *centipoise* (cP) is used:

- $1cP = 10^{-3}$Pas

Another frequently quoted unit of viscosity is the **kinematic viscosity,** ν. If ρ is the lubricant density, then:

$$\nu = \frac{\eta}{\rho}, \tag{5.2}$$

ρ (dimensions of M/L^3), is usually about 850 kg/m^3 (0.85 gm/cm^3). The dimensions of ν are therefore: L^2/T. In SI units they are (m^2/s), but are rarely quoted as such in the commercial literature.

In the cgs system, the unit of ν is the *Stoke* (S) in (cm^2/s). More often, the *Centistoke* (cSt) in (mm^2/s) is used. [Remember, 1(S) = 100(cSt)]

Finally, converting ν in cSt to the SI system:

- $\left(\frac{\text{m}^2}{\text{s}}\right) = \text{cSt} \times 10^{-6}$

The value of ν for a liquid lubricant (identified here as an oil) is determined experimentally with a **viscometer**. Traditionally, a common type measures the time the lubricant takes to travel under its own weight down a capillary tube of diameter appropriate to the viscosity of the lubricant being tested. Measurements are given as flow times. One type of viscometer gives the flow time, 't', in '**Saybolt Universal Seconds**' (SUS). The formulae below then expresses, according to the SUS range, the kinematic viscosity in cS directly:

$$\begin{aligned}
\nu &= 0.224t - 185/t && \text{for } 34\theta < 115\ SUS \\
\nu &= 0.233\theta - 1.55 && \text{for } 115 < \theta < 215\ SUS \\
\nu &= 0.2158\theta && \text{for } \theta > 215\ SUS
\end{aligned}$$

A discussion on some simple ways of measuring viscosity, and various applications involving it, is given by Middleman.[1]

5.3. Effect of Temperature on Viscosity

Unlike gases, where the viscosity increases with temperature, the viscosity of oil decreases strongly with temperature. The standard experiment to find the effect of temperature is to use a capillary tube viscometer under controlled temperatures, and then curve fit the results.

The simplest fit, called **Reynolds viscosity equation**,[2] is:

$$\eta = \eta_0 \exp(-\beta\Delta\theta). \tag{5.3}$$

Here, η_0 is the viscosity at some representative temperature, θ_i in degrees Celsius, $\Delta\theta$ is the temperature rise from $\theta_i : \theta = \theta_i + \Delta\theta$, (the temperature at a point in the oil film in degrees Celsius) and β is the viscosity-temperature coefficient. Equation (5.2) is not accurate except over a very low temperature range. Another more accurate fit over a wider temperature range is by Vogel[3] and discussed in Ref. 2:

$$\eta = a \exp\left(\frac{b}{\Theta - c}\right), \tag{5.4}$$

where the constants a, b and c must be determined for each oil from 3 sets of data supplied by the manufacturer, and Θ is the required absolute temperature at a point in the film in degrees K.

5.4. The American Society for Testing Materials (A.S.T.M.) Chart

A.S.T.M. supplies a data sheet based on which a formula by Walther[4] and discussed in Ref. 2, is based upon:

$$\log\log(\nu + 0.6) = d - e\log\Theta. \tag{5.5}$$

Equation (5.5) can, therefore, be conveniently plotted on special ASTM (log)-(loglog) paper as a straight line. The solution is kinematic viscosity, ν, expressed in cs, at supplied temperature Θ in degrees K. In order to plot the line, its viscosities at two other temperatures must be known. These are generally supplied by the manufacturer.

5.5. Viscosity Index of Lubricants (VI)

This index is supposed to give an indication of the '*goodness*' of an oil on the basis of its variation with temperature, an oil with a low viscosity variation over a wide range of temperature being considered a good oil. The VI compares the kinematic viscosities of the test oil, at 40°C and 100°C , to the viscosities of two chosen standard oils, both having the *same* viscosity as the test oil at 100°C and *different* viscosities at 40°C. The test oil has a viscosity, U_t at 40°C, while the standard ones respectivly have viscosities, H_{st} (the VI = 100 oil) and L_{st} the VI = 0 oil) also at 40°C. The VI of the

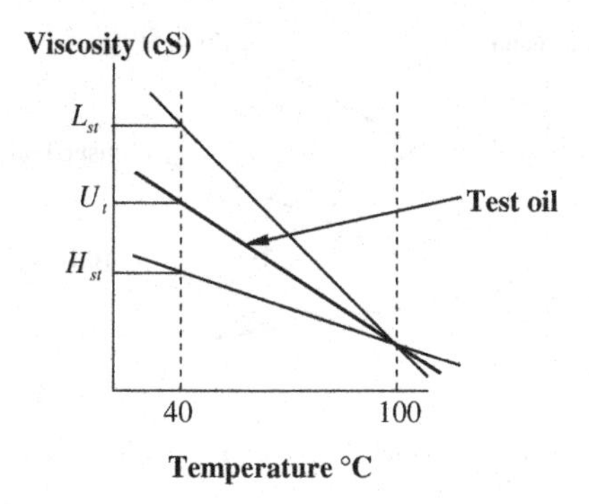

Fig. 5.2. Viscosity index of oils.

test oil is, therefore:

$$\mathrm{VI} = \frac{L_{st} - U_t}{L_{st} - H_{st}} \times 100 \tag{5.6}$$

L_{st}: VI=0 oil (cS) at 40°C

H_{st}: VI=100 oil (cS) at 40°C.

Figure 5.2 explains the procedure, which uses the ASTM chart.

5.6. Polymer Thickened Oils

Generally, a thick *straight** oil has a large viscosity-temperature slope, $d\eta/d\theta$. This means that if it has a high viscosity at the low temperatures end of its operating range, at the higher temperature end the viscosity might be unacceptably low. The opposite is true with thinner oils. They have a lower value of $d\eta/d\theta$ throughout the same operating range.

The addition of polymer additives, to a thin *base*† oil, has the effect of raising its viscosity *with little change* in its $d\eta/d\theta$. Thus, the *finished*‡ oil is more acceptable, having a higher viscosity than the base oil throughout the temperature range. This is one way of creating what is known as a **multi-grade oil**, the effect of the process being illustrated in Fig. 5.3. A consequence of employing a polymer additive is to improve the VI of the finished oil, that can sometimes reach a value of 140.[2]

*A *straight* oil means one employed with no additives.

†A *base* oil has not *yet* any additives.

‡A *finished* oil is the base oil plus the additives.

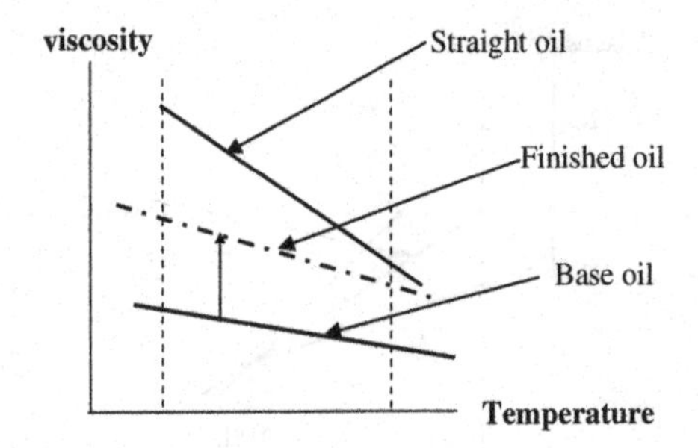

Fig. 5.3. Polymer thickened oils. The finished oil has a low viscosity-temperature slope but a higher viscosity than its base oil.

However, at high shear rates, the polymer can lose its effectiveness, with the viscosity dropping back to that of the base oil, but recovering when the shear rate effect is removed.

5.7. Blends of Oils

The aim of mixing two lubricants of different viscosities is to obtain a blend that has a required intermediate viscosity. How can we find approximately the desired proportions of the original lubricants in the blend? **ASTM**[§] **D341** suggests a procedure using their ASTM viscosity-temperature chart based on Eq. (5.5).

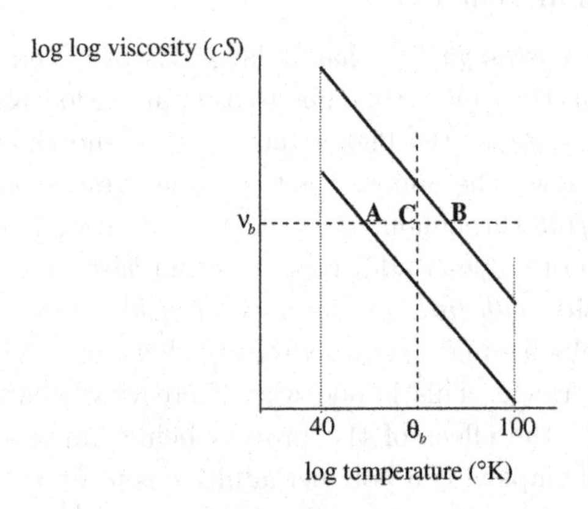

Fig. 5.4. Blends of oils.

§ASTM: American Society of Testing Materials.

- Plot the known viscosity-temperature lines, of the two base oils, say between 40°C and 100°C on standard ASTM paper.
- Referring to Fig. 5.4, draw a line parallel to the temperature axis at the wanted viscosity, η_b and intersecting the two viscosity-temperature lines.
- With a ruler, measure the distance AB between the intersection points.
- Draw a vertical line through the required operating temperature of the blend. It will cross the line AB at C in Fig. 5.4.
- The proportion of the total volume of the blend that constitutes the base oil with the *higher* viscosity, is given by the ratio of measured distances

$$X_b = \frac{AC}{AB}.$$

- It follows that the base oil with the *lower* viscosity has a proportion of the total volume of $1 - X_b$.

5.8. Grades of Oil

The American Society of Automotive Engineers (SAE) divides oils into grades that cover different viscosity ranges. A selection of results in Table 5.1 for engine oils come from their handbook. **W** denotes Winter, **VI** must be >60, Summer grades have no W and only specify one operating temperature (no VI implied).

5.9. Effect of Pressure on Viscosity

The behavior of lubricants under pressure is most significant when there are concentrated contacts, such as in a ball bearing or between a pair of

Table 5.1. SAE J300 viscosity grades for engine oils.

SAE viscosity grade	Low temp (°C) cranking viscos. cP Max.	Kinematic viscosity (cSt) at 100°C (Min.)	Kinematic viscosity (cSt) at 100°C (Max.)	HTHS viscosity (Cp) at 150°C and $10^6\,s^{-1}$
10 W	3500 at −20	4.1		2.9
20 W	4500 at −10	5.6		3.7
25 W	6000 at −5	9.3		3.7
30		9.3	<12.5	
40		12.5	<16.3	
60		21.9	<26.1	

gear teeth. We have seen in Chapter 3, for dry contacts, how high these pressures can reach. The **Barus law**[5] gives a relationship that shows how the lubricant viscosity varies with pressure at constant temperature:

$$\eta_s = \eta_0 \exp(\alpha p), \tag{5.7}$$

where η_s is the oil viscosity at gauge pressure, p, η_0 is the viscosity at $p = 0$, and α is a constant, depending on the oil, called the **pressure viscosity coefficient** with units of $\mathrm{m}^2 N^{-1}$. Equation (5.7) can be quite inaccurate, one reason being because α itself can vary with both temperature and pressure. Equation (5.7) is, however, satisfactory for calculation purposes, where there is a rolling contact (for example, in a moderately loaded ball bearing). It is the *sliding* between the contacting surfaces that is the main cause of temperature rise (for example, between a pair of involute gear teeth). To give us some idea of scale, the peak pressure in a ball bearing can reach 4 GPa, where at such pressures the lubricant appears to have nearly solidified. We will see in Chapter 10, that the **piezoviscous** property of oil is important when we deal with Elastohydrodynamic lubrication.

Table 5.2 gives viscosity and pressure viscosity coefficients of some lubricants, showing the effect of temperature on them.

We have already alluded to the inaccurate nature of Barus's equation at very high pressures. This can be demonstrated by way of an example. Take a typical oil of dynamic viscosity of 0.03 Pas at atmospheric pressure, and with a pressure viscosity coefficient of 10^{-8} Pa^{-1}. Using Eq. (5.7): $\bar{\eta} = \frac{\eta}{\eta_0} = \exp(\alpha p)$. At $p = 1\,\mathrm{GPa}$, $\bar{\eta} = \exp(10) \approx 21365$. This means that the viscosity of the lubricant appears to have increased by more than 4

Table 5.2. Dynamic viscosity and pressure coefficients of some lubricants.

	Dynamic viscosity η_0 at atmospheric pressure (Pas $\times 10^{-2}$)			Pressure-viscosity exponent α ($\mathrm{Pa}^{-1}\mathrm{s} \times 10^{-9}$)	
Temperature	30°C	60°C	100°C	30°C	60°C
High VI oils					
Light machine oil	38	12.1	5.3		18.4
Cylinder oil				34	28
Low viscosity oils					
Spindle oil	30.7	8.6	3.1	25.7	20.3
Heavy machine oil	310	44.2	9.4	34.6	26.3
Other lubricants					
Castor oil	360	80	18	15.9	14.4
Sanotrac 40	30			34.5	

orders of magnitude. At such high pressures the lubricant becomes similar to an amorphous solid, but as we shall see below, predicting this level of increase in viscosity is erroneous.

A more accurate expression, found by Roelands[6] and developed further by Houpert,[7] is often used in computing. It includes the effects of both temperature and pressure on the viscosity and can be expressed in the same form as Eq. (5.7), but with a **modified pressure viscosity coefficient**:

$$\eta_R = \eta_0 \exp(\alpha^* p), \tag{5.8}$$

where α^* is a function of both p and θ:

$$\alpha^* = \frac{1}{p}[\ln(\eta_0) + 9.67]\left\{\left(\frac{\Theta - 138}{\Theta_0 - 138}\right)^{-S_0}\left[\left(1 + \frac{p}{1.98 \times 10^8}\right)^Z - 1\right]\right\}. \tag{5.9}$$

Z and S_0 are constants, independent of temperature and pressure, defined below, p is in Pa, Θ and Θ_0 are in °K. ($\Theta_0 = \theta_0 + 273$ and $\Theta = \theta + 273$). The constants are obtained from:

$$Z = \frac{\alpha_0}{5.1 \times 10^{-9}[\ln(\eta_0) + 9.67]}, \quad S_0 = \frac{\beta_0(\Theta_0 - 138)}{\ln(\eta_0) + 9.67},$$

where β_0, and α_0 are at atmospheric temperature and pressure.

Let us compare the viscosity based on Eq. (5.7) with that obtained from Eq. (5.8). We find that in this case $\eta_{\text{Barus}}/\eta_{\text{Roelands}} = 188$, a huge reduction from the value given above to a more realistic value.

5.10. Lubricant Density

Knowing how the lubricant density varies with pressure is necessary when the pressures are excessive, as they will sometimes be in Chapter 10, when we will discuss numerical solutions to EHL problems. A widely used formula that shows this variation is:

$$\rho = \rho_0\left(1 + \frac{0.6 \times 10^{-9} p}{1 + 1.7 \times 10^{-9} p}\right), \tag{5.10}$$

where p is in Nm^{-2} and ρ_0 for oils is typically 0.87 at 20°C.

The variation of density with temperature is found to be negligible in lubrication problems.

5.11. Effect of Shear Rate on Viscosity

Fluid behavior is described as **non-Newtonian** if the viscosity depends on shear rate. Lubricants like grease suffer from **thixotropic** behavior if there is a steady fall in viscosity with the duration of the shear action. The normal viscosity is slowly regained after the shear action ends.

A liquid lubricant can suffer from **shear thinning**, which makes the viscosity reduce at high shear rates ($>10^9\,\mathrm{s}^{-1}$), the normal viscosity being regained as soon as the shear action ends. The shear rate action itself does not normally affect liquid lubricants. In the case of **Elastohydrodynamic (EHL)** conditions, although the shear rate is usually below $10^9\,\mathrm{s}^{-1}$, the shear stress within the lubricant can itself cause shear thinning. We will see in Chapter 10 that shear thinning is the primary cause of friction levels found in EHL contacts.

5.12. Worked Example*

A lubricating oil has temperature-viscosity characteristics shown in Table 5.3, and a density, independent of temperature, of 850 Kg/m.

(a) Use these viscosities to determine the constants in a Vogel temperature-viscosity expression (Eq. (5.4))

$$\ln \eta = \ln(a) + \frac{b}{(\Theta - c)}.$$

(b) Use the accompanying part of a data sheet to find the oil VI, from standard oil ASTM data, (Table 5.4). Hence, calculate the viscosities at 50°C and −5°C.

Table 5.3. Temperature-viscosity characteristics in lubricating oil.

Temperature °C (θ)	Viscosity cP (η)
20	387
60	35.1
100	9.9

*Adapted from Imperial College, London undergraduate tutorials.

Table 5.4. ASTM data.

Kinematic viscosity at 100°C (cSt)	(40°C) (cSt)	H(40°C) (cSt)
9.90	145.2	81.67

Table 5.5. Temperature-viscosity characteristics in lubricating oil.

Temperature (Θ)°K	Viscosity cP
293	387 (455cSt)
333	35.1 (41.3cSt)
373	9.9 (11.64cSt)

Solution

(a) We use the corresponding temperatures in °K, ($\Theta = \theta + 273$) as shown in Table 5.5

Put Vogel's Eq. (5.4) into ln form, to get the constants a, b and c:

$$\ln(\eta) = \ln(a) + \frac{b}{(\Theta - c)}.$$

Then, use Table 5.5 to get three simultaneous equations:

(1) $\ln(387) = \ln(a) + b/(293 - c)$,
(2) $\ln(35.1) = \ln(a) + b/(333 - c)$,
(3) $\ln(9.9) = \ln(a) + b/(293 - c)$.

Solving equations (1) to (3) we obtain the constants for Vogel's equation as: $a = 0.1597$ cP, $b = 700.81$°K, $c = 203.09$°K (**Answer**). Using these constants in Eq. (5.4) for $\theta = -5$°C (= 268°K) and $\theta = 50$°C (= 323°K) we obtain: $\eta(-5°\text{C}) = 7759$ cP and $\eta(50°\text{C}) = 54.8$ cP (**Answer**).

(b) Using Eq. (5.5) and the data from Table 5.4 at 20°C, and 100°C, we need to find the two constants, d and e, in the Walther Eq. (5.5), reproduced below, which is based on the ASTM chart:

$$\log\log(\nu + 0.6) = d - e\log\Theta. \qquad (5.5)$$

Choose the 293°K and 373°K values and their corresponding viscosities from Table 5.5, by solving the two equations formed from Eq. (5.5). The constants come to: $d = 9.715$ and $e = 3.77$.

Then, using Eq. (5.5), find ν at 40°C. It comes to $\nu = 106$ cSt, this being the required value of U_t in Eq. (5.6). The VI of the test oil can then be found from Eq. (5.6), reproduced below, with the values inserted.

$$\mathrm{VI} = \frac{145.2 - 106}{145.2 - 81.67} \times 100 = 0.617\ (\textbf{Answer}).$$

The VI of the test oil falls in between the standard oil's values, with an intermediate slope. Nowadays, modern lubricants are of better quality with VI's often exceeding 100.

5.13. Closure

This chapter has discussed how some properties of lubricating oils, such as viscosity, change with temperature and pressure and how these properties can also be altered by suitable blending of different oils. In the next chapter we will insert a lubricating oil between two solid sliding surfaces and show how a coherent film may be formed. Another important property of a lubricant is surface tension, which is discussed in Chapter 13.

References

1. Middleman, S. *An Introduction to Fluid Dynamics*, John Wiley and Sons Inc NY (1998).
2. Cameron, A. *Basic Lubrication Theory*, Longman Ltd. (1970).
3. Vogel, H. *Physik Z.* **22** (1921) 645.
4. Walther, A. and Saseffeld, H., *VDI, Forschungshaft* (1954) **441**.
5. Barus, C. Isothermals isopietics and isometrics in relation to viscosity, *American Journal of Science* 3rd *Series* **45** (1893) 87–96.
6. Roelands, C. J. A. Correlation aspects of the viscosity-temperature-pressure relationships of lubricating oils, *Druk VRB Kleine der A3-4 Groningen* (1966).
7. Houpert, L. New results of traction force calculations in elastohydrodynamic contacts, *ASME Journal of Tribology* **107** (1985) 241–248.

CHAPTER 6

THE REYNOLDS AND ENERGY EQUATIONS

6.1. Introduction

In Chapter 5, we discussed the properties of lubricants used in viscous film bearings. Below, we will state, and sometimes derive, some of the equations needed to design these bearings. Where appropriate, we will give some examples of their application. Our approach will be as follows:

- explain the principle of the **hydrodynamic wedge**, both by analogy and by dimensional analysis
- derive Reynolds equation from first principles
- define the energy equation, which accounts for thermal effects in bearings. In subsequent chapters, in order to suit local conditions, modified versions will be employed.

6.2. Reynolds Equation

If you stand by the parallel of a straight part of a steadily flowing river, you will note that the stream velocity is highest in the middle, reducing to zero at the sides. In other words the velocity distribution, of the water, a real fluid, roughly takes a curved shape depicted in Fig. 6.1. An identical velocity distribution would be found whatever position along the bank is chosen, as long as the river geometry is the same.

Additionally, in order for the river to flow at all, its source, on the left of Fig. 6.1, must be at a higher altitude than at its mouth, which is at sea level. The difference in level is called the *head*. It is associated with falling pressures (a negative pressure gradient) driving the stream towards the mouth.

Noting this behavior, let us turn to the next model in Fig. 6.2(a): a simple arrangement of two parallel impervious surfaces, one above the other, long into the paper and separated by a narrow parallel gap. The top surface is stationary and the bottom surface, moving at constant velocity, causes a viscous liquid lubricant to be entrained steadily from left to right.

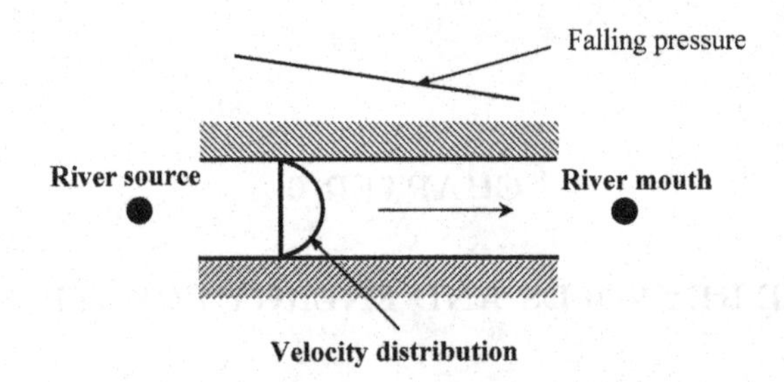

Fig. 6.1. Flowing river.

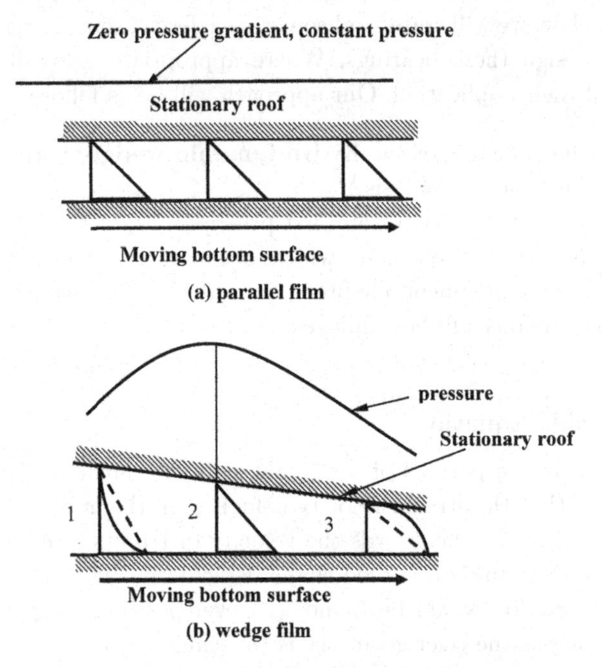

Fig. 6.2. Principle of the wedge film.

The following points should be noted in Fig. 6.2(a):

- Just as in Fig. 6.1, as the flow is constant, the velocity of distribution of the oil is the same whatever position along the gap is chosen.
- As pointed out in Chapter 5, because oil is a Newtonian viscous fluid, it must have the same local velocity as the boundaries wherever it contacts them.

- Here, the shape of the velocity distribution can only be triangular because, unlike in Fig. 6.1, there is no pressure source, making the pressure gradient zero throughout.

Let us now change the gap shape into a narrowing wedge of small angle, with the same boundary velocity as for the parallel case (Fig. 6.2(b)). Clearly, maintaining the original triangular velocity distributions of Fig. 6.2(a), would infringe continuity. Therefore, the local velocities at the three positions shown must alter so as to maintain the same area under each distribution. This can only be achieved by modifying the oil velocity distributions. In position 1; the widest part of the gap, the distribution bulges backwards while it bulges forwards in position 3, the narrowest part. Note that the *forward* bulge of the velocities in position 3 must be associated with *falling* pressures, as is the distribution in Fig. 6.1. The reverse is true in position 1. Therefore, there must be one point where the velocity distribution is triangular. This is the position 2 and, from Fig. 6.2, there must be a pressure maximum at that point.

The presence of a rising and falling pressure distribution, along the hydrodynamic wedge width, suggests that it must possess **load capacity**. This is a characteristic of all hydrodynamic bearings, because a load capacity is their *raison d'etre*. Like the *head* in a flowing river, an external power source is still needed for driving the moving surface against the retarding friction force of the oil. Incidentally, the same behavior results if the narrowing gap is curved, as long as the inclination angles remain small.

Alternatively, a bearing with *parallel and stationary* surfaces could have been designed if an oil pump was available. Try sketching a possible axisymmetric design that can accommodate a vertical load.

In the next section we will put some algebra into the **hydrodynamic wedge** concept, solely from knowledge we have acquired so far. Also, from now onwards we will often use the general words, **hydrodynamic bearing** or **wedge pair** to explain the wedge concept.

6.3. Reynolds Equation by Dimensional Analysis

Having found the forms of the velocity disribution profiles and the pressure distribution over a hydrodynamic bearing. We will go one step further by finding a mathematical relationship between the variables, solely by dimensional analysis. This is a useful technique, because it helps you to delve into the physics of the problem.

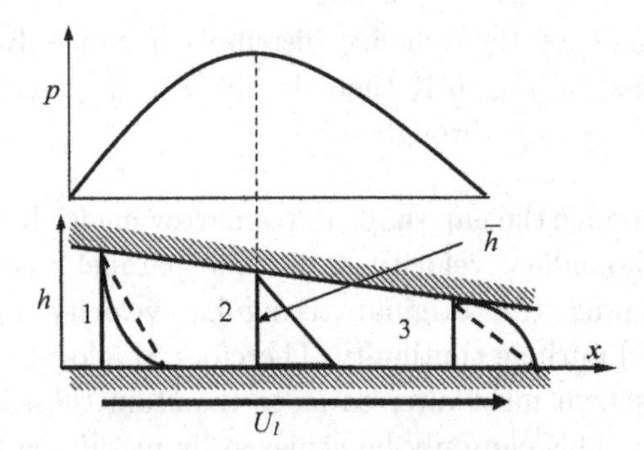

Fig. 6.3. Reynolds equation by dimensional analysis.

Referring to Fig. 6.3, let the volume flow per unit length,* q_x, be of constant density and constant viscosity, η. Recall that, from Sec. 6.2, we discovered that the velocity distribution in a hydrodynamic bearing has two components causing, the total flow, q_x, also to have two parts:

- one due to the velocity of the bottom surface, U_1 (called the **Couette flow**)
- the other from the rising or falling pressure distribution. (called the **Poiseuille flow**, $f(p)$).

The aim of the dimensional analysis is to obtain $f(p)$ and q_x in terms of the other variables: η, U_1 and $h(x)$. This procedure will go part of the way towards a full solution. The area under a velocity profile in Fig. 6.3 is q_x. It can be written as:

$$q_x = \frac{U_1 h}{2} - f(p), \tag{6.1}$$

where $q_x = \frac{\text{total flow rate}}{\text{wedge length into the paper}}$, with dimensions of $\frac{L^3}{TL} = \frac{L^2}{T}$.

For dimensional homegenity in Eq. (6.1), $f(p)$ also has dimensions L^2/T.

As we have reasoned in Sec. 6.2, $f(p)$ is a function of the **pressure gradient** $(\frac{dp}{dx})$, (not magnitude). In addition, it must also depend on the

*Throughout the book, bearing *length* means here the bearing dimension into the paper. This is in keeping with traditional journal bearing terminology, its length being the axial length into the paper.

local h and η. Therefore, let:

$$f(p) = k\left(\frac{dp}{dx}\right)^a h^b \eta^c, \tag{6.2}$$

where a, b and c are unknown indices and k is an unknown constant. Employing a dimensional analysis with units of *force, length and time*, we can write:

$$\left(\frac{dp}{dx}\right)^a (h)^b (\eta)^c \equiv \left(\frac{F}{L^2 L}\right)^a L^b \left(\frac{F}{L^2}T\right)^c = \frac{L^2}{T}.$$

Equate indices of force:	$F^a F^c = 0$	$\therefore a = -c$	
and of time:	$T^c = T^{-1}$	$\therefore c = -1,$	giving $a = 1$
and of length:	$\frac{1}{L^3} L^b L^2 = L^2$	$\therefore b = 3$	

Hence:

$$f(p) = k\frac{h^3}{\eta}\frac{dp}{dx}, \quad \text{so} \quad q_x = \frac{U_1 h}{2} - k\frac{h^3}{\eta}\frac{dp}{dx}$$

We can also state that at some point along the bearing, as yet unknown, $\frac{dp}{dx} = 0$. Let this be at $h = h_c$ (station 2, Fig. 6.3).

There, $q_x = \frac{U_1 h_c}{2}$ and as q_x is constant:

$$\frac{U_1 h_c}{2} = \frac{U_1 h}{2} - k\frac{h^3}{\eta}\frac{dp}{dx}, \tag{6.3}$$

or, rearranging Eq. (6.3):

$$\frac{dp}{dx} = \frac{U_1 \eta}{2k}\left(\frac{h - h_c}{h^3}\right). \tag{6.4}$$

A full solution gives k as $1/12$. One integration of Eq. (6.4), yields:

$$p = 6U_1\eta \int \frac{h - h_c}{h^3} dx + const. \tag{6.5}$$

h_c and *const.* are integration constants that need 2 boundary conditions along the film to find their values. Equation (6.4) is called the one dimensional Reynolds equation. A full three dimensional solution will now be given below.

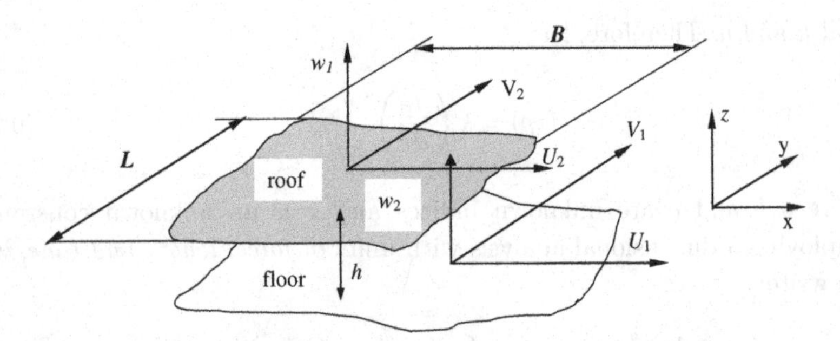

Fig. 6.4. Two generalized surfaces, in relative motion, bounding an oil film.

6.4. Derivation of Reynolds Equation in Three Dimensions

As we did in Sec. 6.3, some realistic assumptions are needed to derive Reynolds equation. Even though we have not yet designed a hydrodynamic bearing, lets assume firstly that the film is thin, so that in profile it has a large aspect ratio (width†/mean film thickness) is typically 1000:1). As an example, for a thrust bearing, the minimum film thickness can be 25 μm with a wedge angle 0.075°. This implies that the film thickness dimensions, $h(x)$, are much less than those defining its top and bottom surfaces, B and L, as illustrated in Fig. 6.4. We may also neglect the film weight (gravity forces).

The differential equation governing pressure distribution in a Newtonian lubricant film was first obtained by Reynolds.[1] It can be derived from the full Navier Stokes (NS) equations[2] by making simplifying assumptions at appropriate points in the analysis. The finished article is a perfectly generalized version of Reynolds equation. However, in our case we will use the more direct engineeing approach: by making simplifying assumptions at the start of the analysis. This approach offers more insight into the physics of the equation.

Assumptions made in the direct derivation of Reynolds equation:

(1) the oil film has negligible mass (gravity forces neglected),
(2) because its so thin, assume pressure is constant across the film (z direction),
(3) no slip at the boundaries (Newtonian fluid),
(4) lubricant flow is laminar (low Reynolds numbers),

†Remember, *width* is measured n the x direction.

(5) Inertia and surface tension forces are negligible compared with viscous forces,
(6) Because it is thin, shear stresses and velocity gradients are only significant across the film (z direction),
(7) The lubricant is Newtonian (high shear rates are not present),
(8) The lubricant viscosity is constant across the film (z direction),
(9) The boundary surfaces (roof and floor in Fig. 6.4) follow some designated geometry but are always at low angles to each other.

6.4.1. *Equilibrium of forces on a lubricant element*

Referring to Fig. 6.5, let the two bounding surfaces have perfectly general motion defined by their velocity vectors. Consider the local forces acting on a lubricant element, defined at x, y, z in a column of thickness, h. The element center has velocity components u, v, w.

Neglecting shear stress and velocity gradients in the x and y directions (Assumption 6) we have:

$$\sum F_x = 0,$$

$$\therefore \frac{\partial \tau_{xz}}{\partial z} = \frac{\partial p}{\partial x}. \tag{6.6}$$

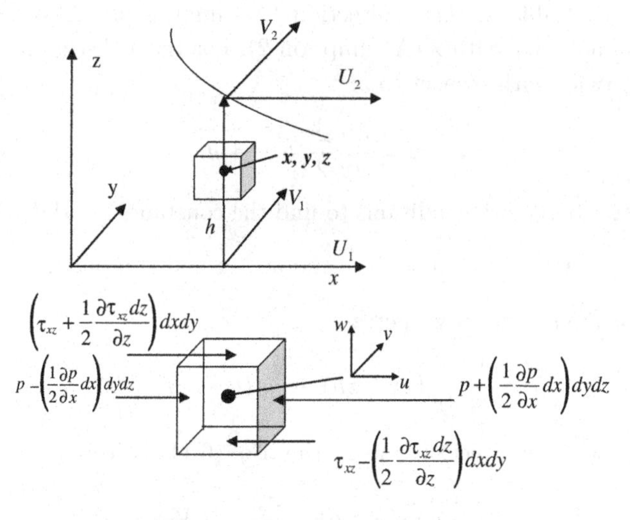

Fig. 6.5. Forces on an element.

Here, τ_{xz} means shear stress in the x direction in a plane having z as its normal.

Similarly:

$$\sum F_y = 0$$

$$\therefore \frac{\partial \tau_{yz}}{\partial z} = \frac{\partial p}{\partial y}. \tag{6.7}$$

Also, from Eq. (5.1):

$$\tau_{xz} = \eta \frac{\partial u}{\partial z}, \tag{6.8}$$

$$\tau_{yz} = \eta \frac{\partial v}{\partial z}. \tag{6.9}$$

Combining Eq. (6.6) with (6.8), and Eq. (6.7) with (6.9):

$$\frac{\partial}{\partial z}\left[\eta \frac{\partial u}{\partial z}\right] = \frac{\partial p}{\partial x}, \tag{6.10}$$

$$\frac{\partial}{\partial z}\left[\eta \frac{\partial v}{\partial z}\right] = \frac{\partial p}{\partial y}. \tag{6.11}$$

6.4.2. *Velocity distribution*

Let η be invariable in the z direction (Assumption 8). Also, let $\partial p/\partial x$ and $\partial p/\partial y$ not vary with z (Assumption 2). For the x direction, integrate Eq. (6.10) twice with respect to z:

$$\eta u = \frac{\partial p}{\partial x}\frac{z^2}{2} + cz + d.$$

We need two boundary conditions to find the constants c and d. These are:

$$\text{At } z = h, \quad u = U_2 \quad \text{and at} \quad z = 0, \quad u = U_1.$$

Solving for the constants we get:

$$u = \frac{1}{2\eta}\frac{\partial p}{\partial x}\left(z^2 - zh\right) + \frac{z}{h}\left(U_2 - U_1\right) + U_1. \tag{6.12}$$

Similarly, for the y direction, integrating Eq. (6.11) twice:

$$v = \frac{1}{2\eta}\frac{\partial p}{\partial y}\left(z^2 - zh\right) + \frac{z}{h}\left(V_2 - V_1\right) + V_1. \tag{6.13}$$

Just as we surmised from the wedge-shape study above and found also by dimensional analysis, both Eqs. (6.12) and (6.13) describe velocity distributions composed of two parts. There is a parabolic part due to the pressure gradient (Poisseuille flow) and a linear part due to the boundary surface velocities (Couette flow).

6.4.3. *Mass continuity*

To complete our full derivation of Reynolds equation, we must invoke **mass continuity**. This states that there is the same mass of fluid per second entering a column of oil height, h, as leaving it. Here, we do not have to assume that the fluid is of constant density in the x and y directions. Referring to Fig. 6.6, let m_x and m_y be the mass flows per unit width in the x and y directions at the column center.

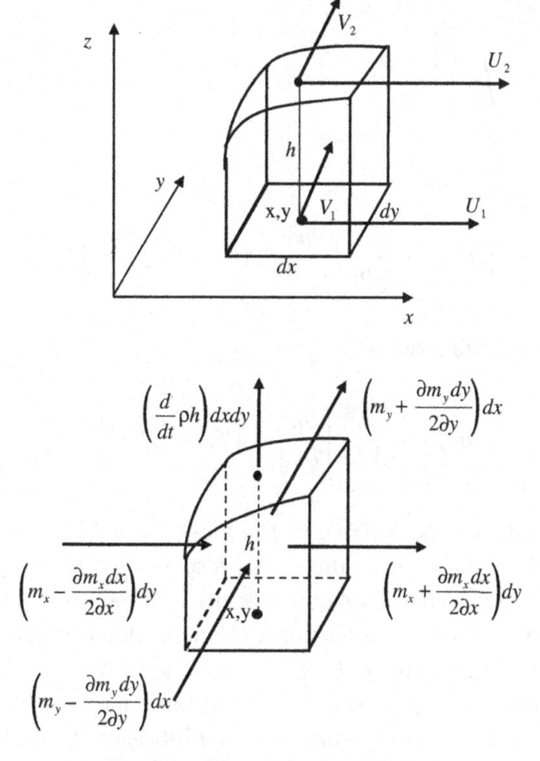

Fig. 6.6. Flow through a column.

The net mass flow out of the column in the z direction is $(d/dt)\rho h dx dy$ and the mass flows in the x and y directions are, by definition:

$$m_x = \rho q_x = \rho \int_0^h u dz, \tag{6.14}$$

$$m_x = \rho q_x = \rho \int_0^h u dz. \tag{6.15}$$

Also, from Fig. 6.6, equating the flows entering the column to those leaving it:

$$\frac{\partial m_x}{\partial x} + \frac{\partial m_y}{\partial y} + \frac{\partial}{\partial t}(\rho h) = 0. \tag{6.16}$$

Additionally, from Eqs. (6.12)–(6.14) and (6.15):

$$m_x = -\rho \int_0^h \left[\frac{1}{2}\eta \frac{\partial p}{\partial x}(z^2 - zh) + \frac{z}{h}(U_2 - U_1) + U_1\right] dz. \tag{6.17}$$

Therefore:

$$m_x = -\frac{\rho h^3}{12\eta}\left[\frac{\partial p}{\partial x}\right] + (U_1 + U_2)\left[\frac{\rho h}{2}\right]. \tag{6.18}$$

Similarly, in the y direction:

$$m_y = -\frac{\rho h^3}{12\eta}\left[\frac{\partial p}{\partial y}\right] + (V_1 + V_2)\left[\frac{\rho h}{2}\right] \tag{6.19}$$

Equations (6.18) and (6.19) give the mass throughputs of the lubricant per unit length in the x and y directions respectively. Like Eqs. (6.12) and (6.13) each of these is composed of a pressure induced term (Poiseuille flow) and a boundary velocity induced term (Couette flow). If there is only one surface velocity, say in the x direction, then $V_1 = V_2 = 0$. In Eq. (6.19), however, a mass flow, m_y, still occurs in that direction. This now is due only to the pressure gradient, $\frac{\partial p}{\partial y}$, causing some lubricant to flow transversely from a high pressure to a low pressure region in that direction. This action in a bearing is called *side leakage*.

Substitute Eqs. (6.18) and (6.19) into Eq. (6.16) and re-arrange the order so that the pressure induced terms only are on the left hand side. The result is the full **Reynolds equation**:

$$\frac{\partial}{\partial x}\left[\frac{\rho h^3}{\eta}\frac{\partial p}{\partial x}\right]+\frac{\partial}{\partial y}\left[\frac{\rho h^3}{\eta}\frac{\partial p}{\partial y}\right]$$

$$=6\left\{\frac{\partial}{\partial x}[\rho h(U_1+U_2)]+\frac{\partial}{\partial y}[\rho h(V_1+V_2)]+2\frac{d}{dt}(\rho h)\right\}. \qquad (6.20)$$

Equation (6.20) is the fundamental equation of fluid film lubrication theory, with units of $\text{kg m}^{-2}\text{s}^{-1}$. If the density, ρ, is constant, it cancels out leaving terms of dimension m s^{-1}.

It is the full Reynolds equation in three dimensions for compressible or incompressible flow of a Newtonian fluid. On the left hand side are the Poiseuille pressure induced terms. The right hand side is composed of Couette terms which, from left to right, are divided into **wedge** and **squeeze** components. Equation (6.20) accounts for flow components in the x and y directions. Neither boundary surface need have a uniform velocity vector parallel to the xy plane because the components are respectively situated within the differentials on the right hand side, $\frac{\partial}{\partial x}$ and $\frac{\partial}{\partial y}$. The same applies to the right hand side, where ρ and η are again within the differential operators. Both may vary with x and y because of their sensitivities to pressure (we have seen that it rises and falls) and/or temperature. Finally, the squeeze term, $\frac{\partial}{\partial t}(\rho h)$, need not be uniform over the region of pressure. If the squeeze term is expanded, it can be written as: $\rho(w_1-w_2)+h\frac{d\rho}{dt}$ where $(w_1-w_2)=\frac{dh}{dt}$ and w_1 and w_2, are respectively the roof and floor velocities, from whatever cause.

6.5. Simplifications of Reynolds Equation

Generally, we will be dealing with simplified versions of Eq. (6.20). For the two examples below, assume:

- the oil density and viscosity are constant.
- the roof and floor of the film are non-porous and have no normal velocity components.
- the surface velocities are in the x direction only.

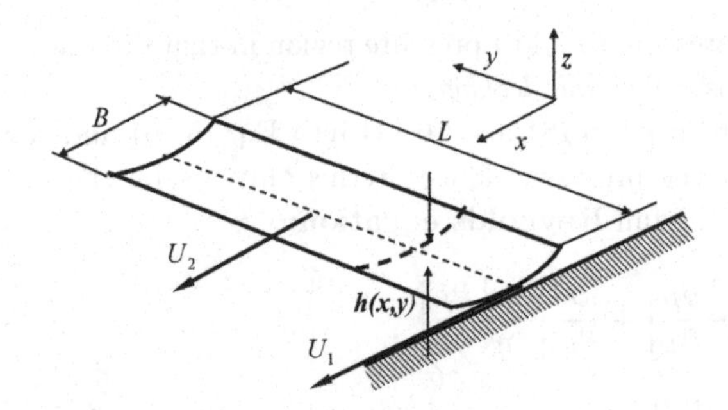

Fig. 6.7. Long bearing $L \gg B$.

6.5.1. *Long bearing*

As a simplification we assume that the transverse length of the bearing, L, (y direction) is effectively infinite. In practice we call this a **long bearing** (L much greater than the width, B, in the x direction). The long transverse length assumption makes the pressure distribution uniform in the y direction, except close to the edges, where the pressures must drop to zero there. The shape is illustrated in Fig. 6.7.

In this case, dp/dy can be neglected compared with dp/dx, so Eq. (6.20) becomes:

$$\frac{d}{dx}\left(\frac{h^3}{\eta}\frac{dp}{dx}\right) = 6(U_1 + U_2)\frac{dh}{dx}. \tag{6.21}$$

Equation (6.21) can be integrated with respect to x:

$$h^3\frac{dp}{dx} = 6(U_1 + U_2)\eta h + C,$$

where C is an integration constant.
Let $h = h_c$ (as yet unknown) at $dp/dx = 0$.
Equation (6.21) modifies to:

$$\frac{dp}{dx} = 6(U_1 + U_2)\eta\left[\frac{h - h_c}{h^3}\right]. \tag{6.22}$$

Equation (6.22) is the differential form of Eq. (6.5) we obtained by dimensional analysis. It also confirms the multiplying constant as $k = 1/12$. This equation will be used frequently throughout the book. The **long bearing assumption** is reasonably accurate if $L/B > 3$. Below is an example.

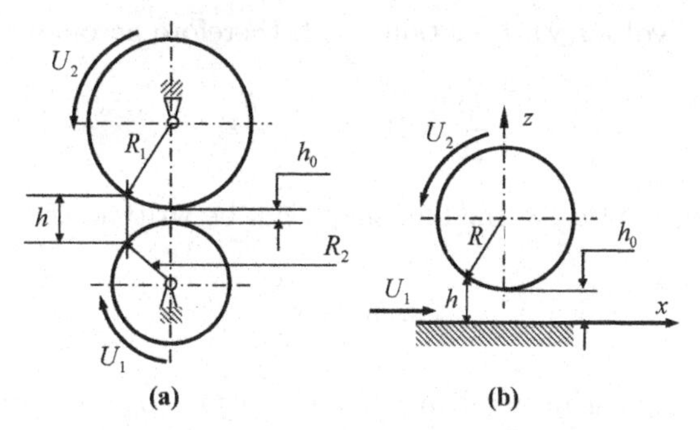

Fig. 6.8. Line contact geometry (a) two discs (b) equivalent contact pressure distribution.

6.5.2. *Long bearing approximation for rigid cylinders*

An important application of Eq. (6.22) is the lubrication of long rigid cylinders where we will need to use additional boundary conditions in Reynolds equation for a complete solution. The model below is a useful precursor to the analysis in Chapter 10, where the bounding surfaces are assumed to be elastic.

Figure 6.8(a) shows two rigid cylinders, loaded on their circumferences, in rolling-sliding lubricated contact, producing a **convergent-divergent wedge film**. Just as the one in Fig. 3, for static contacts, this arrangement is called a **hydrodynamic line contact**, even if the surfaces are distorted. Let us first find the pressure distribution created by the wedging action of the surfaces.

6.5.3. *Line contact pressure distribution*

We will make the following assumptions:

(1) The discs are considered long in the y direction, so that there is no transverse flow except near their ends,
(2) $R \gg h$,
(3) Conditions are isoviscous throughout, so that $\eta = \eta_0. =$ constant.

Referring to Fig. 6.8 (b) and Eq. (3.8), in order to simplify the coordinate system let the discs be replaced by an equivalent disc of reduced radius, R, in contact with a plane, where $1/R = 1/R_1 + 1/R_2$. In addition, in keeping with normal practice for lubricated concentrated contacts, let the mean velocity of the Couette flow component be $U = \frac{U_1+U_2}{2}$ (called the

entrainment velocity). Equation (6.22) therefore becomes:

$$\frac{dp}{dx} = 12U\eta_0 \frac{h - h_c}{h^3}. \tag{6.23}$$

Also, from Assumption 3, the film shape can be written as:

$$h \approx h_0 + \frac{x^2}{2R}. \tag{6.24}$$

Equation (6.24) will be used in Chapter 10 for counterformal EHL line contacts.

It is usual practice to employ dimensionless groups, so let:

$$\bar{x} = \frac{x}{\sqrt{2Rh_0}}, \quad \bar{h} = \frac{h}{h_0},$$

Making this substitution, Eq. (6.24) becomes:

$$\bar{h} = 1 + \bar{x}^2,$$

and letting $\bar{p} = \frac{h_0^{3/2} p}{12U\eta_0 \sqrt{2R}}$, with $\bar{x}_c$ the $\bar{x}$ coordinate when $\bar{h} = \bar{h}_c$, we get:

$$\frac{d\bar{p}}{d\bar{x}} = \frac{\bar{x}^2 - \bar{x}_c^2}{(1 + \bar{x}^2)^3}. \tag{6.25}$$

Integrating Eq. (6.25) with respect to $\bar{x}$:

$$\bar{p} = \int \frac{\bar{x}^2 d\bar{x}}{(1 + \bar{x}^2)^3} - \bar{x}_c^2 \int \frac{d\bar{x}}{(1 + \bar{x}^2)^3}. \tag{6.26}$$

These are standard integrals can be found using Math CAD or from Ref. 3. The solution comes to:

$$\bar{p} = \left[\frac{-\bar{x}}{4(1 + \bar{x}^2)^2} + \frac{\bar{x}}{8(1 + \bar{x}^2)} + \frac{1}{8} tg^{-1}(\bar{x})\right]$$

$$- \bar{x}_c^2 \left[\frac{x}{4(1 + \bar{x}^2)^2} + \frac{3}{8} \frac{\bar{x}}{(1 + \bar{x}^2)} + \frac{3}{8} tg^{-1} \bar{x}\right] + C_1. \tag{6.27}$$

We will need two additional boundary conditions to determine $\bar{x}_c$ and C_1. (Remember in deriving Eq. (6.23) we started by using boundary conditions at the top and bottom of the film (z direction). By this stage the boundary values are in the x direction. One pair, called the **Full Sommerfeld**[4] condition, discussed in Ref. 3, is:

$$p = 0 \quad \text{at} \quad \bar{x} = -\infty.$$

The condition at $\bar{x} = -\infty$ is called a **fully flooded** or **drowned inlet**. Inserting the above two boundary conditions into Eq. (6.27) we get after some manipulation and reference to standard integrals:

$$\bar{p} = \frac{-\bar{x}}{3(1+\bar{x}^2)^2}. \tag{6.28}$$

Equation (6.28) has an antisymmetric shape producing zero load capacity because of the positive and pressure loops of equal area. One alternative model is to ignore the negative pressures. It is called the **Half Sommerfeld boundary condition**[4] and discussed in Ref. 3. This is curve (b) in Fig. 6.9, obtained from Eq. (6.28) with $p = 0$ when $\bar{x} \leq 0$. Although it gives reasonable approximate answers for the pressure distribution, the abrupt change of pressure gradient to zero at $x = 0$, cannot occur because flow continuity is contravened (The Poiseuille flow component has suddenly

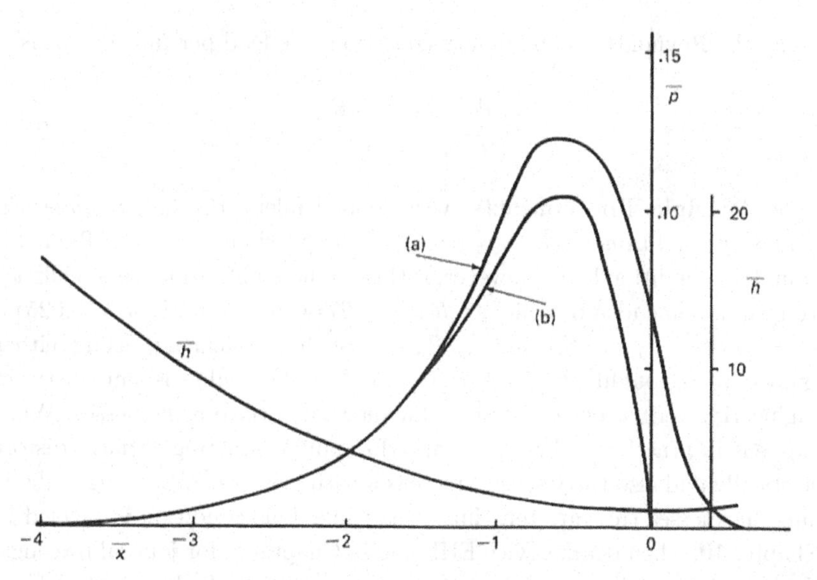

Fig. 6.9. Line contact pressure distribution for a flooded inlet distance. (a) Reynolds exit condition: (b) Half Sommerfeld exit condition.

vanished, although the Couette flow component continues across the z axis). Swift and Steiber[5] have suggested a more realistic boundary condition that does not contravene flow continuity, known as the **Reynolds exit boundary condition** (curve (a) in Fig. 6.9). It is often the one of choice in numerical solutions, especially when there are high loads, such as under EHL conditions (see Chapter 10). The Reynolds exit boundary condition states that the pressure exit boundary is where $p = dp/dx = 0$ (a little beyond $x = 0$). Experiments show that with Reynolds condition where film rupture occurs, the remaining Couette flow component has to expand into a widening gap, causing the flow to break up into oil carrying partitions (fingers) separated by air gaps (cavities). There are some illustrations of these in Ref. 6.

6.5.4. *Line contact load*

Using the half Sommerfeld boundary condition, integration of Eq. (6.28) between $x = -\infty$ and 0, produces a load per unit length of[6]:

$$\frac{W}{L} = \frac{4U\eta_0 R}{h_0}. \tag{6.29}$$

Using the Reynolds exit boundary condition, the load per unit length is[6]:

$$\frac{W}{L} = \frac{4.9U\eta_0 R}{h_0}. \tag{6.30}$$

Thus, for **rigid line contacts** with flooded inlets, the half Sommerfeld boundary condition produces a reasonable approximation to the Reynolds boundary condition load. However, is this predicted film thickness realistic? To give us some idea of scale, let $R = 0.0127\,\text{m}$, $W = 5000\,\text{N}$, $L = 0.025\,\text{m}$, $U = 1\,\text{m/s}$, $\eta_0 = 0.1\,\text{Nm}^{-2}\text{s}$. The least film thickness, using either expression is roughly $h_0 = 3 \times 10^{-8}\,\text{m}$. This low value is unrealistic in engineering terms, because it is below normal surface roughnesses. What happens in practice is that the counterformally contacting surfaces distort elastically and also the viscosity increases with pressure, producing realistic film thicknesses that are ten times this value This condition is tackled in Chapter 10, when dealing with EHL and in Chapter 8, for journal bearings, the boundary conditions discussed above are again applied to rigid surfaces that are in highly conformal contact.

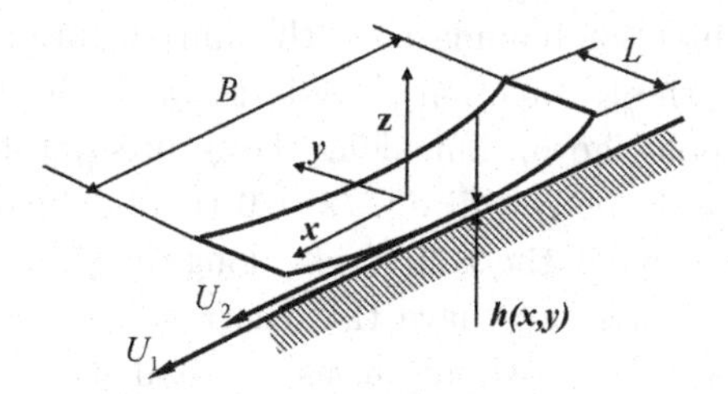

Fig. 6.10. Narrow bearing.

6.5.5. *Narrow bearings*

At the other extreme, if the bearing is assumed to be narrow, as in Fig. 6.10, $B << L$, we can assume now that $\partial p/\partial x$ can be neglected compared with $\partial p/\partial y$ With the above assumption, Eq. (6.20) becomes:

$$\frac{d}{dx}\left\{h^3\frac{dp}{dy}\right\} = 6(U_1 + U_2)\eta\frac{dh}{dx}. \tag{6.31}$$

Another assumption we must make for this simple bearing, is that $h \neq f(y)$. We can, therefore, remove h from the bracket to give:

$$\frac{d^2p}{dy^2} = 6(U_1 + U_2)\eta\frac{dh}{dx}\frac{1}{h^3}. \tag{6.32}$$

Integrating Eq. (6.31) with respect to y and noting that dh/dx does not vary with y:

$$\frac{dp}{dy} = 6\left\{(U_1 + U_2)\eta\frac{dh/dx}{h^3}\right\}y + C_1. \tag{6.33}$$

Integrating again with respect to y:

$$p = 6\left\{(U_1 + U_2)\eta\frac{dh/dx}{h^3}\right\}\frac{y^2}{2} + C_1y + C_2. \tag{6.34}$$

For the long bearing, we have generated two integration constants that can be obtained in the following way: Because of symmetry about the x axis, at the bearing ends: $p = 0$ along $y = \pm L/2$ and also at the center $dp/dy = 0$ along $y = 0$, yielding: $C_1 = 0, C_2 = -3(U_1 + U_2)\eta\frac{dh/dxL^2}{h^3}\frac{L^2}{4}$.
Thus:

$$p = 3(U_1 + U_2)\eta\frac{dh/dx}{h^3}(y^2 - (L^2/4)). \tag{6.35}$$

Note that in Eq. (6.35), dh/dx has not been stipulated, because only certain film shapes are eligible for short bearing solutions. The reason is that, along both the x and y direction boundaries of the film, the pressure must be zero. This is fine along the x direction boundaries with the rising and falling symmetrical parabola shown, but along the y direction boundaries of the film pressure can only be zero if $dh/dx = 0$ there. Thus, Eq. (6.35) must always have a zero slope in the x direction along the y direction boundaries. Fortunately, journal bearings have this characteristic so that Eq. (6.27) offers a simple design approximation, as we shall see later. On the other hand, for a straight wedge shaped bearing (flat roof and floor) Eq. (6.27) cannot be used. Finally, the narrow bearing assumption for certain bearing geometries can be used in practice with reasonable accuracy only if $L/B < 0.5$.

6.5.6. *Squeeze film bearings*

Equation (6.20) also applies to another type of bearing behavior depending on the **squeeze film effect**. If we place a flat plate onto a uniform thin film of oil, it will sink down slowly. The more viscous the oil, the more slowly it will sink. This viscous resistance is governed by the squeeze film effect. It is accommodated by the last term on the right hand side of Eq. (6.20). Noting that at any instant h is constant for a flat plate, and assuming ρ is invariable, Eq. (6.21) becomes:

$$\frac{\partial}{\partial x}\left[\frac{h^3}{\eta}\frac{\partial p}{\partial x}\right] + \frac{\partial}{\partial y}\left[\frac{h^3}{\eta}\frac{\partial p}{\partial y}\right] = 12\frac{dh}{dt}. \tag{6.36}$$

For a circular flat plate or a journal bearing, Reynolds equation is better expressed in polar coordinates. We will tackle a similar problem in Chapter 12 when we deal with alternating loads on journal bearings.

6.6. Rolling Contacts

Another important modification of Eq. (6.20) applies to any surface of revolution rolling without slip on a *stationary* plane covered by a lubricant film, such as the cylinder in Fig. 6.11(a). This is the same as the example above for the line contact of rigid discs in Sec. 6.5.3. However, the values to be assigned to U_1, are not immediately obvious. A full discussion of how this is done is given by Gohar in Ref. 6, but a quick solution can be found. In Fig. 6.11(a), if we put a backward velocity, $-U_1$, onto the cylinder center,

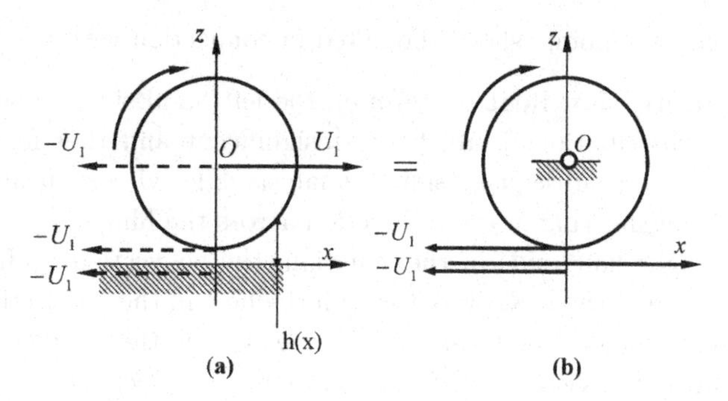

Fig. 6.11. Rolling cylinder.

the film roof and the floor (all dotted), we freeze point O, thus making the film roof and floor velocities at A both $-U_1$, as in (b).

Equation (6.22) therefore becomes for a rolling contact:

$$\frac{dp}{dx} = 12U_1\eta\left[\frac{h-h_c}{h^3}\right]. \tag{6.37}$$

Note again that Fig. 6.11 represents a **convergent-divergent wedge**, with the inlet part on the right hand side in both (a) and (b). We will frequently encounter this form of fluid film kinematics, when we deal with lubricated rolling element bearings.

6.7. The Energy Equation

The other fundamental equation needed is the **Energy equation**, where a full proof is given in Ref. 3. We will present it here in two dimensional form without proof, at a level suitable for our purposes. It is, for a Newtonian fluid:

$$\underbrace{\nu u\theta\frac{\partial p}{\partial x}}_{\textbf{compressive heating}} + \underbrace{\eta\left(\frac{\partial u}{\partial z}\right)^2}_{\textbf{viscous heating}} = \underbrace{\rho u c_p\frac{\partial \theta}{\partial x}}_{\textbf{convection cooling}} - \underbrace{k_t\frac{\partial^2\theta}{\partial z^2}}_{\textbf{conduction cooling}}, \tag{6.38}$$

where $\theta(x)$ is the temperature rise of the oil from inlet, ν is its coefficient of thermal expansion, c_p its specific heat (at constant pressure), and k_t its thermal conductivity.

The following points should be noted in connection with Eq. (6.38):

- The **compression heating** term on the left hand side, caused by the pressure distribution, is relatively insignificant and therefore will be ignored in our subsequent simple analysis. The viscous heating term comes from shearing in the x direction across the film.
- On the right hand side of the equation, the convection cooling term, carries some of the heat away through the flow in the x direction, while the conduction cooling term carries some away in the z direction across the solid boundaries. This term was discussed in Sec. 4.4.
- When later we deal with heat transfer in hydrodynamic bearings, the **viscous heating** and **convection cooling** terms are the most important because of the relatively thick films encountered.
- On the other hand, when we deal with **Elastohydrodynamic** contacts **(EHL)**, the viscous heating and conduction cooling terms are the most important because the oil films are much thinner.
- For our purposes, Eq. (6.38) is simplified considerably if we assume that the bearing film is sensibly parallel and long in the y direction, with the dominant heating from Couette flow (caused only from the x direction velocities of the boundaries).

With the above assumptions Eq. (6.38) becomes:

$$\eta\left(\frac{\partial u}{\partial z}\right)^2 = \rho u c_p \frac{\partial \theta}{\partial x} + k_t \frac{\partial^2 \theta}{\partial z^2}. \tag{6.39}$$

6.7.1. *Significance of terms in the energy equation*

Firstly, we must determine the relative significance of the right hand side terms of Eq. (6.39), in relation to the film thickness. With the above assumptions, we have Fig. 6.12.

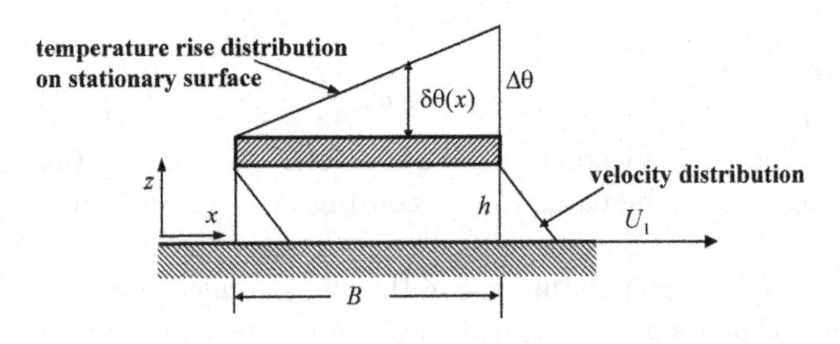

Fig. 6.12. Thermal effects.

6.7.2. Convected heat only

Dealing with the **convection cooling** term only in Eq. (6.26) we will use that term to derive an order of magnitude expression for the heat convected away. Assume that in this case the bounding surfaces are completely insulated so that all the heat is carried away by convection along the film. Let the average fluid velocity be $U_1/2$ and the maximum average temperature rise in the film be $\Delta\theta/2$. Then, at position x, the convected heat flow out across distance dx is:

$$\rho c_p \frac{d\theta}{dx} dx \int_0^h u dz = \rho c_p \left(\frac{d\theta}{dx}\right)\left(\frac{U_1 h}{2}\right) dx.$$

But at $x = B$, $d\theta/dx = \Delta\theta/2B$. Therefore, total convected heat flow out of the bearing end is

$$\frac{\rho c_p U_1 h}{2}\left(\frac{\Delta\theta}{2B}\right)\int_0^B dx = \frac{\rho c_p U_1 h \Delta\theta}{4} = \left(\frac{U_1 \rho h}{2}\right) \times c_p \times \frac{\Delta\theta}{2}. \tag{6.40}$$

The last term in Eq. (6.40) is an easy way of remembering that the convective heat flow rate is: mass flow per second × specific heat × temperature rise.

6.7.3. Conducted heat only

If all heat is removed by **conduction**, an order of magnitude solution, assumes that the temperature gradient *across* the film varies linearly for each value of x, rising from zero on the moving surface (along $z = 0$) to $\delta\theta/h$ on the stationary top surface, giving a parabolic temperature distribution across the film. Therefore, integrating with respect to z, the heat flow rate into the top surface through a column of width dx and height, h, is:

$$k_t dx \int_0^h (d^2\theta/dz^2) dz = k_t dx(\delta\theta/h). \tag{6.41}$$

Substituting $\delta\theta = x\Delta\theta/B$ into Eq. (6.41), the total conducted heat flow through the whole top surface is:

$$\frac{k_t \Delta\theta}{Bh}\int_0^B x dx = \frac{k_t \Delta\theta B}{2h}. \tag{6.42}$$

Table 6.1. Diffusivity and Peclet numbers.

Fluid	Oil	Water	Air (STP)
Diffusivity (κ) (m^2/s)	8.96×10^{-8}	1.4×10^{-7}	2.180×10^{-5}
Peclet number (Pe)	14.92	9.346	0.061

6.7.4. *Heat flow ratio*

Therefore, from Eqs. (6.40) and (6.42), the ratio of convected to conducted heat (defined here as the fluid **Peclet number**) is given by the relationship:

$$Pe = \frac{\left(\frac{U_1 h}{4} c_p \rho \Delta\theta\right)}{\left(\frac{\Delta\theta k_t B}{2h}\right)} = \frac{\frac{U_1 h^2}{2B}}{\frac{k_t}{\rho c_p}}. \tag{6.43}$$

The bracketed group of material constants on the right hand side of Eq. (6.43) is called the **thermal diffusivity** (κ) of the fluid. For various fluids, Table 6.1 gives values of thermal diffusivity and Peclet Numbers for a typical high speed long journal of diameter $D = 0.0076\,\text{m}$ ($B = \pi D$), surface speed $U_1 = 61\,\text{ms}^{-1}$ and radial clearance $h = 10^{-4}\,\text{m}$ bearing (see Chapter 8).

Note that k_t has units of: $\text{kJ}\,\text{m}^{-1}\text{s}^{-1}(^\circ\text{K})^{-1}$ and σ has units of: $\text{kJ}\,\text{Kg}^{-1}(^\circ\text{K})^{-1}$.

For the three fluids shown, oil is the best medium for carrying some of the heat away by convection, while air is best for transfer by conduction.

Had the model been two involute gear teeth meshing in oil, typically of contact width 0.002 m, total surface speed 10 m/s and an average film thickness of 1 micron, then Pe = 0.0279, which is about 1/500 of that of the journal bearing above.

6.8. Closure

In this chapter we have discussed the significance of the hydrodynamic wedge in Lubrication theory. We then derived Reynolds equation, applying it to both long and narrow geometries, together with the various boundary conditions employed. Finally, the relative significance of the terms composing the Energy equation was analyzed in relation to heat transfer. Some of these initial results will be applied to the design of bearings in subsequent chapters. The next chapter will analyse the thrust bearing and its practical applications.

References

1. Reynolds, O., *Phil. Trans.* **177** (1886) 157–244.
2. Middleman, S., *An Introduction to Fluid Dynamics*, John Wiley & Sons NY (1998).
3. Cameron, A., *Principles of Lubrication* Longmans (1964).
4. Sommerfeld, A., *Zeits. f. Math. U. Phys.* **40** (1904) 97–155.
5. Steiber, Das-Schwimmlager, *VDI* (1933).
6. Gohar, R., *Elastohydrodynamics*, 2nd ed., Imperial College Press (2001).

CHAPTER 7

THRUST BEARINGS

7.1. Introduction

The purpose of the hydrodynamic thrust bearing is to accommodate axial loads. It is often used in heavy equipment, such as on ships, to take the horizontal thrust component along the propeller shaft. Alternatively, the rotor may have attached to it the blades of a vertical axis axial flow water turbine in a hydroelectric scheme.

Thrust bearings follow the general principle of the hydrodynamic wedge discussed in Chapter 6. There, we gave you some worked examples and introduced you to the boundary conditions necessary for a solution. A pressure distribution is created by oil being dragged into a converging wedge formed from an inclined stationary bearing pad (the stator) and the motion of a *runner* (sometimes called a *collar*). The pad is either machined to a fixed inclination or pivoted, allowing it to take up an optimum inclination according to the external conditions (a **pivoted pad bearing**). The pivot can either be designed to make a line contact with the supporting housing or it can be a spherical button making a point contact. We will deal with fixed inclination and pivoted pad bearings later.

Because the motion of the runner in a bearing is normally rotary, the stator is split into several sector shaped pads (6–18), forming a number of wedges with the runner. Figure 7.1(a) shows the pad shape, together with the resulting oil film pressure contours and lubricant flow lines resulting from the runner's rotary motion. The arrangement gives identical converging films over each of the pads. Figure 7.1(b) illustrates the arrangement of a pivoted pad type bearing.

The part section through X–X shows a pivoted pad resting on the housing. In the end view the equispaced pads are shown. Between each pair is a *supply chamber* with a lubricant supply hole. The pressure contours in Fig. 7.1(a) suggest that there are transverse pressure gradients, causing some of the entrained lubricant to leak out radially. It is this *side leakage* lubricant that is externally pumped back into the film wedges via these supply holes.

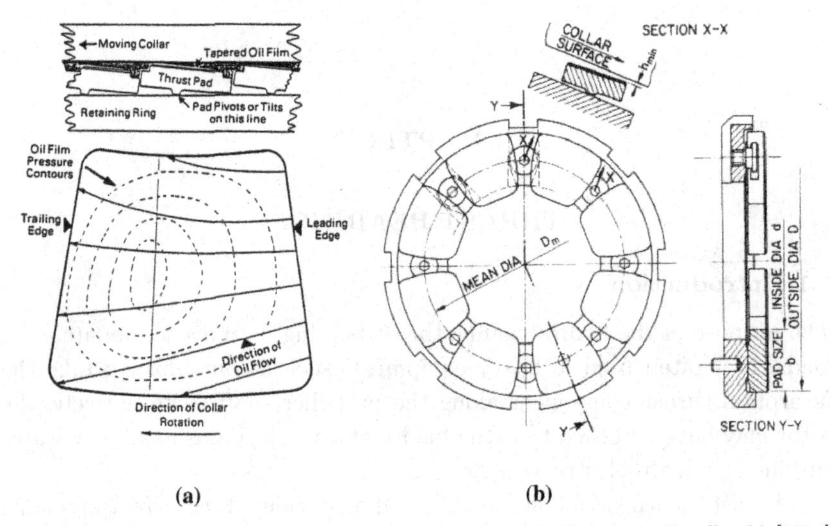

Fig. 7.1. Pivoted pad thrust bearing. From: M J Neal 'Tribology Handbook' [1995] reproduced by permission of Elsevier Ltd.

7.2. Analysis of Thrust Bearings

7.2.1. *Geometry*

As we did in Sec. 6.4.1, a simplified analysis assumes that the rectangular pads are rigid and long transverse to the direction of the runner velocity. For simplicity, the origin of coordinates for the pair is taken at the wedge start, as in Fig. 7.2.

With this geometry, as the inlet is on the right, the runner velocity is $-U_1$ and the minimum film thickness position is at $x = d_0$. The *inclination factor* (*or ratio*) is defined as:

$$K = \frac{h_1 - h_0}{h_0} = \frac{t_p}{h_0}.$$

Dimension t_p is called the taper of the **wedge pair**.

$$\text{Also: } h = \frac{xh_0}{d}, \quad d = B\frac{h_0}{(h_1 - h_0)} = \frac{B}{K}. \text{ Therefore: } x = \frac{hB}{h_0 K}.$$

7.2.2. *Pressure distribution*

Equation (6.23) now becomes:

$$\frac{dp}{dx} = -6U_1\eta\frac{h - h_c}{h^3}, \tag{7.1}$$

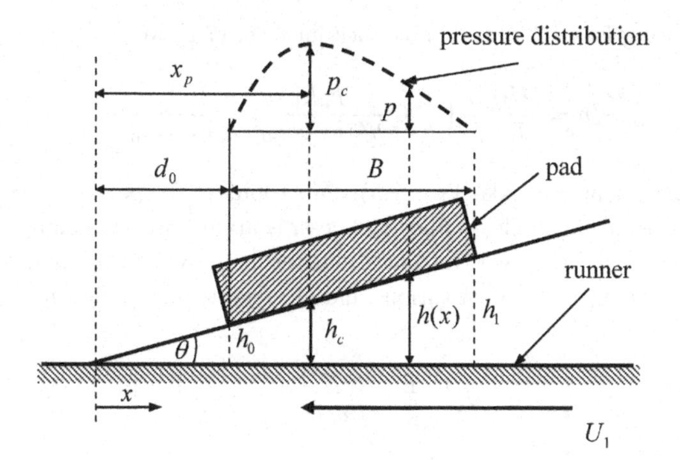

Fig. 7.2. A single plane thrust bearing pair.

where h_c is the film thickness at the maximum pressure.

$$\text{As} : x = \frac{hB}{h_0 K}, \quad dx = \frac{dhB}{h_0 K},$$

Eq. (7.1) can, therefore, be cast in terms of h to give:

$$-dp = \frac{6U_1 \eta B}{K h_0} \left(\frac{dh}{h^2} - \frac{h_c dh}{h^3} \right).$$

Integrating with respect to h:

$$p = \frac{-6U_1 \eta B}{K h_0} \left(-\frac{1}{h} + \frac{h_c}{2h^2} + C \right). \tag{7.2}$$

To obtain the integration constant, C, and the film thickness, h_c, we know that $p = 0$ at $h = h_0$ and $h = h_1$, giving:

$$-\frac{1}{h_0} + \frac{h_c}{2h_0^2} + C = 0,$$

$$-\frac{1}{h_1} + \frac{h_c}{2h_1^2} + C = 0.$$

Solving for C and h_c, after some manipulation we get:

$$h_c = \frac{2h_0 h_1}{h_1 + h_0}, \tag{7.3}$$

$$C = \frac{1}{(h_1 + h_0)}. \tag{7.4}$$

Substituting Eqs. (7.3) and (7.4) back into Eq. (7.2) we get:

$$p = \frac{6U_1\eta B}{Kh_0}\left(\frac{1}{h} - \frac{h_0h_1}{h^2(h_1+h_0)} - \frac{1}{h_1+h_0}\right). \tag{7.5}$$

It is convenient to make Eq. (7.5) dimensionless. Inspection of Fig. 7.2 indicates that h_0, the least film thickness, is an important design variable, to be retained. Also, we can put h_1 in terms of K and h_0, and let $\bar{h} = h/h_0$. Inserting these substitutions and after some manipulation, Eq. (7.5) becomes:

$$p = \frac{6U_1\eta B}{Kh_0^2}\left[\frac{1}{\bar{h}} - \frac{K+1}{\bar{h}^2(K+2)} - \frac{1}{K+2}\right]. \tag{7.6}$$

The terms in the square bracket are dimensionless, so if we transpose the remaining dimensional terms to the left hand side of Eq. (7.6) we have a dimensionless pressure as:

$$\bar{p} = \frac{h_0^2 p}{6U_1\eta B} = \frac{1}{K}\left[\frac{1}{\bar{h}} - \frac{K+1}{\bar{h}^2(K+2)} - \frac{1}{K+2}\right]. \tag{7.7}$$

Equation (7.7) gives the dimensionless pressure distribution in terms of the local dimensionless film thickness, $\bar{h}$. In order to find the maximum pressure, p_c (at $h = h_c$), substitute Eq. (7.3) into Eq. (7.5) to get:

$$p_c = \frac{6U_1\eta B}{Kh_0^2}\left[\frac{(h_1-h_0)^2}{4h_1(h_1+h_0)}\right]. \tag{7.8}$$

Finally, by using the definition of K to remove h_1, Eq. (7.8) also can be made non-dimensional to yield $\bar{p}_c$:

$$\bar{p}_c = \frac{K}{4(K+2)(K+1)}. \tag{7.9}$$

7.2.3. *Load capacity*

Knowing the pressure distribution, we can find the load per unit length, W/L as:

$$\frac{W}{L} = \int_{d_0}^{d_0+B} p\,dx = \frac{B}{h_0K}\int_{h_0}^{h_1} p\,dh.$$

Substituting for p from Eq. (7.6) and integrating:

$$\frac{W}{L} = \frac{6U_1\eta B^2}{K^2h_0^2}.$$

Putting the dimensional constants on the left hand side and integrating:

$$\frac{Wh_0^2}{6U_1\eta LB^2} = \frac{1}{K^2}\left|\ln(h) + \frac{h_0h_1}{h\,(h_0+h_1)} - \frac{h}{(h_0+h_1)}\right|_{h_0}^{h_1}$$

$$= \frac{1}{K^2}\left[\ln\left(\frac{h_1}{h_0}\right) - \frac{2\,(h_1-h_0)}{(h_1+h_0)}\right].$$

Let $W^* = \frac{Wh_0^2}{6U_1\eta LB^2}$, the dimensionless load, and substitute $h_1/h_0 = K+1$, to give:

$$W^* = \frac{1}{K^2}\left[\ln(K+1) - \frac{2K}{K+2}\right]. \tag{7.10}$$

From the definition of the dimensionless load, we can define a dimensionless film thickness as:

$$\frac{h_0}{B} = \sqrt{6W^*}\sqrt{\frac{U_1\eta L}{W}}. \tag{7.11}$$

Finally, if the ln term in Eq. (7.10) is expanded as an infinite series, as: $h_1 \to h_0, K \to 0, W^* \to 0$, a condition that defines a parallel surface hydrodynamic bearing and thus one with zero load capacity. A plot of W^* against K, for different pad length-to-width ratios, is shown in Fig. 7.3 (The $L/B = 10$ curve represents a long bearing). Observe that there always exists an optimum value of K for maximum load capacity, being highest for the long bearing. This property will be exploited in the following section.

7.2.4. *Pivot point location*

The center of pressure of the pressure distribution curve (distance x_p in Fig. 7.2) defines the required pivot position for **tilting pad bearings.**

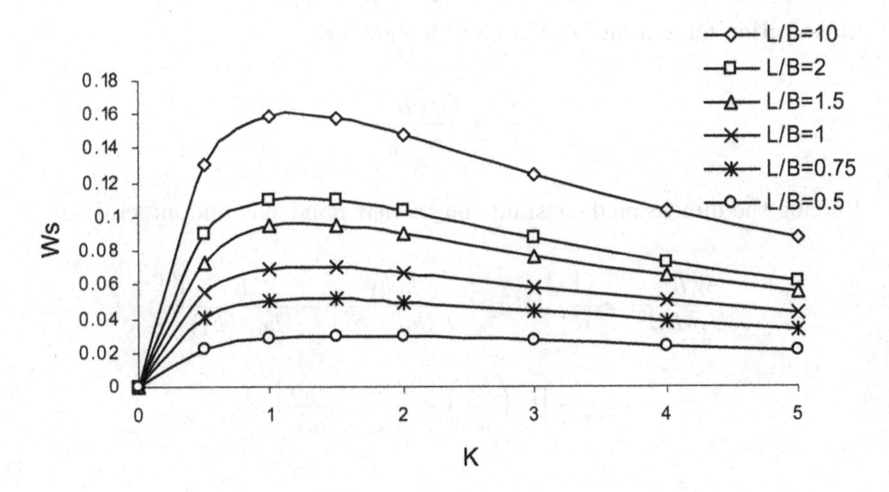

Fig. 7.3. Plane thrust bearings. Inclination ratio against dimensionless load.

Thus:

$$\frac{x_p W}{L} = \int_c^{c+B} px\,dx.$$

Substituting for p from Eq. (7.5):

$$\frac{x_p W}{L} = \frac{6U_1\eta B^3}{K^3 h_0^3}\int_{h_0}^{h_1}\left[1 - \frac{h_0 h_1}{h\,(h_1+h_0)} - \frac{h}{(h_1+h_0)}\right]dh,$$

$$\therefore \quad \frac{x_p W}{L} = \left| h - \frac{h_0 h_1}{h_0+h_1}\ln h - \frac{h^2}{2\,(h_0+h_1)}\right|_{h_0}^{h_1}.$$

Substituting $W^* = \frac{h_0^2 W}{6U_1\eta B^2 L}$ we get, after some rearrangement:

$$\frac{x_p}{B} = \frac{1}{W^* K^3}\left(\frac{K}{2} - \frac{K+1}{K+2}\ln(K+1)\right)$$

It is more usual to define the position of the center of pressure, X, from the leading edge of the bearing, where $X = (B + x_p) - d$ in Fig. 7.2. We can also replace W^* above by a function of K using Eq. (7.10). Finally, letting $X^* = X/B$, we obtain, after considerable

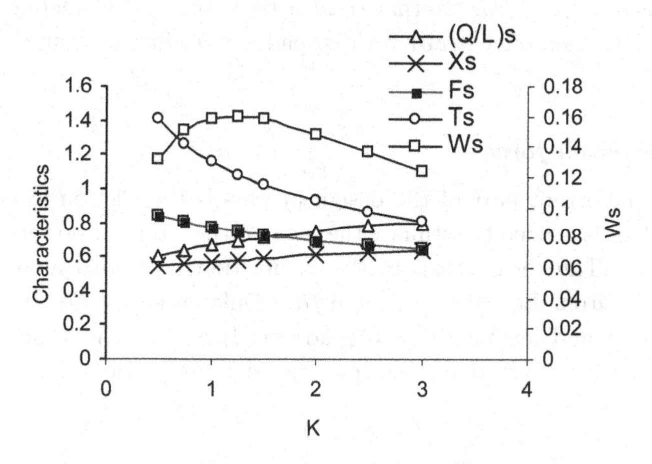

Fig. 7.4. Characteristics of a long pad bearing.

manipulation:

$$X^* = \frac{2(K+3)(K+1)\ln(K+1) - K(6+5K)}{2K[(K+2)\ln(1+K) - 2K]}. \tag{7.12}$$

Figure 7.4 shows how K varies withX^* for an infinitely long tilting pad bearing (crosses). The variation in X^* over the range of K is relatively small.

As $K \rightarrow 0$ (a parallel surface bearing) the vanishing pressure distribution becomes symmetrical, with $X^* \rightarrow 0.5$ (not shown).

The K values in Fig. 7.4, define different inclinations of the pad. As the maximum value of W^* is at $K = 1.2$, the optimum pivot position for highest load capacity in a long tilting pad bearing is there. Note also from Eq. (7.12) that X^* is a function only of K. Thus, once the value of K is chosen, a rigid pivoted pad bearing will always operate at this film configuration whatever the actual load, and with h_1/h_0 being always constant. A fixed inclination pad is not accorded this luxury.

It is worth pointing out here, that in our simple analysis, we have assumed that the pads and runner are both rigid. In practice they may not be, either by accident or design. It is possible to design the pads with their pivot positions at the center ($K = 0$). This would give zero load capacity for rigid surfaces. If, however, the pads can flex under the pressure distribution to produce a wedge film geometry, they can accommodate either sense of runner rotation, a useful design feature. Hall and de Guerin discuss this design.[1] An alternative idea by Ettles and Cameron[2] utilises thermal distortion of centrally pivoted pads to produce a similar effect.

7.2.5. *Friction force*

Another important part of the design process is the friction force on the runner. This is needed to estimate the temperature rise and power absorbed in the film. The shear stress in the x direction developed at a point in the film is, from Eq. (6.8) $\tau = \eta \partial u / \partial z$. Differentiating Eq. (6.13) with respect to z and letting $U_2 = 0$ (pad speed) and remembering that U_1 (runner speed) is negative because of the sign convention chosen for this chapter, then:

$$\frac{\partial u}{\partial z} = \frac{1}{\eta}\frac{\partial p}{\partial x}\left(z - \frac{h}{2}\right) + \frac{U_1}{h}.$$

The shear stress at the runner ($z = 0$) is therefore:

$$\tau_0 = -\frac{dp}{dx}\frac{h}{2} + \frac{\eta U_1}{h}.$$

The shear force in the x direction at the runner surface (of length L) follows as:

$$F_0 = \int_0^L \int_{d_0}^{d_0+B} \tau_0 \, dx dy = L \int_{d_0}^{d_0+B} \left(\underbrace{-\frac{dp}{dx}\frac{h}{2}}_{(a)} + \underbrace{\frac{\eta U_1}{h}}_{(b)} \right) dx \qquad (7.13)$$

Integrating by parts term (a) on the right hand side of Eq. (7.13):

$$\int_{d_0}^{d_0+B} \left(-\frac{dp}{dx}\frac{h}{2} \right) = - \left| p \frac{h}{2} \right|_{d_0}^{d_0+B} + \int_{d_0}^{d_0+B} p \frac{dh}{2}.$$

The first term of (a) above is zero, because p is zero at each end of the bearing.

For the second term of (a) above, multiply top and bottom by dx to get $\int_c^{c+B} \frac{p}{2}\frac{dh}{dx} dx$. Noting that $dh/dx = \tan\theta =$ constant, and $L \int_c^{c+B} \frac{p}{2} dx = \frac{W}{2}$, we can write the complete first term of (a) in Eq. (7.13) as $(W/2)\tan\theta$.

Turning our attention now to term (b) in Eq. (7.13), from the bearing geometry $h = xh_0/c = xh_0K/B$. Therefore, term (b) is:

$$\int_{d_0}^{d_0+B} \frac{\eta U_2}{h} dx = \frac{\eta U_2 B}{h_0 K} \int_{d_0}^{d_0+B} \frac{dx}{x} = \frac{\eta U_1 B}{h_0 K} \ln\left(\frac{d_0 + B}{d_0}\right).$$

Adding terms (a) and (b) the complete expression for the friction force per unit length at the runner is:

$$\frac{F_0}{L} = \frac{W \tan\theta}{2L} + \frac{\eta U_1 B}{h_0 K} \ln\left(\frac{d_0 + B}{d_0}\right).$$

Alternatively, using the definition of K and putting W in terms of W^*, we can write in general terms:

$$\frac{F_0 h_0}{LB\eta U_1} = F^* = \left[\frac{4\ln(1+K)}{K} - \frac{6}{2+K}\right]. \tag{7.14}$$

The dimensionless term in brackets in Eq. (7.14) applies to long bearings only, while the definition of F^*on its left hand side, is perfectly general. The long bearing friction characteristic, F^*, is also shown in Fig. 7.4.

Note that as, $K \to 0, W \to 0$, Eq. (7.14) becomes:

$$\frac{F_0}{L} = \frac{\eta U_1 B}{h_0}. \tag{7.15}$$

Equation (7.15) leaves the friction force only with the **Couette flow** term because of the absence of a pressure distribution. In all cases, the magnitude of the friction force on the pad (F_h) will be the same as on the runner when its inclination is taken into account in the analysis[1] and see Chapter 8 for journal bearings.

7.2.6. *Mass flow*

Another feature of the design process is mass flow through the bearing. Like the friction force, this is needed to estimate the power absorbed by the film. From Eq. (6.19), making $U_2 = 0$ and U_1 negative, the mass flow per unit length is

$$m_x = -\frac{\rho h^3}{12\eta}\left[\frac{\partial p}{\partial x}\right] - U_1\left[\frac{\rho h}{2}\right].$$

Also, $dp/dx = 0$ at $h = h_c$, so:

$$m_x = \frac{-\rho U_1 h_c}{2}. \tag{7.16}$$

Moreover, using the definition of K, Eq. (7.3) can be written as $h_c = \frac{2h_0(1+K)}{2+K}$ giving, for long bearings:

$$m_x = -\rho U_1 h_0 \frac{(1+K)}{2+K} \tag{7.17}$$

In general terms, and omitting the negative sign, the total volumetric flow rate is:

$$\frac{Q}{LU_1 h_0} = Q^* = \frac{1+K}{2+K}. \tag{7.18}$$

Remember that the film shape function of K, in Eq. (7.18) applies *only* to long bearings. For finite length bearings, the function of K on the right hand side would be replaced by a value obtained from a numerical solution. Long bearing flow characteristics are shown in Fig. 7.4.

7.2.7. *Temperature effect*

As we will see later when designing journal bearings, it is important to consider the effect of temperature on the viscosity of the lubricant. For thrust bearings, the mean lubricant viscosity can be 20 to 40% of the entry viscosity to the bearing housing, before distribution to the wedge pairs. As shown in Fig. 7.1, the full design procedure for a bearing with a plurality of sector shaped pads requires a complex numerical analysis involving the Reynolds and Energy equations together with the lubricant temperature/viscosity equation. Also taken into account in a complete solution, is pad distortion caused both by conductive heating, as well as from the individual pressure distributions.[2,3] There is insufficient space here to discuss the numerical procedure. All we will do is to explain an approximate design process that can be used on a bearing with rigid plane pads, which estimates a single representative lubricant film temperature and its corresponding effective viscosity. The method is based on data from existing isothermal numerical solutions.

Referring to the Energy Eq. (6.38), for an element in a single parallel film of thickness h and assuming convection only (as discussed in Sec. 6.7), we have:

$$\eta\left(\frac{\partial u}{\partial z}\right)^2 = \rho u c_p \frac{\partial \theta}{\partial x}. \tag{7.19}$$

The left hand side of Eq. (7.19) is from viscous heating and the right hand side is from convective cooling. Integrating across the film of thickness h and assuming that η is constant *throughout* the lubricant let it be called now the **effective viscosity**, η_e. Using this *single value* to represent the variable viscosity, we now have:

$$\eta_e \int_0^h \left(\frac{\partial u}{\partial z}\right)^2 dz = \rho c_p \int_0^h u \frac{\partial \theta}{\partial z} dz. \tag{7.20}$$

Assuming, additionally a linear mean film temperature rise over width B of $\Delta\theta$, and ignoring pressure effects, if U_1 is the runner velocity:

$$\frac{\partial u}{\partial z} = \frac{U_1}{h} \quad \frac{\partial \theta}{\partial x} = \frac{\Delta\theta}{B}.$$

We can now take both terms outside the integrals in Eq. (7.20). With these substitutions we get, after integrating and rearranging:

$$\frac{\Delta\theta}{\eta_e} = \frac{2BU_1}{\rho c_p h^2}. \tag{7.21}$$

We could have also obtained Eq. (7.21) directly by writing: mass flow/s $\times$ specific heat $\times$ temperature rise $=$ work done per second (power), that is:

$$\frac{\rho U_1 h}{2} \times c_p \times \Delta\theta = \left(\frac{\eta_e U_1 B}{h}\right) U_1.$$

Moreover, using this power balance, we can write in general terms for a long bearing and *any* film geometry, including the pressure effects:

$$(Q\rho) c_p \Delta\theta = F U_1. \tag{7.22}$$

where Q is the inlet flow into the film.

We already have general expressions for F and Q. Substituting these from Eqs. (7.14) and (7.17) into Eq. (7.22) yields:

$$F^*\left(\frac{\eta_e U_1 BL}{h_0}\right) = Q^*\left(LU_1 h_0\right)\rho c_p \Delta\theta.$$

Then, letting $\theta^* = F^*/Q^*$ we can now write:

$$\Delta\theta = k_1\theta^*\left(\frac{\eta_e U_1 B}{\rho c_p h_0^2}\right). \tag{7.23}$$

The empirical constant, k_1, accounts for the possibility that not all the heat is removed by convection from the bearing end, some escaping by conduction through the pad and runner. For a plane thrust bearing pair this is generally taken as $k_1 = 0.6$. All the above expressions for load, pivot position, friction, flow, temperature rise and the corresponding dimensionless parameters for a long bearing are summarized in Appendix Table 7.1 at the end of the chapter. Appendix Table 7.2 shows the numerical results for long bearings and Fig. 7.4 has graphed their behavior.

7.2.8. *Effective temperature*

Assuming h_0, the inlet temperature to the film, θ_1 and U_1 are all stipulated, the effective viscosity, η_e, and total temperature rise *along the pad*, $\Delta\theta$, are unknowns in Eq. (7.23). We, therefore, need another equation for a solution. The method is to seek a value of η_e that can represent the temperature dependent variable viscosity throughout the film. We say that η_e must have a corresponding **effective temperature** θ_e lying between the film inlet and exit temperatures. Thus, we can state that:

$$\theta_e = \left(\theta_1 + k_2\Delta\theta\right). \tag{7.24}$$

Normally, for thrust bearings, θ_e is taken as the mean temperature of the pad, so that $k_2 = 0.5$. Equation (7.24) can also be used for finite length bearings, discussed in Sec. 7.3. It also is valid for journal bearings in Chapter 8. A single viscosity, η_e, for the film at θ_e enables us to employ the isothermal expressions we have already derived above for determining the bearing performance characteristics. How these quantities are obtained will be explained later by some examples.

7.3. Finite Length Plane Thrust Bearings

7.3.1. *Introduction*

These bearings are the ones used in practice. Their pads are generally segmental in plan-form with a number of them creating wedge pairs supporting the annular runner, as we discussed in Sec. 7.1 and illustrated in Fig. 7.1. When analyzing a single wedge pair, the square pad is a reasonable approximation to a segmental shape. Because each pad is now of finite length, the pressure distribution over it is no longer uniform in the y direction because it must fall to zero along the pad sides. The isobars in Fig. 7.1 illustrate this behavior. Besides reducing the load capacity, these falling pressures cause side leakage of the oil, with the make up flow being supplied through feedholes into the chambers between the wedges (Fig. 7.5). The steady state 2 dimensional Reynolds equation (Eq. (6.20)) is, therefore, required to analyze a finite length bearing, a numerical solution usually being necessary. The complete solution requires, in addition, the full energy equation, as well as accounting for the possibility of thermal and mechanical flexure of the pads. There are also complicated thermal mixing processes in the chambers. It is important to understand this process if a simple design procedure is to be used. We will explain this in the next section.

7.3.2. *Thermal design of finite length bearings*

As we have demonstrated above, for a single *long* pad bearing pair, a convective flow analysis needs to only consider the x direction flow entering and leaving the wedge film. In the case of a finite length bearing pair, the situation is more complicated on account of the additional side flow. Pinkus[4] discusses the resulting flow processes. The side flow causes some of the heat to be carried away by convection from the pad sides as well, in addition to the flow from its end. Furthermore, as we have discussed in Sec. 7.1, there is normally a plurality of pairs arranged circumferentially below an annular runner (Fig. 7.1). Consider now Fig. 7.5. It depicts a supply chamber formed by the inlet and outlet profiles of two adjacent pads in the bearing, with the runner, moving at constant speed U_1, entraining the lubricant.

The finite lengths of the pads and their transverse pressure gradients (y direction) cause a proportion of the lubricant to leak out from the film sides. This oil side flow leaves the bearing and travels to a common

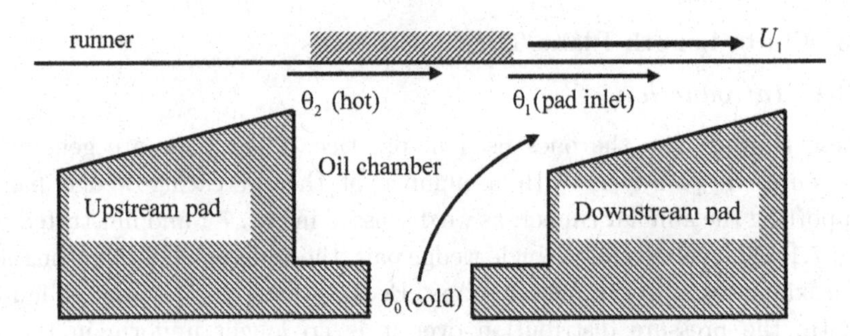

Fig. 7.5. Lubricant mixing in the supply groove.

sump from where it is pumped back and distributed via feed holes to the chambers situated between the pads. When it re-enters the bearing films, the make up oil has been cooled down to bulk temperature θ_0. In each chamber, this cold entering oil, at temperature θ_0, mixes with the hot oil leaving the upstream pad film at temperature θ_2 and heats up to temperature θ_1, the entry temperature to the adjacent downstream film. The mixing of the hot and cold oil in the chamber is a complex process, both thermally and hydrodynamically. Nevertheless, if we now consider one of these bearing pairs from a design perspective, the problem can be simplified.

7.3.3. *Power balance for the effective viscosity*

Figure 7.6 shows a pad film surrounded by a control volume. The power supplied to drive the runner is P, the externally pumped lubricant make-up power is $\rho c_p Q_0 \theta_0$ and the power possessed by the side leakage lubricant is $\rho c_p Q_s \theta_s$.

Considering the external power balance, power going into the control volume equals power leaving it, that is:

$$P + \rho c_p Q_0 \theta_0 = \rho c_p Q_s \theta_s.$$

Since: $Q_0 = Q_s$, then:

$$P = \rho c_p Q_s (\theta_s - \theta_0). \tag{7.25}$$

Remember, $\theta_s - \theta_0$ represents the temperature rise from the inlet *supply chamber* to the equivalent temperature of the side flow.

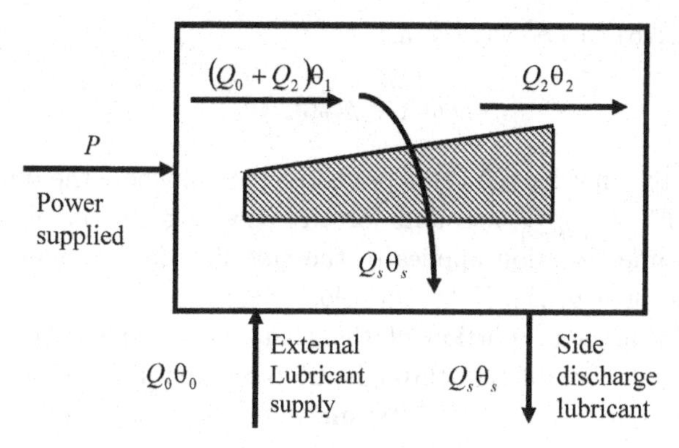

Fig. 7.6. Control volume round a thrust bearing pair. (The ρc_p terms have been omitted.)

If we now consider the power flow *within* the bearing film, assuming perfect thermal mixing in the supply chamber, we can write:

Friction power inputted to the film equals total power carried away from the film sides and its end, that is:

$$P = \rho c_p[Q_s(\theta_s - \theta_1) + Q_2(\theta_2 - \theta_1)]. \tag{7.26}$$

Equation (7.26) is not of direct use in our simple design procedure, because the pad inlet and outlet temperatures are unknown. They can, however, be found if the empirical Eq. (7.24), describing the assumed position of the effective temperature along the film, is used in conjunction with Eq. (7.26). One of the Chapter 7 questions invites the reader to try this procedure.

However, Eq. (7.25) is adequate for our approximate solution, because $\theta_s - \theta_0$ includes this local temperature rise there. Another advantage of using Eq. (7.25) is that θ_0 can be chosen as input data, it being easily measured experimentally. Therefore, the procedure discussed below attempts to find θ_s but cannot alone determine θ_1 or the maximum temperature, θ_2, at exit to the film.

7.3.4. *Effective temperature design coefficients*

From the definition of P and letting:

$$\Delta\theta_s = \theta_s - \theta_0. \tag{7.27}$$

Equation (7.25) can be written as:

$$F_0 U_1 = \rho c_p Q_s \Delta\theta_s. \tag{7.28}$$

We also stated that Eq. (7.14) is perfectly general where the dimensionless coefficient $F^* = \frac{F_0 h_0}{LB\eta U_1}$ accounts for any chosen geometry of the bearing film. The same situation applies to the side flow factor defined as $Q_s^* = \frac{Q_s}{LU_1 h_0}$, equivalent to Eq. (7.18) for a long bearing.

In addition, a proportion of the heat generated in the film, k_1, is assumed lost by conduction through the bounding solid surfaces. Putting these substitutions into Eq. (7.28) and letting $\theta_s^* = F^*/Q_s^*$ we can now write:

$$\Delta\theta_s = \left(\frac{k_1}{\rho c_p}\right)\left(\frac{B\eta U_1}{h_0^2}\right)\theta_s^*. \tag{7.29}$$

The coefficients $W^*, F^*, Q^*, Q_t^*, Q_s^*$ and θ_s^* are tabulated in Appendix Table 7.2. There is now sufficient information for some design examples to be tackled.

In addition, an empirical formula equivalent to Eq. (7.24) for long bearings, can be written as:

$$\theta_S = \theta_1 + k_2(\theta_2 - \theta_1), \tag{7.30}$$

where $k_2 = 0.5$ for thrust bearings.

Equation (7.30) states that the *position* of the equivalent side leakage temperature along the pad is assumed to be midway between the pad inlet and outlet temperatures. In conjunction with Eq. (7.26), an estimate of the *maximum* temperature rise in the bearing may be found.

7.4. Examples of Thermal Design

7.4.1. *Long bearing*

A single plane pivoted pad *long* bearing has $K = 1$, $B = 40\,\text{mm}$, $L = 160\,\text{mm}$, $U_1 = 10\,\text{ms}^{-1}$ and $W = 30\,\text{KN}$. The oil temperature at entry to the bearing film is 40°C, and a viscosity there of 95.85 cSt. The oil specific heat is 1880 $\text{Jkg}^{-1}\,{}^\circ\text{C}^{-1}$ and its density is $850\,\text{kgm}^{-3}$. Find the maximum oil temperature rise in the bearing, its effective temperature, and viscosity, and the least film thickness. Hence find the flow, friction force, coefficient of friction and power consumed in the bearing.

Solution

As $K = 1$ always, we can use Appendix Table 7.1 to obtain the necessary information on the dimensionless coefficients. Assuming $k_1 = 0.8$, the temperature rise can be obtained from Appendix Table 7.1:

$$\Delta\theta = \frac{k_1}{\rho c_p}\left[\frac{F^*}{W^*}\left(\frac{2+K}{1+K}\right)\frac{W}{BL}\right]$$
$$= \frac{0.8}{850 \times 1880}\left[\frac{0.7726}{0.1589}\left(\frac{3}{2}\right)\left(\frac{30 \times 10^3}{0.04 \times 0.16}\right)\right] = 17.11°\text{C } (\textbf{Answer}).$$

We can now find the effective temperature from Eq. (7.24):

$$\theta_e = \theta_1 + 0.5\Delta\theta = 40 + 8.55 = 48.55°\text{C}(= 321.55°\text{K})\ (\textbf{Answer})$$

To obtain the effective viscosity, we will use Vogel's formula from Example 5.4 for this particular oil (remember that the viscosity comes out in cp):

$$\ln(\eta_e) = -1.845 + \frac{700.81}{321.55 - 203.1} = 4.099$$
$$\therefore \quad \eta_e = 58.674cp = 0.0587PaS\ (\textbf{Answer}).$$

From Appendix Table 7.1, the relevant expression is then used to obtain the minimum film thickness:

$$\Delta\theta = k_1\theta^*\left(\frac{\eta_e U_1 B}{\rho c_p h_0^2}\right).$$

Thus, $h_0 = \left(\sqrt{\frac{0.8\times1.75\times0.0587\times10\times0.04}{850\times1880\times16}}\right) = 0.036\,\text{mm}$ (**Answer**).
The other parameters can now be found:

$$Q = Q^*(U_1 L h_0) = 0.667 \times 10 \times 0.16 \times 3.6 \times 10^{-5}$$
$$= 3.842 \times 10^{-5}\,\text{l/s} \rightarrow 2.3\,\text{l/m}\ (\textbf{Answer}).$$
$$F = F^*\left(\frac{LB\eta U_1}{h_0}\right) = 0.7726 \times \left(\frac{0.16 \times 0.04 \times 0.0587 \times 10}{3.6 \times 10^{-5}}\right)$$
$$= 80.62\text{N}\ (\textbf{Answer}).$$
$$\mu = \frac{F}{W} = \frac{80.62}{30 \times 10^3} = 0.0027\ (\textbf{Answer}).$$
$$P = FU_1 = 80.26 \times 10 = 803\,\text{Nm/s}\ (\textbf{Answer}).$$

7.4.2. *Finite length bearing*

A fixed inclination plane thrust bearing has a square plan form with $B = L = 50\,\text{mm}$, and a fixed taper, t, of 0.03 mm. The bearing is required to support a maximum load of 7 kN at a runner speed of 10 m/s. The same mineral oil as in Example 7.1 is to be used at a supply temperature of 30°C. Assuming that 60% of the heat flow in the film is carried away by convection, find the minimum film thickness and temperature rise under these conditions.

Solution

Unlike the pivoted pad bearing in Example 7.1, K is *not* constant for a fixed pad but varies with the dimensionless design groups. The **effective viscosity**, temperature rise and least film thickness are all unknown, so there must be a different approach to the problem. Appendix Table 7.2 for square pads has K as the independent variable. Because h_0 is a wanted variable, we can write:

$$K = \frac{h_1 - h_0}{h_0} = \frac{t}{h_0} = \frac{3 \times 10^{-5}}{h_0}, \tag{a}$$

allowing $h_0$to be found for each value of K.

Assuming that the supplied values of K, given in Appendix Table 7.2, cover an adequate range for a solution, start a new Table (7.1) reproducing the dimensionless groups we will need and then add a new column of h_0 values, as shown below. We can estimate the additional data we need in the following way:

- **In column 3**, $W^*$is needed because the load, W, is supplied.
- **In column 4**, θ_s^* is needed because the problem involves a temperature rise. It can be found from Appendix Table 7.2.
- **For column 5**, use Appendix Table 7.1, for W, to obtain η for each value of h_0:

$$\eta = \frac{W h_0^2}{W^* U_1 L B^2} = 5.6 \times 10^6 \left(\frac{h_0^2}{W^*}\right) Pas. \tag{b}$$

Table 7.1. Tabulated results.

1	2	3	4	5	6	7	8
K	$h_0 \times 10^{-5}$m	W^*	θ_s^*	η(Pas)	$\Delta\theta_s$(°C)	θ_{s1}(°C) = $30 + \Delta\theta_s$	θ_{s2}(°C)
0.5	6	0.0575	6.76	0.501	123.44	153.4	20.96
1	3	0.0689	2.959	0.104	45.19	75.1	44.18
1.5	2	0.67	1.787	0.048	127.69	57.69	60.88
2	1.5	0.0671	1.242	0.027	19.43	49.35	76.68
3	1	0.0583	0.732	0.014	13.18	43.18	100.62
4	0.75	0.0503	0.461	0.00892	9.66	39.66	120.25

- **For column 6**, use Appendix Table 7.1, for $\Delta\theta_s$ where $k_1 = 0.6$. Thus:

$$\Delta\theta_s = k_1\theta_s^*\left(\frac{\eta U_1 B}{\rho c_p h_0^2}\right) = \frac{0.6\theta_s^*\eta_e \times 10 \times 0.05}{850 \times 1880 h_0^2}$$
$$= \theta_s^* \times 1.877 \times 10^{-7}\left(\frac{\eta}{h_0^2}\right).$$

Two other temperature equations are needed to complete Table 7.1.

- **Column 7** produces θ_{s1} by using column 6 in Eq. (7.27):

$$\theta_{s1} = \theta_0 + \Delta\theta_s.$$

- **Column 8** produces θ_{s2} by using η from column 5 in **Vogel's equation** (Eq. (5.4) and tutorial Example 5.1).

$$\ln(\eta) = -1.845 + \left(\frac{700.81}{\Theta_{s2} - 273}\right).$$

Note that in Vogel's Equation, η must be inputted in *cp*. Moreover, Θ_{s2} is outputted in °K and must be converted to C for Table 7.1.

For the correct solution, the values of θ_s from columns 7 and 8 must be the same at one particular value of h_0, it being found by plotting both versions of θ_s against h_0. The intersection of the two curves gives the solution. The graphs are shown in Fig. 7.7

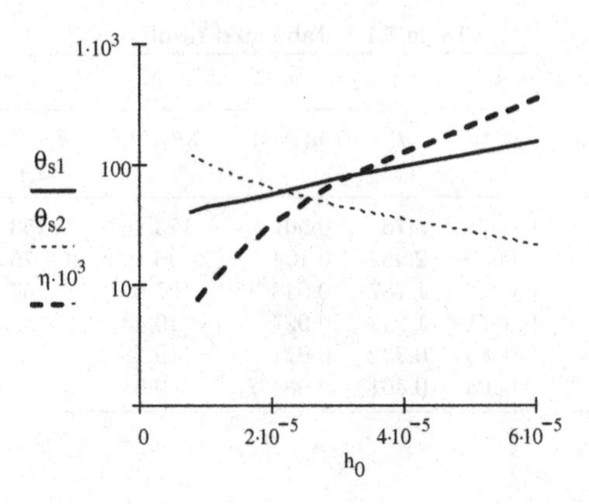

Fig. 7.7. Solution curves for temperature and viscosity.

The answers read off are: $h_0 = 2.1 \times 10^{-5}$ m and $\theta_{se} = 63°\text{C}$, $\eta_e = 38 \times 10^{-3}$ Pas.

Comments

The bearing is running fairly hot because of the relatively thin film. Knowing the correct values of θ_s, η_e and h_0, the other unknown design variables can then be obtained from the design coefficients.

This type of problem is rapidly solved with a software package like Mathcad using its 'Vectorize' operator. It means, for example, that having entered K as a vector from the data table, a vector of h_0 can be found from equation (a) above. Having entered W^* from the data table, Eq. (b) above allows the vector η to be found, and so on building up Table 7.1. Finally, Mathcad allows Fig. 7.10 to be plotted with the intersection point determined, using its 'Trace' facility.

If other design variables are required, a similar procedure is used. For example, to find the side flow Q_{se} we must firstly graph the vector h_0 against the vector Q_s^*. Then find the corresponding value of Q_s^* where $h_0 = 2.1 \times 10^{-5}$m. Hence determine Q_{se} from the definition of Q_s^* in Appendix Table 7.1.

7.5. Other Thrust Bearing Geometries

There are alternative designs of thrust bearing with fixed inclination wedge pairs in use. Here are two examples of long bearings.

7.5.1. *The taper land bearing*

This design is more efficient than the fixed inclination plane wedge pair. In this design the wedge only occupies a part of the total width followed by a parallel land* for the remaining part, as in Fig. 7.8. This arrangement reduces wear when the bearing stops. In keeping with all wedge geometries, there need not be a continuous inclination. As long as there is some slight narrowing of the film, the bearing will operate hydrodynamically.

The geometry is defined as:

$$h = h_0, \quad \text{for } x \leq 0,$$

$$h = h_0\left(1 + \frac{Kx}{B_1}\right), \quad \text{for } x > 0.$$

The optimum configuration for a long bearing is $\frac{B_1}{B_1+B_2} = 0.8, h_1/h_0 = 2.25, W^* = 0.192$. A numerical thermal analysis of this type of bearing type is given by Advani and Gohar.[5]

7.5.2. *The Raleigh step bearing*

Referring to Fig. 7.9, the geometry of this pair is defined as:

$$h = h_0 \quad \text{for } x \leq B_2$$

$$h = h_1 \quad \text{for } B_2 < x \leq B_1 + B_2$$

An analysis of a long Raleigh step bearing[6] follows:

Because both films are constant, from Eq. (7.1), dp/dx must be constant along both films. This causes the pressure distribution to be triangular, reaching a peak at the step.

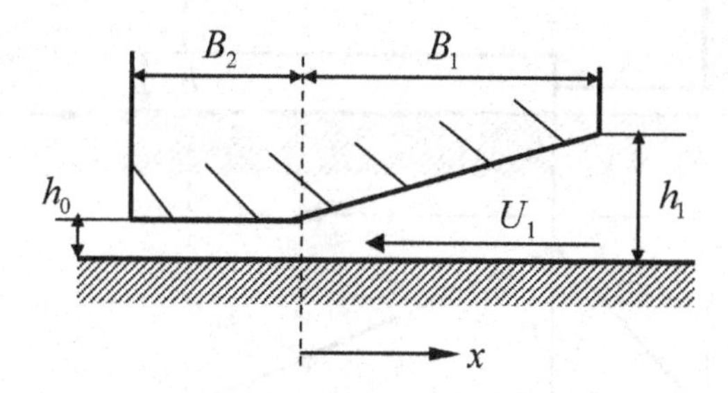

Fig. 7.8. The taper land bearing.

*By lands, we mean the parallel regions where the film is thinnest.

Thus:

$$\left(\frac{dp}{dx}\right)_1 = -\frac{p_c}{B_1} \quad \text{and} \quad \left(\frac{dp}{dx}\right)_2 = \frac{p_c}{B_2}.$$

As the bearing is assumed to be long, the flow per unit width is constant so that:

$$q_x = -\frac{U_2 h_1}{2} - \frac{h_1^3}{12\eta}\left(\frac{dp}{dx}\right)_1 = -\frac{U_1 h_0}{2} - \frac{h_0^3}{12\eta}\left(\frac{dp}{dx}\right)_2.$$

Substituting for $\left(\frac{dp}{dx}\right)_{1,2}$ above, we get:

$$\frac{U_1}{2}(h_h - h_0) = \frac{p_c}{12\eta}\left(\frac{h_1^3}{B_1} + \frac{h_0^3}{B_2}\right) \tag{7.31}$$

Thus, $p_c$can be found from Eq. (7.31). The load capacity is the area of the pressure triangle in Fig. 7.9. The maximum value occurs when $h_1/h_0 = 1.87$ and $B_1/B_2 = 2.588$, the load becoming:

$$\frac{W}{L} = 0.206\eta U_1 \frac{(B_1 + B_2)^2}{h_0^2}. \tag{7.32}$$

An equivalent long plane pivoted pad bearing has a load coefficient of 0.1602. However, an advantage of the **pivoted pad bearing** is that, if there is an overload, h_1/h_0 does not alter because K is set by the pivot position (Eq. (7.12)). On the other hand, the fixed geometry Raleigh step bearing has h_1/h_0 varying with the load. Another problem with the

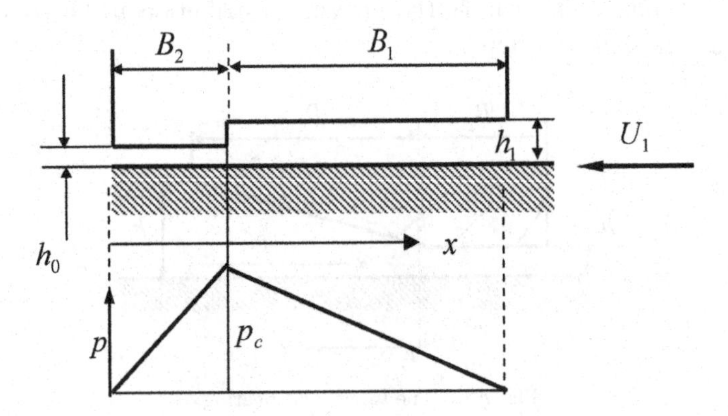

Fig. 7.9. The Raleigh step bearing.

step bearing is one of manufacture. Because of the small step size it is liable to wear away under continuous usage, thus losing its discrete step property.

7.5.3. *The pocket step bearing*

Some attempts have been made to reduce the side leakage of lubricant in finite length bearings by creating side horns to keep the pressure as uniformly high as possible within the bearing. Figure 7.10 uses such arrangement, shown in plan form for a square bearing (Kettleborough[7]). The enclosed inlet area, A, is deepest at constant height, h_1, while the area B, partially surrounding it, is thinner at constant height, h_0. The entrained lubricant is thus entering an inverted channel that has the effect of inhibiting side leakage. This arrangement maintains a more uniform high pressure in the y direction thus increasing the load capacity. One reason for discussing this design here is that, in Chapter 10, when we discuss Elastohydrodynamic lubrication, where the surfaces are allowed to distort elastically under the high pressures, while seeking their optimum load capacity, a similar 'natural' film shape results (see Fig. 10.10).

7.6. Closure

This chapter has dealt with the hydrodynamic lubrication of thrust bearings. By creating a single effective viscosity at an effective temperature for the bearing, computed isothermal dimensionless coefficients were employed in a thermal design procedure. The chapter has covered both long and finite length plane bearings as well as explaining some other designs in current use. In the next chapter we will discuss the design of journal bearings.

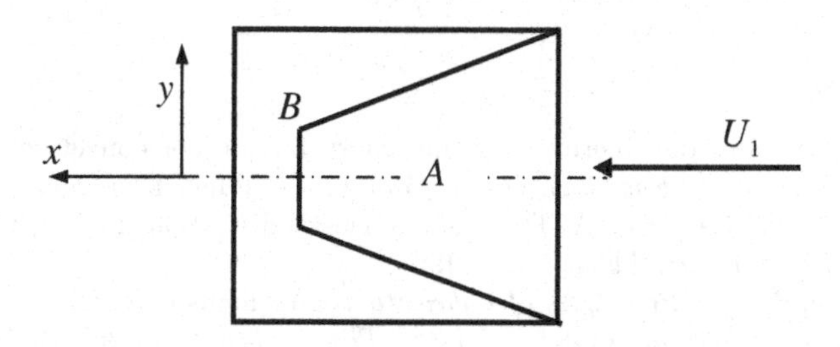

Fig. 7.10. Horned Raleigh step bearing.

Appendices

Appendix Table 7.1. Design coefficients for plane thrust bearing pairs.

(1) All plane thrust bearings	(2) Dimensionless group expressions for long plane thrust bearings ($L/B \geq 3$)
$W = W^* \left(\frac{U_1 \eta L B^2}{h_0^2}\right)$	$W^* = \frac{6}{K^2}\left[\ln(K+1) - \frac{2K}{K+2}\right]$
$X = X^* B$	$X^* = \frac{2(K+3)(K+1)\ln(K+1) - K(6+5K)}{2K[(K+2)\ln(1+K) - 2K]}$
$F = F^* \left(\frac{LB\eta U_1}{h_0}\right)$	$F^* = \left[\frac{4\ln(1+K)}{K} - \frac{6}{2+K}\right]$
$Q_s = Q_s^* (LU_1 h_0)$ (finite length)	
$Q_1 = Q_1^* (LU_1 h_0)$ (finite length)	$Q^* = \frac{Q}{LU_1 h_0} = \frac{1+K}{2+K}$
$Q_2^* = Q_1^* - Q_s^*$ (finite length)	
$\theta_s^* = F^*/Q_s^*$ (finite length)	$\theta^* = \left[\frac{4\ln(1+K)}{K} - \frac{6}{2+K}\right]\left[\frac{2+K}{1+K}\right]$
$\Delta\theta_s = k_1 \theta_s^* \left(\frac{\eta U_1 B}{\rho c_p h_0^2}\right)$ (finite length)	$\Delta\theta = \frac{k_1}{\rho c_p}\left\{\frac{F^*}{W^*}\left(\frac{2+K}{1+K}\right)\right\}\frac{W}{BL}$

Definitions in column (1) apply to all plane thrust bearings, but note that the values of the dimensionless groups for finite length bearings are obtained from separate computed results (Appendix Table 7.2).

Appendix Table 7.2. Results for plane finite length thrust bearing pairs, $L/B = 1$ (after Jackobsen and Floberg[8]).

K	0	0.5	1	1.5	2	3	4
W^*	0.0556	0.05575	0.0689	0.07	0.067	0.0584	0.0503
X^*	0.5	0.548	0.582	0.607	0.627	0.657	0.678
Q_1^*	0.5	0.68	0.847	1.008	1.165	1.470	1.769
Q_t^*	0	0.122	0.246	0.371	0.496	0.75	1.01
F^*	1	0.825	0.728	0.663	0.616	0.549	0.466
θ_s^*	∞	6.76	2.859	1.787	1.242	0.732	0.461

References

1. Hall, L. F. and de Guerin, D. Some characteristics of conventional tilting bearings, *I. Mech. Eng. Lubrication Conf.* (1957) Paper 82 142–6.
2. Ettles, C. and Cameron, A. Thermal and elastic distortions in thrust bearings, *Proc I Mech. Engrs. Lubn. Conf.* (1963).
3. Cameron, A. *The Principles of Lubrication*, Longmans (1966).
4. Pinkus, O. *Thermal Aspects of Fluid Film Lubrication*, ASME Press NY (1990).

5. Advani and Gohar. The taper-land sector shaped thrust bearing, *J. Sci. Tech.* **38**(2) (1971) 83–90.
6. Lord, R. Notes on the theory of lubrication, *Phil Mag.* (1918) 1–12.
7. Kettleborough, C. F. An approximate analytical solution for a stepped thrust bearing, *Trans ASME* **77** (1955) 311–320.
8. Jacobsen, B. and Floberg, L. The rectangular plane thrust bearing. *Trans Chalmers University Tech. Gothenburg* **203** (1958) Appendix 10 page 42.

CHAPTER 8

JOURNAL BEARINGS

8.1. Introduction

Journal bearings are one of the most common types of hydrodynamic bearing, their primary purpose being to support a rotating shaft. Examples of their use are for radially supported automotive engine crankshafts, steam turbine rotors in power stations and propeller shafts in ships. The aim of the procedure, outlined below, is to design a bearing that maintains a coherent lubricant film, despite the associated high temperatures and avoid any vibration problems. Like the thrust bearing in Chapter 7, part of the power needed to drive the shaft is used in pumping the oil through a wedge shaped film and overcome the friction forces generated. We will commence by describing the 360 degree journal bearing film geometry.

8.2. Film Geometry

Figure 8.1 shows the bearing **journal** (the rotating member, center C, attached to the shaft) and surrounded by the **bearing bush** (the normally stationary member, center O). In order to support the load, the journal center must take up an eccentric position defined by eccentricity, e.

In Fig. 8.1(a), O is the bearing center, C the journal center and distance OC is called the **eccentricity**, e. If CO is produced to the bearing surface at E, the maximum film thickness h_1 is there (sometimes called the maximum clearance). Diametrically opposite at F, there is the minimum film thickness (h_0). Diameter, FE is called the *line of centers*, from which the bearing polar coordinate angle, ϕ, is measured. We will now find an expression for the film thickness $h(\phi)$.

If $OA = R_2$ is the bearing bush radius, R_1 is the journal radius, then $CA = R_1 + h$.

If $e = 0$, then $R_2 - R_1$ is the radial clearance, c. It follows that $h_1 = c + e$ and $h_0 = c - e$. Defining ϕ and α as shown in Fig. 8.1(b), $R_2 + h = CD + DA = e\cos\phi + R_1\cos\alpha$ and from $\Delta OCA, e/\sin\alpha = R_1/\sin\phi$. Also, $\cos^2\alpha + \sin^2\alpha = 1$. Using the two latter relationships, we find that

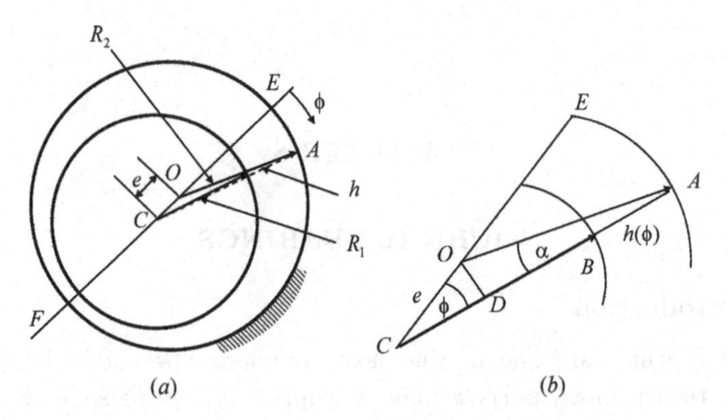

Fig. 8.1. Journal bearing geometry.

$\cos\alpha = \sqrt{1 - \frac{e^2 \sin^2\phi}{R_2^2}}$. But for the formation of a hydrodynamic film, its thickness must be small compared with the other dimensions, such that $e/R_2 \ll 1$. We can, therefore, use the binomial theorem, allowing us to write $\cos\alpha = 1 - \frac{e^2 \sin^2\phi}{2R_1^2}$, so that $R_1 + h = e\cos\phi + R_2(1 - \frac{e^2 \sin^2\phi}{2R_2^2})$. Again, because $e/R_2 \ll 1$ the film thickness can now be written as: $h = (R_2 - R_1) + e\cos\phi$. Thus, we can define the film thickness at position ϕ from the line of centers as:

$$h = c + e\cos\phi. \tag{8.1}$$

Alternatively, if the **eccentricity ratio** is defined by $\varepsilon = e/c$, hence:

$$h = c(1 - \varepsilon\cos\phi). \tag{8.2}$$

The eccentricity ratio (ε) of a journal bearing defines its film shape with all other design variables being related to it. For the thrust bearing (Chapter 7) its equivalent is the geometry factor, K. From Fig. 8.1 and Eq. (8.2), we can deduce that when $\phi = 0, h_0 = c(1 - \varepsilon))$ and when $\phi = \pi, h_1 = c(1 + \varepsilon)$.

8.3. The Pressure Equation for a Narrow Bearing

We will confine ourselves here to the narrow bearing because this is the simplest to analyze. The narrow bearing pressure is given by Eq. (6.27), reproduced below with $U_2 = 0$:

$$p = 3(U_1)\eta\frac{dh/dx}{h^3}(y^2 - (L^2/4)). \tag{8.3}$$

In order to apply Eq. (8.3) to journal bearings, we must make the substitution: $x = R\phi$, where R will from now on represent the bearing journal radius, R_1 (which approximately equals R_2).

Using $dx = Rd\phi$. If Eq. (8.2) is differentiated with respect to ϕ : $\frac{dh}{d\theta} = -(c\varepsilon \sin\phi)R$. With this substitution in Eq. (8.3) we get:

$$p = \frac{3U_1\eta c\varepsilon \sin\phi}{Rc^3(1+\varepsilon\cos\phi)^3}\left(\frac{L^2}{4} - y^2\right). \tag{8.4}$$

A study of Eq. (8.4) shows that the pressure distribution is antisymmetrical about $\phi = \pi$, producing the **Full Sommerfeld**[1] solution described in Sec. 6.5. The **Half Sommerfeld**[1] solution, suppresses the negative pressure loop from $\phi = \pi$ back to $\phi = 0$. This is preferable, again being closer to more accurate numerical solutions for finite length bearings that employ the Swift-Steiber boundary condition.[2] These different boundary conditions for journal bearings are demonstrated in Fig. 8.2.

In (a) the pressure distribution is anti-symmetric about $\phi = \pi$. Such a solution is unlikely in practice, because oil cannot stand high negative pressures. In (b), the negative loop has been suppressed. The problem here is that, as discussed in Sec. 6.5.3, at $\phi = \pi$ the finite slope of the pressure distribution indicates both Poisseuille and Couette flow components. However, just beyond $\phi = \pi$ there is zero pressure gradient and hence there remains only Couette flow, thus causing a discontinuity of flow at $\phi = \pi$. Nevertheless, the Half Sommerfeld solution is reasonably accurate for narrow bearings. In (c), the pressure and its slope are both made zero at some angle, α, a little beyond $\phi = \pi$ (about 8°). This is the **Reynolds exit boundary condition** that is often employed in numerical solutions for journal bearings.

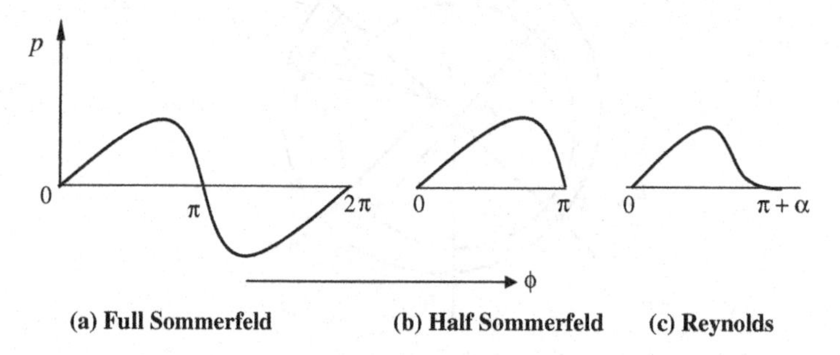

Fig. 8.2. Journal bearing boundary conditions.

8.4. Load

The reaction force on a narrow journal bearing, under Half Sommerfeld boundary conditions, is found by integrating the pressure from $\phi = 0$ to π as well as in the y direction along the bearing length. Referring to Fig. 8.1, these boundaries define a crescent shaped converging wedge from A to F. In Fig. 8.3 component pressure forces are shown, together with the gravity force on the journal that is rotating at surface speed U_1.

$$\text{Component along the line of centers} \quad W_x = \int_0^{\pi} \int_{-L/2}^{+L/2} pR \cos\phi d\phi dy.$$

$$\text{Component normal to the line of centers} \quad W_z = \int_0^{\pi} \int_{-L/2}^{+L/2} pR \sin\phi d\phi dy.$$

The eccentricity $CO = e$, does not concern us at present as we are only equating forces. The parts of the above integrals involving θ are quite involved, their derivations being found in the appendix to Chapter 12 of Cameron.[3] The solutions to the x and z direction load components come to:

$$W_z = \frac{\pi U_1 \eta L^3 \varepsilon}{4c^2(1-\varepsilon^2)^{3/2}}, \quad W_x = \frac{-U_1\eta L^3}{c^2}\frac{\varepsilon^2}{(1-\varepsilon^2)^2}.$$

The total load is the sum of these components, being $W = (W_x^2 + W_z^2)^{1/2}$ After some arranging it comes to:

$$W = \frac{U_1\eta L^3}{c^2}\frac{\pi\varepsilon}{4(1-\varepsilon^2)^2}\left[\left(\frac{16}{\pi^2}-1\right)\varepsilon^2+1\right]^{1/2}. \tag{8.5}$$

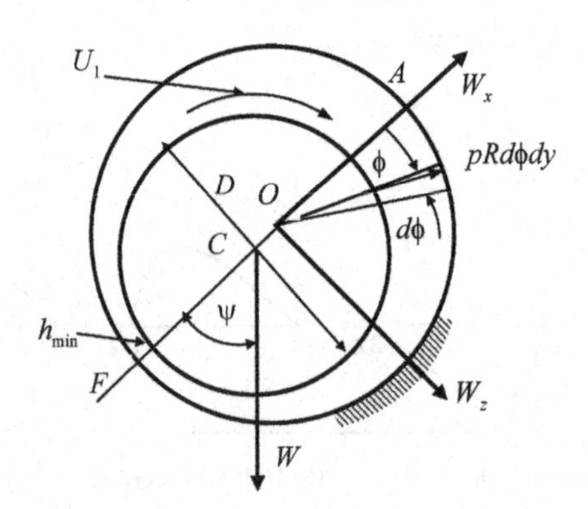

Fig. 8.3. Forces on a journal bearing.

We see now that for a given bearing, Eq. (8.5) is a relationship between load and eccentricity. A more common form is to express Eq. (8.5) in terms of a dimensionless group called the **Sommerfeld Number** (S).

If the journal diameter is $D = 2R$, surface speed, $U_1 = \pi DN$, where N is in rev/s and we define S as:

$$S = \frac{W}{NDL\eta}\frac{c^2}{R^2}. \tag{8.6}$$

In terms of S, Eq. (8.5) for narrow bearings can be rearranged further to give:

$$S = \left(\frac{L}{D}\right)^2 \frac{\pi^2\varepsilon}{(1-\varepsilon^2)^2}(0.62\varepsilon^2 + 1)^{1/2}. \tag{8.7}$$

In the case of finite length bearings:

$$S = f\left(\varepsilon\frac{L}{D}\right), \tag{8.8}$$

where f is evaluated numerically (see Appendix Table 8.2 for some solutions).

From its definition in Eq. (8.6) $S \propto W$ and as $h_m = c(1-\varepsilon)$, Eq. (8.7) allows us to plot S against h_m, which is effectively the typical load-deflection curve shown in Fig. 8.4.

Because stiffness is defined by dW/dh_m, Fig. 8.4 shows that the bearing is most stiff somewhere in the middle part of the range of $h_{\min}$, and is softer at either end. This load-deflection shape also occurs in finite length hydrodynamic, as well as externally pressurized bearings.

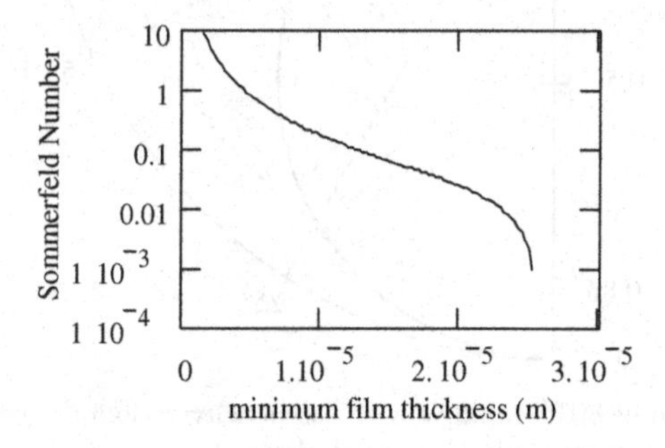

Fig. 8.4. Relation between S number and minimum film thickness.

8.5. Attitude Angle

From Eq. (8.6), having found the magnitude of $h_{\min}$ from a known value of ε, the inclination, ψ, of the line of centers with respect to the resultant load W needs to be determined (*FA* in Fig. 8.2). It is called the **attitude angle**, ψ, and can be found from the definitions of W_x and W_y to give, after some manipulation:

$$\tan\psi = -\frac{W_x}{W_z} = \frac{\pi}{4}\frac{\sqrt{1-\varepsilon^2}}{\varepsilon}. \tag{8.9}$$

This relationship, which is approximately a semi-circle, is illustrated in Fig. 8.5.

Although not drawn to scale, Fig. 8.5 illustrates how the orientation of the line of centers, *OC*, changes as the eccentricity ratio varies. For example, if $\varepsilon = 0.5, \psi = 52.4°$. In the limiting positions, from Eq. (8.8), as $\varepsilon \to 0, \psi \to 90°, h_{\min} \to c$, the journal taking up a nearly concentric position in the bearing. This situation occurs when the journal is unloaded such as when in the vertical position. At the other extreme, $\psi \to 0°$, $\varepsilon \to 1$, corresponds to a heavily loaded journal with $h_{\min} \to 0$ near the $\Psi = 0°$ position in Fig. 8.4. Of course, at zero speed (start up) *OC* is vertical. The journal center of finite length bearings also follows approximately this locus.[3]

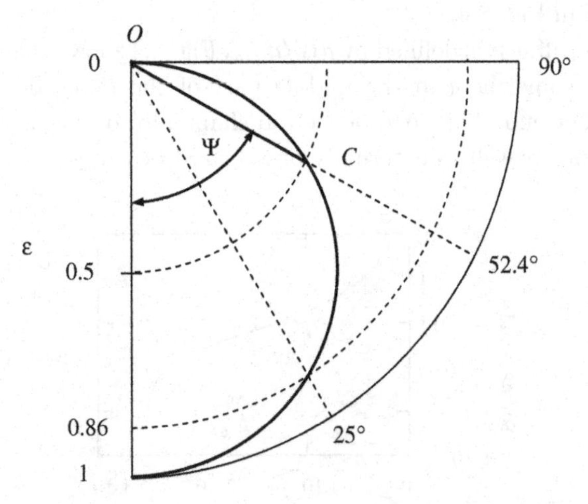

Fig. 8.5. Variation of attitude angle with eccentricity ratio for narrow bearings. (Note that as drawn, the journal is rotating anticlockwise.)

8.6. Lubricant Flow

8.6.1. *Side flow from hydrodynamic action*

A finite length or **narrow journal bearing** loses oil from side leakage (in the y direction) which must be replenished from a supply pump. This flow has two components, the first being from the Poiseuille flow pressure gradient and the second coming directly due to the supply pump pressure. Dealing initially with the first component, it is the difference between the total circumferential hydrodynamic flow at the start of the pressure curve (point A in Fig. 8.3) and at that at the finish (point F in Fig. 8.3). We are interested in this loss of oil along $y = \pm L/2$, it being usually fed back into the film under pressure from the supply pump somewhere upstream of point A in the unloaded part of the bearing.

Letting $x = R\phi$, from Eq. (6.18) $q_\theta = \frac{-h^3}{12\eta R}\frac{dp}{d\phi} + U_1\frac{h}{2}$. However, we know that narrow bearing theory neglects the Poiseuille flow caused by the ϕ direction pressure gradient, when compared with the Couette term. Under these conditions the narrow bearing side leakage flow rate over the pressurized region AF must be:

$$\begin{aligned}Q_s = q_\theta L = \frac{U_1}{2}(h_A - h_F)L &= \frac{U_1 L}{2}[c(1+\varepsilon) - c(1-\varepsilon)]\\ = \left(\frac{U_1 c L}{2}\right)2\varepsilon &= (\pi R N L c)2\varepsilon.\end{aligned}$$

At this stage we can define a flow factor that applies to *all* types of finite length journal bearing as:

$$Q_s^* = \frac{Q_s}{(\pi R N L c)}. \tag{8.10}$$

For narrow bearings we see that the **side flow factor** is simply:

$$Q_s^* = 2\varepsilon. \tag{8.11}$$

Apart from narrow bearings, Q_s^* must be evaluated numerically. In fact, for 360° finite length bearings, Q_s^* also varies approximately linearly with ε (Appendix Table 8.2). This is not the case for both finite and narrow 180° bearings where Q_s^* peaks near $\varepsilon = 0.8$ before reducing.[3]

8.6.2. *Pressurized flow from a single supply hole*

Let us now find an expression for the second component of side flow, Q_a, which is caused solely by the pump pressure gradient. The entry flow from

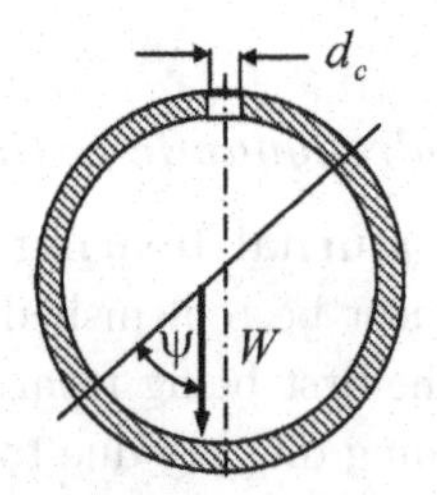

Fig. 8.6. Single entry hole for a 360° bearing.

an external pump depends partly on the bearing design. In our case we will employ a 360° bearing with a single circular feedhole, of diameter d_c, sited near the gravity load line at the top of the bearing bush mid plane ($y = 0$) as in Fig. 8.6.

The film thickness where the hole is sited is given by:

$$h_f = c(1 + \varepsilon \cos(-\psi)) = c(1 + \cos\psi). \tag{8.12}$$

It is assumed that the axial pressure drop for the pump flow, from mid plane to either edge ($y = \pm L/2, p = 0$), is linear. It is also assumed that the thickness of the film at the supply hole, h_f, is where all the pump axial leakage component (y direction) occurs, none leaving over the rest of the bearing arc. If Q_a^* is the axial flow factor for a single hole, the pump axial flow is:

$$Q_a = \left(\frac{h_f^3 p_0 d_c}{12\eta L}\right) Q_a^*. \tag{8.13}$$

The straight line law defining the axial flow factor for a hole is[3]:

$$Q_a^* \approx 1.2 + 11\frac{d_c}{L}. \tag{8.14}$$

8.6.3. *Total pump delivery flow*

The total side flow, Q_p, the supply pump must deliver is the sum of the two flow components defined by Eqs. (8.10) and (8.13), that is:

$$Q_p = Q_s + Q_a = (\pi RNLc)Q_s^* + \left(\frac{h_f^3 p_0 d_c}{12\eta L}\right) Q_a^*. \tag{8.15}$$

Observe that once the bearing geometry has been determined, including a choice of supply hole diameter, the only unknowns are η at the supply temperature and p_0, both values of which must be selected.

8.6.4. *Double 180° journal bearings*

In order to accommodate vertical loads in both directions, a more common bearing design composes two finite length 180° bearings joined together at $\phi = \pi/2$ and $\phi = 3\pi/2$. At each joint an oil entry port or hole is placed, receiving half the total flow from the supply pump. This arrangement, illustrated in Fig. 8.6, has a bearing performance little different from the 360° bearing where, in practice, most of the significant pressures occur anyway over its bottom half. The double 180° bearing compels this to happen because of the positions of the entry ports. An illustration of the arrangement is shown in Fig. 8.7.

Note that, as in the 360° journal bearing, the end of the pressure distribution occurs at a small distance beyond the line of centers on account of the assumed Reynolds boundary condition of zero pressure and slope. From the zero pressure point back to pressure commencement, continuity dictates that in order to fill the widening gap, the flow must break into streamers alternating between air and oil. This has some significance when we later consider friction.

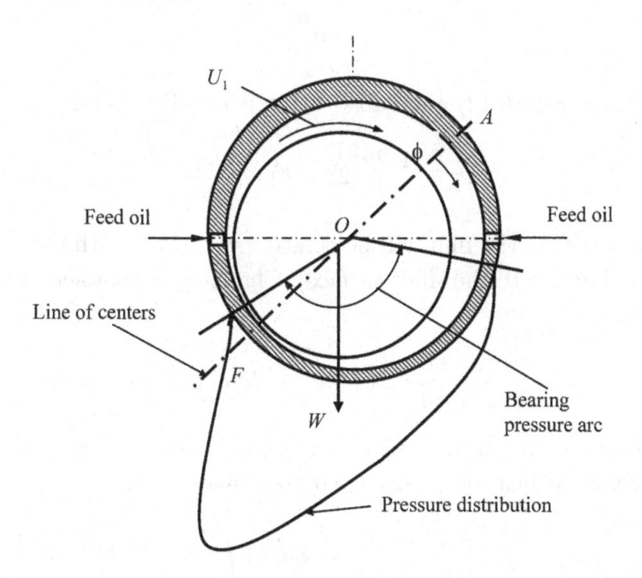

Fig. 8.7. Double 180° bearing.

8.6.5. *Circumferential groove bearings*

For a bearing subjected to a load of varying magnitude and direction, in order to maintain a constant supply of oil to the film, a midplane circumferential feeding groove right round the bearing is used. On account of the groove width, this design has the effect of dividing the bearing into two parts, each with an L/D ratio less than that of the whole bearing. Applications of these bearings will be discussed in Chapter 12, which deals with alternating loads and squeeze films.

8.7. Friction in Journal Bearings

In order to measure the friction force in a journal bearing (**viscous friction**), we can start with Eq. (6.12) for the velocity at a point x, z in the film where, for journal bearings, U_1 is the journal surface speed and $U_2 = 0$.

$$u = \frac{1}{2\eta}\frac{\partial p}{\partial x}(z^2 - zh) + \frac{z}{h}U_1. \tag{6.12}$$

We need to now find the shear stress at this point because, in general terms, friction force is shear stress times the area. Shear stress is given by Eq. (5.1) reproduced below:

$$\tau = \eta\frac{\partial u}{\partial z}. \tag{5.1}$$

Differentiating Eq. (6.12) and substituting it into Eq. (5.1):

$$\tau = \frac{1}{2}\frac{\partial p}{\partial x}(2z - h) + \frac{\eta U_1}{h}.$$

The shear stress in the film will be considered at $z = h$ (the bearing bush surface) and at $z = 0$ (the shaft surface). Changing to complete derivatives, then:

$$\tau = \pm\frac{dp}{dx}\frac{h}{2} + \frac{\eta U_1}{h} \quad (+ \text{ at } z = h \text{ and } - \text{ at } z = 0).$$

An integration of the shear stress over the film boundaries yields the friction force. For narrow bearings using polar coordinates it is:

$$F_{h,0} = \int_0^L \left(\int_0^\pi \pm\frac{dp}{Rd\phi}\frac{h}{2}Rd\theta + \int_0^{2\pi}\frac{\eta U_1}{h}Rd\phi\right)dy.$$

The first double integral is the friction force due to the Half Sommerfeld pressure distribution. Assuming the bearing runs full, the second, right round the bearing, comes from the Couette flow. Taking the first double integral, integrating by parts and recalling that $h = c(1 - \varepsilon\cos\phi)$ so that $dh/d\theta = -c\varepsilon\sin\theta$, we get:

$$\int_0^L \int_0^\pi \pm\frac{h}{2}\frac{dp}{d\phi}d\phi = \pm\frac{1}{2}\int_0^L \left(|hp|_0^\pi - \int_0^\pi \frac{dh}{d\phi}pd\phi \right)$$
$$= \pm\frac{1}{2}\int_0^L \left(0 + \int_0^\pi c\varepsilon p\sin\phi d\phi \right) dy.$$

It is unnecessary to solve the above second integral because we have shown in Sec. 8.4 that $\int_0^L \int_0^\pi pR\sin\phi d\phi dy = W\sin\psi$, so a substitution in terms of W can be made there to give:

$$\int_0^L \int_0^\pi \pm\frac{h}{2}\frac{dp}{d\phi}d\phi = \pm\frac{1}{2R}c\varepsilon W\sin\psi.$$

Turning now to the Couette component of friction and substituting for h:

$$\int_0^L \int_0^{2\pi} \frac{\eta U_2}{h}Rd\phi dy = \frac{\eta U_2 R}{2c}\int_0^L \int_0^{2\pi} \frac{d\phi}{1+\varepsilon\cos\phi}dy = \frac{2\pi\eta U_1 RL}{c(1-\varepsilon^2)^{1/2}}.$$

The solution details for the above integral can be found in Ref. 3. Thus, the complete solution for friction force is:

$$F_{h,0} = \pm\frac{c\varepsilon W}{2R}\sin\psi + \frac{2\pi\eta U_1 RL}{c(1-\varepsilon^2)^{1/2}}, \tag{8.16}$$

or

$$F_{h,0} = \pm F_p + F_c.$$

Equation (8.16) is also accurate for other L/D ratios where the oil film is assumed to extend right round the bearing without **cavitation**.* The $\pm$ sign in Eq. (8.16) seems to suggest that the film, which is assumed to be weightless and having constant pressure across it, has different friction forces on its respective boundary surfaces. We should therefore investigate whether another force may be present to rectify this anomaly. Consider the

*The presence of air in the wake forming air cavity patterns in the form of streamers at the end of the pressure distribution.

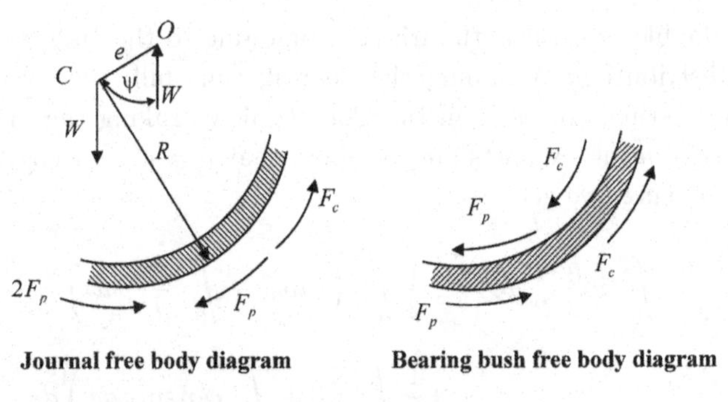

Fig. 8.8. Forces on the journal and bearing.

friction forces on the journal shown in Fig. 8.8 below together with the pressure reaction force through O and the journal load through C. From the first term of Eq. (8.17) below, the external force on the journal surface due to eccentricity, e, only, can be written as $2F_pR$.

We see, therefore, that the friction forces on the journal and bearing surfaces, including the effect of the couple on the journal, are now equal and opposite:

$$F_{0,h} = \frac{c\varepsilon W}{2R}\sin\psi + \frac{2\pi\eta U_1 RL}{c(1-\varepsilon^2)^{1/2}}. \tag{8.17}$$

8.7.1. *Effect of the Reynolds boundary condition on friction*

Besides the Half Sommerfeld narrow bearing solution, Eq. (8.17) can also apply to all L/D ratios when the bearing is assumed to run full. More accurate numerical solutions for finite length bearings employ the **Reynolds boundary condition** at the end of the pressure distribution. We discussed this briefly with the other boundary conditions in Sec. 8.3. As shown in Fig. 8.2, the Reynolds boundary condition determines the end of the pressure loop by stating that when $p = dp/dx = 0, p = 0$. It occurs at a small angle, α, beyond the minimum film thickness, that is at $\theta = \pi + \alpha$. The Reynolds boundary condition is more realistic than the assumption for the narrow bearing that simply ignores the negative pressures generated by the Full Sommerfeld solution after $\theta = \pi$, hence contravening continuity of flow there. The effect of employing the Reynolds condition is to reduce the Couette friction in the unloaded part of the bearing film because, in order to maintain the flow continuity, the oil must break into streamers of air and oil, the significant contribution coming from the oil.

8.7.2. *Coefficient of friction*

By definition, the coefficient of friction is:

$$\mu = \frac{F}{W} = \frac{c\varepsilon}{2R}\sin\psi + \frac{2\pi\eta U_1 RL}{Wc(1-\varepsilon^2)^{1/2}}. \tag{8.18}$$

Equation (8.18) can be written in terms of the **Sommerfeld Number**, defined in Eq. (8.5) by $S = \frac{W}{\eta NDL}(\frac{c}{R})^2$. Substituting for W in Eq. (8.19) we obtain:

$$\mu\frac{R}{c} = \mu^* = \frac{\varepsilon}{2}\sin\psi + \frac{2\pi^2}{(1-\varepsilon^2)^{1/2}}\left(\frac{I}{S}\right). \tag{8.19}$$

When using the Reynolds boundary condition, the factor I in Eq. (8.19) accounts for the reduced friction in the cavitated region that occurs downstream of the zero pressure point. Tabulated values of μ^* for a 360° bearing with cavitation, for various L/D ratios, are given in Appendix Table 8.2 at the end of the chapter.

8.8. Thermal Design of Journal Bearings

8.8.1. *Introduction*

As we noted in Chapter 7 when designing thrust bearings, the effect of oil heating on oil viscosity must be included in order to obtain a realistic result. A complete solution of the finite length bearing problem, uses the Reynolds and the full energy equations solved together numerically in two dimensions. Fortunately, accurate isothermal numerical solutions for finite length bearings are available, allowing us to use these results in an approximate thermal design procedure By isothermal we mean that the lubricant is assumed to be isoviscous at some effective viscosity, initially of unknown value. The aim of this approximate method is to find this value using a simplified version of the energy equation together with the available numerical results from the isothermal solution and, most importantly, the thermal viscosity equation of state of the oil.

The simplifying assumptions we will make are similar to those used for the thrust bearing design. They are:

(1) The oil carries away by convection all the heat generated with any loss through conduction to the surfaces accounted for by an empirical factor (and see Sec. 6.7).

(2) The oil viscosity (now called the **effective viscosity**) is constant through the film thickness, allowing us to employ Reynolds equation.
(3) Additionally, this effective viscosity is assumed constant circumferentially as well as transversely.
(4) The pressurized lubricant flow to the film from the supply hole or groove is dominantly circumferential at the hydrodynamic pressure zone start, most of the axial flow component being near the groove exit before pressure commencement.[3]

8.8.2. *Finding the effective viscosity*

Just as in the case of a line of thrust bearing pads (Fig. 7.6), in Fig. 8.9 a control volume for power can be drawn round the effective bearing arc (omitting here the ρc_p terms).

The power flows in Fig. 8.9 have the same meaning as in Chapter 7: The external powers are:

- P, generated by the friction forces.
- $Q_s\theta_s\rho c_p$[†] removed by side leakage of the oil at a unique temperature θ_s.
- $Q_s\theta_0\rho c_p$, needed to return this side leakage oil back to the bearing, via the inlet supply hole, at some lower temperature, T_0.

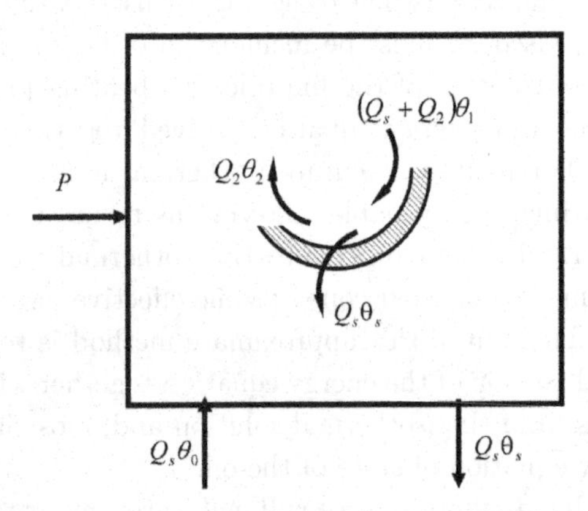

Fig. 8.9. Convective heat transfer in a journal bearing. (The lubricant properties are omitted.)

[†]Note that Q_s is the side flow leaving the bearing in the pressurized region. Any additional side flow is caused by the inlet pump pressure (see Sec. 8.6.2). On its return to the bearing, most of this entering pump flow leaves axially at the start of the pressurized region and so is considered not to affect the power balance.[3]

The internal powers are:

- $(Q_s + Q_2)\theta_1 \rho c_p$ entering the pressurized film arc
- $Q_2\theta_2\rho c_p$ leaving the film pressurized arc.

As in Sec. 7.7, we need not consider here the internal power balance in the simple design method discussed below (but see Question 8.1).

Thus, for the external powers to balance:

$$P = Q_s(\theta_s - \theta_0)\rho c_p$$

If $\Delta\theta_s = (\theta_s - \theta_0)$ = temperature rise from the supply hole to the effective temperature of the side leakage oil, then:

$$P = Q_s \Delta\theta_s \rho c_p. \tag{8.20}$$

Again, just as in Sec. 7.7, we can state that for the lubricant leaving by the sides and re-entering via the supply hole:

$$\begin{aligned}\text{Heat generated} = \text{mass flow per second} \times \text{specific heat}\\ \times \text{temperature rise}\end{aligned}$$

or

$$k_1 F U_1 = Q_s \rho c_p \Delta\theta_s, \tag{8.21}$$

where k_1 recognizes that not all this power has been lost by convection. Some is lost by conduction through the bounding surfaces (Sec. 6.7).

From Secs. 8.6 and 8.7:

$$Q_s = Q_s^* \pi NRLc, \quad F = \mu W, \quad \mu = \mu^* \frac{c}{R}.$$

Also, let $K_1 = \frac{k_1}{\rho c_p}$. We can also let the flow entering the pressure region of the film, $Q_1 = Q_1^* \pi NRLc$, and that leaving equal $Q_2 = Q_2^* \pi NRLc$ (needed in Example 8.1).

With the substitution for the side flow, Eq. (8.21) can be written as:

$$\Delta\theta_s = \left(\frac{2K_1 W}{RL}\right)\frac{\mu^*}{Q_s^*}. \tag{8.22}$$

We can also write, the friction power generated by the film as:

$$P = \mu^* \left(W \frac{c}{R} \pi DN \right). \tag{8.23}$$

Finally, the **Reciprocal Sommerfeld Number**, $1/S$, yields from Eq. (8.6):

$$\eta = \left(\frac{1}{S} \right) \left(\frac{W}{NDL} \right) \left(\frac{c}{R} \right)^2. \tag{8.24}$$

All the dimensionless factors are displayed in Appendix Table 1. Appendix Table 2 gives their values, taken from Ref. 4 for finite length 360° bearings. Just as in the worked Example 2 in Chapter 7, we will use the same approach for a journal bearing.

8.8.3. *Design example (Some of the data in this example is taken from Ref. 5)*

A steam turbine rotor supports a pair of identical hydrodynamic 360° bearings. Each bearing is required to take a steady radial load, W, of 200 kN, while the rotor revolves at a constant speed of 3000 rev/min. The oil used is the same as in design example 7.2, with a supply temperature, θ_0, of 40°C and a maximum pump pressure of 9×10^5 N/m^2. Because of surface finish constraints, the minimum oil film thickness, h_0, must not be less than 2×10^{-5} m. The specified bearing dimensions are $D = 0.2$ m and $L = 0.1$ m.

Estimate the radial clearance, effective temperature and equivalent viscosity of the side leakage oil Take the conductive heating constant as $k_1 = 0.75$, the lubricant density as 850 Kg/m^2 and its specific heat as 1880 J/Kg°C.

Solution

For the solution, we need to build up Table 8.1, based on the given data supplied and the values of the design coefficients, for a range of radial clearances. We start by entering a range of eccentricity ratios as well as the corresponding values of the design coefficients for a bearing with an L/D ratio of 0.5. This data is in Appendix Table 8.2.

- Columns 1, 2, 3 and 4 can be entered immediately.
  The required minimum oil film thickness, h_0, is given. For this value, from Eq. (8.1) $c = \frac{h_0}{1-\varepsilon}$ enabling us to enter, into column 5, the corresponding c value for each ε.

Table 8.1. Solution to design example.

1	2	3	4	5	6	7	8	9
ε	$1/S$	Q_s^*	μ^*	$c \times 10^{-6}$ m	$\Delta\theta_s(^\circ\text{C})$	ηcP	$\theta_{s1}(^\circ\text{C})$	$\theta_{s2}(^\circ\text{C})$
0.4	0.7855	0.751	17.096	33.33	427.23	17.46	467.2	79.1
0.45	0.6337	0.845	14.2	36.364	315.4	16.76	356	80.4
0.5	0.5093	0.938	11.814	40	236.2	16.30	276	81.3
0.55	0.4059	1.032	9.813	44.44	178.38	16.04	218	82
0.6	0.3192	1.126	8.102	50	135	15.96	175	81.7
0.65	0.2463	1.220	6.626	57.14	101.87	16.08	142	81
0.7	0.185	1.315	5.342	66.67	76.24	16.44	116	81
0.75	0.1338	1.409	4.22	80	56.17	17.13	96.2	79.7
0.8	0.092	1.504	3.233	100	40.33	18.32	80.3	77.6
0.85	0.058	1.599	2.352	133.33	27.59	20.48	67.6	74.2
0.9	0.031	1.695	1.519	200	16.81	24.96	58.8	68.6
0.95	0.0119	1.792	0.892	400	9.34	38.08	49.3	57.9

- The values of $\Delta\theta_s$ are entered into column 6 by using Eq. (8.23).
- We can now use Eq. (8.25) to enter the values of η into column 7.
- For column 8, an effective temperature θ_{s1} can be found from its definition: $\theta_{s1} = \theta_0 + \Delta\theta_s$.
- For column 9, the values of η from column 7 must be converted to cP in order to be used in the **Vogel equation** (5.4) and tutorial example 5.1, to obtain an effective temperature. For the chosen oil, it is:

$$\theta_{s2} = \left(203.09 + \frac{700.81}{\ln(\eta) + 1.843}\right) - 273 \text{ in } ^\circ\text{C}.$$

  Remember that η is entered in cP.
- The values of θ_{s1} and θ_{s2} will be different except at one particular value, which yields the correct solution. Thus, if columns 8 and 9, as ordinates, are plotted against column 5, the intersection point is the wanted value, as in Fig. 8.2 below.

From Fig. 8.10 the required value of c at the intersection of θ_{s1} and θ_{s2} is approximately $c_e = 104$ microns at $\theta_{se} = 80^\circ$C, $\varepsilon_e = 0.8$ and $\eta_e = 19cP$, allowing us to find the remaining design variables from the relevant expressions above.

8.9. Mass Unbalance in Rotors

One problem prevalent in rotating machinery is the **whirling** of rotors. These are vibrations of the rotor geometric center that follow some closed (stable) or open (unstable) path relative to a fixed axis through the support

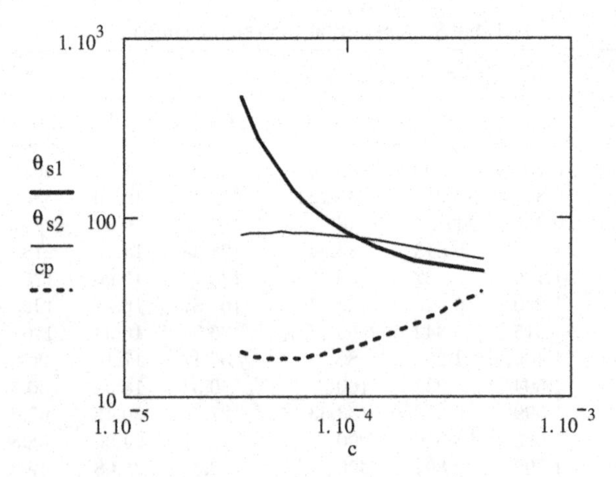

Fig. 8.10. Worked example. Radial clearance c (m), temperatures qs1 and qs2 in degrees C, viscosity cp in (cP).

bearings, which may themselves be vibrating in their oil films. If there is mass unbalance, the rotor mass center will itself be rotating under centrifugal force.

To discuss the effect of mass unbalance, the problem of whirl may be simplified by firstly considering a uniform flexible rotor, with its rotation axis vertical, suspended on freely pivoted *rigid* bearings. Keyed to it at mid span is a horizontal disc with an offset mass center of gravity. Let the combined mass of the disc and rotor equal m. Figure 8.11 shows the arrangement.

Assume for simplicity that there is negligible damping present and a steady rotational speed ω of the disc and rotor about O on the vertical axis through the bearings. If $OP = u$ and the offset of its mass center of gravity

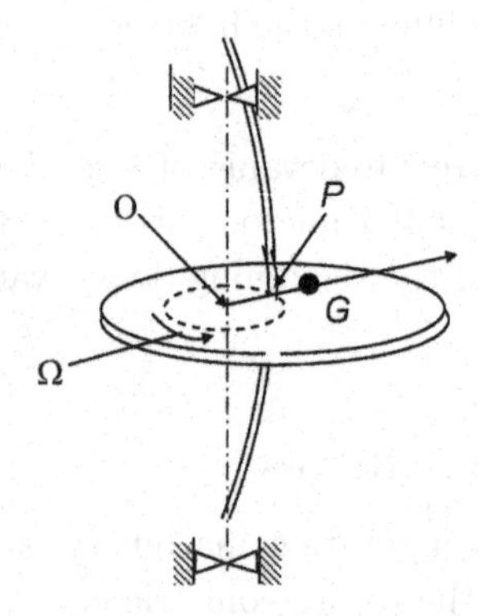

Fig. 8.11. Whirlingdisc.

is $PG = r$, then for zero damping, u and r will always be collinear as they rotate together about O. Had there been damping present, then OP and PG would not be collinear, but at some phase angle ϕ to each other that would vary with the amount of damping present. Now the elastic deflection at the disk geometric center is caused by the centrifugal force at G. If the bending stiffness of the shaft is k_r, a balance of steady state forces along the radius vector OG yields

$$m\Omega^2(r+u) = k_r u. \quad \text{Therefore:} \quad u = \frac{\Omega^2 r}{(k_r/m) - \Omega^2}.$$

As the natural frequency of the system in bending is $\omega_n = \sqrt{k_r/m}$, we can alternatively write the deflection expression as:

$$u = \frac{(\Omega/\omega_n r)}{1-(\Omega/\omega_n)^2}. \tag{8.25}$$

Equation (8.25) tells us that if $\Omega/\omega_n < 1$, G is outside P.

$$\text{If} \quad \Omega/\omega_n = \pm 1, \quad u = \pm\infty.$$

This is the undamped natural frequency condition. In practice, the resonance amplitude is finite, because of the damping being present.

If $\Omega/\omega_n > 1, u$ is negative causing G to be between P and O.

At excessive speeds u, which is now reducing with speed increase, eventually has G coincident with O. In this position $u = r$, producing a deflection equal to r.

The above behavior is called **synchronous whirl**. It can only be caused by a forced excitation, such as unbalance. The safest practice is to keep the operating speeds below the natural frequency by making the rotor bending stiffness correspondingly high. If the rotor and disc are horizontal, point O is now the mid span *static* equilibrium position.

8.9.1. *Natural frequency of journal bearings*

Turning our attention to the journal bearings supporting the rotor at its ends, each oil film is both a spring and a viscous damper. This situation is complicated by the film geometry varying with speed, making the eccentricity ratio and attitude angle also vary. To find a simple expression for the undamped natural frequency of small vibrations, referring to Fig. 8.12, let the rotor be horizontal under a gravity load of W on

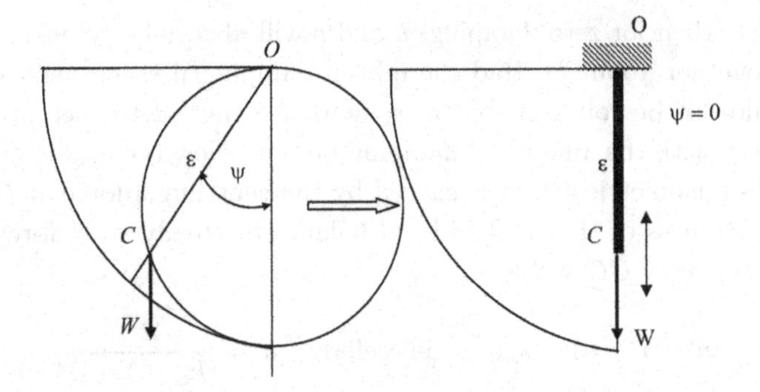

Fig. 8.12. Approximate attitude angle against eccentricity ratio.

each bearing. Also assume that the journal center, C at attitude angle, Ψ, vibrates up and down in one dimension instead of moving along the equilibrium semicircle.

If k_o is the oil film stiffness at one of the bearings, its natural frequency is:

$$f_0 = \frac{1}{2\pi}\sqrt{\frac{\text{stiffness}}{\text{mass}}} = \frac{1}{2\pi}\sqrt{\frac{gk_0}{W}}\text{vib/s.}$$

$$\text{Now} \quad k_0 = \frac{dW}{de} = \frac{dW}{cd\varepsilon}.$$

$$\text{Also,} \quad \frac{dS}{S} = \frac{dW}{W} \quad \text{where the Sommerfeld Number,} \ S = \frac{W/L}{ND\eta}\left(\frac{c}{R}\right)^2.$$

$$\therefore f_0 = \frac{1}{2\pi}\sqrt{\frac{gdS}{cSd\varepsilon}}.$$

For the stiffness we need, $dS/d\epsilon$ which can be obtained approximately from Appendix 8.2 by plotting a graph of S against ε (see Fig. 8.4 and Question 8.2). For example, for a bearing of $L/D = 1$ between $0.35 < \varepsilon < 0.7$, a curve fit in that region can be approximated to:

$$S = a\exp(b\varepsilon), \qquad \text{where } a \text{ and } b \text{ are constants}$$

Differentiating with respect to ε:

$$dS = bSd\varepsilon.$$

Choosing a suitable pair of coordinates in this range $b \approx 4$.

And letting a constant, $n = \sqrt{b}, n = 2$, to give:

$$\omega_0 = n\sqrt{\frac{g}{c}} \text{ rad/s.} \tag{8.26}$$

Equation (8.26) is an approximate expression for the undamped natural frequency of a hydrodynamic journal bearing. We see that it depends only on the radial clearance and the value of $n = 2$ above. A fuller solution to the problem gives $n = 1.3$ between $0.1 < \varepsilon < 0.7$. Above $\varepsilon = 0.7$, the film 'spring' becomes highly nonlinear.

8.9.2. *Half speed cylindrical whirl of journal bearings*

Having found the natural frequency of a hydrodynamic bearing, we should investigate what happens when the journal rotational speed, Ω, is some multiple of ω_0. Just as in the case of a flexible rotor there occurs a synchronous resonance at $\Omega = \omega_0$ that can generally be passed through easily. More seriously, another problem arises at a rotor speed some way beyond ω_0. Consider the hydrodynamic bearing flow analyzed in Fig. 8.1

The journal shown in Fig. 8.13(a) is rotating at speed Ω at an eccentricity, e, while the line of centers through OC is rotating at some speed, ω, such as at its natural frequency, (ω_0). In (b), impart a global anti-clockwise rotation of ω to the journal, bearing bush and the line of centers. Figure 8.13(c) shows the resultant motion, now with the line of centers stationary, while the bearing bush rotates anticlockwise at ω. If we inspect the right hand side of Reynolds Eq. (6.20), it shows that the flow depends on the *sum* of the two surface velocities U_1 and U_2, these being proportional to their angular velocities. Thus, 8.13(d) is equivalent to Figure 8.13(c) if

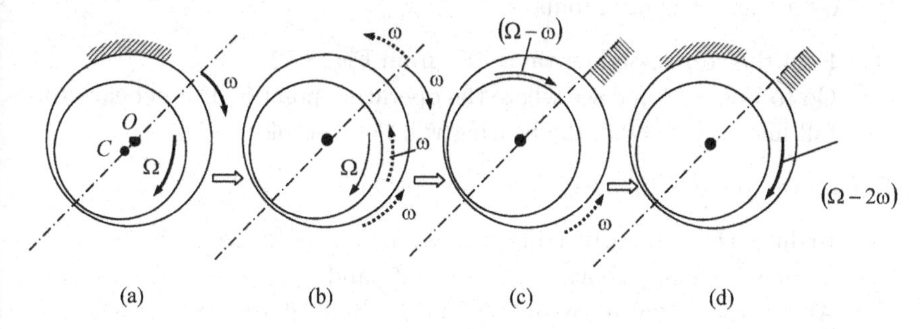

Fig. 8.13. Half-speed whirl.

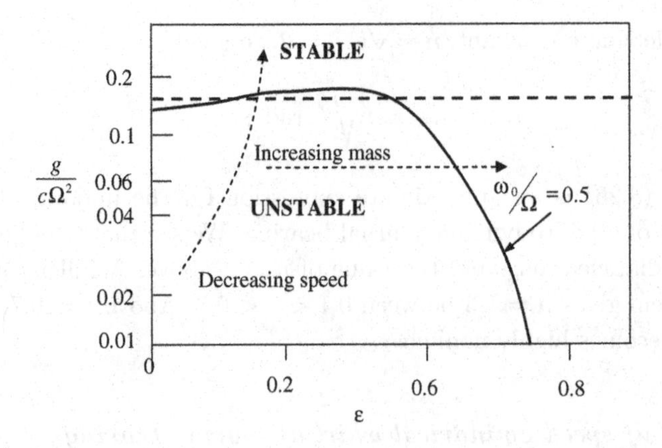

Fig. 8.14. Stability map for a horizontal journal bearing of $L/D = 1$.

both the surface velocities are transferred to the journal surface, making the bush again stationary, while the journal rotates at $\Omega - 2\omega$. The **half-speed cylindrical whirl** condition arises when $\Omega = 2\omega_0$. In this case, the film entraining velocity is effectively zero, causing the journal to spiral outwards into the bearing bush surface. Assuming this condition and letting $\omega = \omega_0$ in Eq. (8.26) with $n = 1.3$

$$\Omega = 2.6\sqrt{\frac{g}{c}}\text{rad/s},$$

or, non-dimensionally for bearings of $L/D = 1$:

$$\frac{g}{c\Omega^2} = 0.148. \tag{8.27}$$

The *undamped* two dimensional numerical solution to this problem, produces the stability map shown in Fig. 8.14, with the one dimensional solution shown as a dotted horizontal straight line of height 0.148.

Assuming S has been found:

(1) Find the corresponding value of ε from Fig. 8.12.
(2) Go to Fig. 8.14 and see where the operating point is. If it is below the full line, $\omega_0/\Omega = 0.5$, the bearing will be unstable.

To cure the problem in an isothermal solution:

(3) Reduce Ω sufficiently (this will alter S and hence ε) to take the operating point above $\omega_0/\Omega = 0.5$ and into the stable region. Alternatively, try increasing the load (this will alter S and hence ε) again taking the operating point into the stable region.

Note that a more rigorous search for a cure should include the thermal procedure as well.

8.9.3. *Natural frequency of the complete rotor/bearing system*

If the flexibility of the rotor and its two bearings are considered together as two springs in series supporting the disc, then if $n = 1.3$, $\varepsilon < 0.65, L/D = 1$, M_{rot} = natural frequency of the rotor in revolutions/min and c = bearing radial clearance in metres, M_{eff}, the vibration frequency of the system in thousands of cycles/min, is given empirically by:

$$\frac{1}{M_{eff}^2} = \frac{1}{M_{rot}^2} + \frac{c}{155} \times 10^5. \tag{8.29}$$

If $M_{eff} = 0.48$ of M_{rot} a severe vibration may occur.

8.10. Journal Bearings with Periodic and Rotating Loads

Particularly with internal combustion engines, their crankshaft and crankpin bearings are subjected to periodic inertia and firing loads from the cylinders via the connecting rods. This important problem is covered in Chapter 12.

8.11. Closure

The theory outlined in this chapter, should enable the reader to produce approximate designs of journal bearings, taking into account both their thermal and dynamic behavior.

Appendices

Appendix Table 8.1. Design coefficients and formulas.

	Journal bearing design coefficients	Closed solutions
1	$h_0 = c(1-\varepsilon)$	
2	$\eta_e = \left(\frac{1}{S}\right)\left(\frac{W}{NDL}\right)\left(\frac{c}{R}\right)^2$	$S = \left(\frac{L}{D}\right)^2 \frac{\pi\varepsilon}{(1-\varepsilon^2)}(0.62\varepsilon^2+1)^{1/2}$ (narrow 360° bearing)
3	$\mu = \frac{F}{W} = \mu^*\left(\frac{c}{R}\right)$	$\mu^* = \frac{2\pi^2}{S}$ (Petroff‡ bearing)
4	$F = \mu^*\left(W\frac{c}{R}\right)$	$F = \frac{\pi^2 ND^2 L\eta}{c}$ (Petroff bearing)
5	$Q_s = Q_s^*(\pi NRLc)$	$Q_s^* = 2\varepsilon$ (narrow bearing)
6	$H = \mu^*\left(W\frac{c}{R}\pi DN\right)$	$H = \frac{\pi^3\eta D^3 N^2 L}{c}(1-\varepsilon^2)^{-1/2}$ (narrow bearing)
7	$\Delta\theta_s = \frac{\mu^*}{Q_s^*}\left(\frac{2K_1 W}{RL}\right)$	

‡A Petroff bearing is one that is always concentric. Thus friction and flow are from the Couette term only, so that the effect of load that is assumed to be negligible.

Appendix Table 8.2. Design coefficients for finite length 360° bearings (Data from Ref. 4).

ε	0.4	0.45	0.5	0.55	0.6	0.65	0.7	0.75	0.8	0.85	0.9	0.95
$\frac{L}{D}=1$												
$1/S$	0.26	0.216	0.179	0.147	0.121	0.098	0.078	0.06	0.044	0.031	019	008
μ^*	5.79	4.962	4.282	3.71	3.216	2.783	2.393	2.037	1.702	1.378	1.051	0.69
Q_s^*	0.631	0.708	0.785	0.861	0.937	0.937	1.012	1.163	1.238	1.313	1.386	1.46
Q_1^*	1.27	1.298	1.326	1.353	1.378	1.403	1.426	1.449	1.47	1.489	1.51	1.528
$\frac{L}{D}=0.5$												
$1/S$	0.785	0.634	0.509	0.406	0.319	0.246	0.185	0.134	0.092	0.058	0.031	0.012
μ^*	17.1	14.2	11.81	9.813	8.102	6.626	5.342	4.22	3.233	2.362	1.592	0.892
Q_s^*	0.751	0.845	0.938	1.032	1.126	1.22	1.315	1.409	1.504	1.599	1.695	1.792
Q_1^*	1.365	1.413	1.455	1.5	1.545	1.6	1.634	1.785	1.723	1.767	1.811	1.855

Note that:

(1) $Q_1^* = Q_s + Q_2^*$.

Q_1^* is the *total* dimensionless flow entering the bearing pressure arc, Q_s^* is the side leakage component and Q_2^* (not shown above) is the circumferential flow component that always circulates through the film.

(2) μ^* includes cavitation friction in the unloaded zone.

References

1. Sommerfeld, A., *Zeits. f. Math. U. Phys.* **40** (1904) 97–155.
2. Steiber, *Das-Schwimmlager, VDI* (1933).
3. Cameron, A., *Principles of Lubrication*, Longmans (1964).
4. Khonsari, M. M. and Booser, E. R., *Applied Tribology*, John Wiley & Sons (2001).
5. Arnell, R. D., Davies, P. B., Halling, J. and Whomes, T. W., *Tribology Principles and Design*, MacMillan (1991).

CHAPTER 9

EXTERNALLY PRESSURIZED (EP) BEARINGS

9.1. Introduction

So far we have considered only the hydrodynamic bearing, where a wedge shaped lubricant film separates two surfaces in relative motion. We had also seen in Chapters 7 and 8, that the total power required for a hydrodynamic bearing must overcome lubricant shear friction as well as driving a small pump that returns the side leakage oil back into the wedge inlet region.

Unlike hydrodynamic bearings, in the case of **externally pressurized bearings** (EP), the need for relative motion between the surfaces is removed. Instead, EP bearings must have an external pump powerful enough to generate sufficient pressures to separate the surfaces as well as returning the leakage oil. When a liquid is being used, they are called **hydrostatic bearings** and when a gas is being used they are often called **aerostatic bearings.** Remember, in Chapter 6 we invited you, in a text example, to discuss the physics of a hydrostatic bearing.

Because of the pump, the pressurized film created (typically $30\,\mu$m) can be maintained at zero speed; hence the word *static* in the name where, for this condition, the friction force must be zero.

A rotating EP thrust bearing, therefore, does not require a wedge shaped film to operate. On the other hand, a *rotating* EP journal bearing becomes a **hybrid bearing**, because the lubricant film generation is created partly from the wedge action and partly from the hydrostatic action. The problems of wear at start up in EP bearings are not present on account of the initial separation of the surfaces before rotation commences. In comparison with the equivalent hydrodynamic bearing, the external control of the supply pressure makes the EP bearing relatively stiff, with negligible deflection under load.

The chief disadvantages of hydrostatic bearings are the relative cost, space and weight requirements of the pump circuit. Also, just like the human heart, pump failure can result in bearing failure.

9.2. Some Applications of EP Bearings

- **Large machine tool slideways**
  Intermittent slideway motion with high direct and side loads, make the hydrostatic oil bearing ideal for this application.
- **Large astronomical and radio telescope supports**
  Hydrostatic oil bearings are used where a slow, vibration free, rotational motion is needed.
- **Dentist drills**
  As extremely high rotational speeds are needed here, the drill can be supported by an aerostatic hybrid journal bearing.
- **Dynamometers**
  Because of their low friction characteristics aerostatic bearings are ideally suited for this application.
- **Metrology**
  The low run out of aerostatic bearings makes them a useful component in a measuring instrument where smooth directed motion is essential.

Some examples of hydrostatic bearing designs are illustrated in Fig. 9.1

9.3. Principles of Hydrostatic Bearings

Figure 9.2 most easily explains the principle of the **hydrostatic bearing**. It represents the profile of a hydrostatic step bearing.

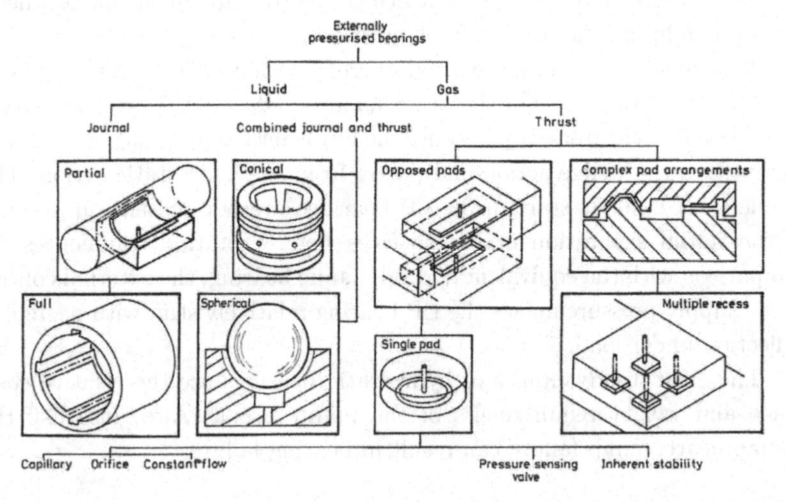

Fig. 9.1. Hierarchy of hydrostatic bearings. From Ref. 1 reproduced by permission of Elsevier Ltd.

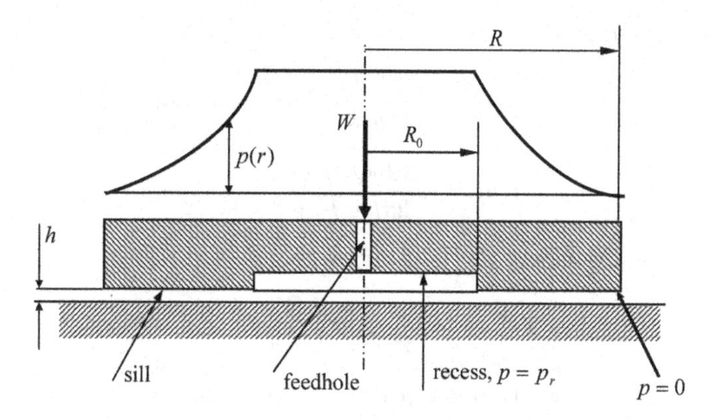

Fig. 9.2. Circular hydrostatic step bearing.

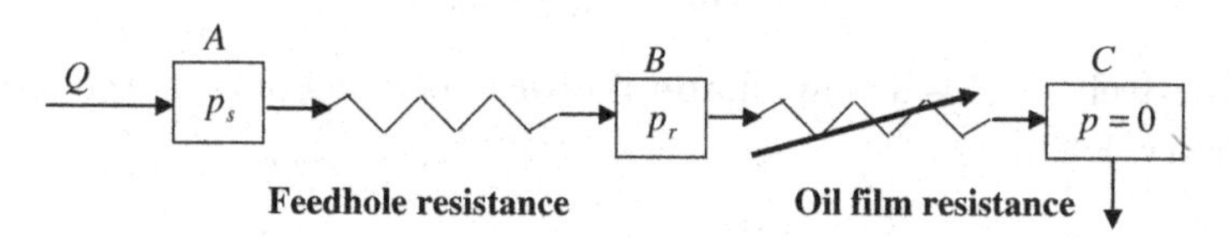

Fig. 9.3. Hydrostatic bearing circuit resistances.

The bearing composes a circular disc capable of rotating about its spin axis under a load, W. There is a deep recess on the bottom face above the film (typically $100\,h$) with oil being fed from a constant flow pump through a feedhole into the recess. The hole is usually of small diameter so that there is some resistance to the pumped oil. The relatively large depth of the recess compared with the film thickness, h, under the sill, makes the recess oil practically stagnant with only **creeping flow**, Q, leaking out under the sill. By continuity, the same flow enters via the feedhole. Effectively, we can define the design as equivalent to the electric circuit shown in Fig. 9.3, with gauge pressures equivalent to voltages and the flow equivalent to current.

Any increase in load will make the oil film under the sill thinner, thus increasing its resistance. For a constant flow pump, this must increase the recess pressure, p_r, thus increasing the film stiffness, a very desirable characteristic. We will see below that feedholes, allow a compound bearing, like that in example (a) in Fig. 9.1, to employ a single pump supplying several independent bearing surfaces. In our demonstration example, referring to Eq. (6.18), because the bearing is axisymmetric, the steady state volumetric flow through an annulus of the film of radius r, length $2\pi r$

and thickness h, is:

$$Q = -\frac{h^3}{12\eta}\frac{dp}{dr}2\pi r = \text{ constant},$$

$$\therefore \quad \int dp = -\frac{12\eta Q}{2\pi h^3}\int\frac{dr}{r}.$$

Boundary conditions are $p = 0$ at $r = R$ giving:

$$p = \frac{6\eta Q}{2\pi h^3}\ln\left(\frac{R}{r}\right). \tag{9.1}$$

Equation (9.1) defines the pressure distribution in the oil film under the sill. To obtain the flow, we see that $p = p_r$ at $r = R_0$, giving a radial flow of:

$$Q = \frac{p_r \pi h^3}{6\eta}\left[\frac{1}{\ln(R/R_0)}\right]. \tag{9.2}$$

Alternatively, if $\bar{B}$ is a **flow shape factor** depending only on the bearing geometry, let:

$$\bar{B} = \frac{\pi}{6\ln(R/R_0)}, \tag{9.3}$$

allowing Eq. (9.2) to be expressed as:

$$Q = \frac{p_r h^3}{\eta}\bar{B}. \tag{9.4}$$

As Eq. (9.4) shows, if h reduces under load, and Q is constant, p_r must increase. Let us now find an expression for the load under steady state conditions. As the flow within the recess is negligible, because of its depth relative to the leakage film below the sill, the recess pressure is nearly constant. Therefore, integrating Eq. (9.1) and substituting Eq. (9.2) for Q:

$$W = \int_{R_0}^{R} 2\pi r p dr + \pi R_0^2 p_r,$$

$$\therefore \quad W = \frac{p_r \pi}{2}\left[\frac{R^2 - R_0^2}{\ln(R/R_0)}\right]. \tag{9.5}$$

Introducing a load factor, $\bar{A}$ and letting $A = \pi R^2$, $\bar{R} = R/R_0$,

$$\bar{A} = \frac{1}{2}\left(\frac{(1/\bar{R}^2)}{\ln(\bar{R})}\right), \tag{9.6}$$

$$W = A p_r \bar{A}. \tag{9.7}$$

The particular geometry of the step bearing is defined by the factors, $\bar{A}$ and $\bar{B}$. Moreover, the same form of the flow and load equations apply to other bearing geometries, for example one that is rectangular. All that alters are the expressions for $\bar{A}$ and $\bar{B}$.

9.3.1. *Optimization*

In order to select the appropriate pump, we must determine the **lifting power**, P, required to support the given load. This is shown to be:

$$P = Qp_r. \tag{9.8}$$

In Fig. 9.3, it is the power absorbed between points B and C in the film that is used to lift the load, called the power to lift ratio, thus:

$$\frac{Power}{W} = \frac{Qp_r}{W} = \frac{p_r h^3 \bar{B}}{\eta W} = \left(\frac{W}{A\bar{A}}\right)^2 \frac{h^3}{\eta}\frac{\bar{B}}{W} = \frac{Wh^3}{A^2\eta}\left(\frac{\bar{B}}{\bar{A}^2}\right).$$

Hence the power needed in terms of the load is:

$$P = \frac{W^2h^3}{A^2\eta}\left(\frac{\bar{B}}{\bar{A}^2}\right), \tag{9.9}$$

the power to **lift ratio** being:

$$\frac{P}{W} = \frac{Wh^3}{A^2\eta}\bar{P}. \tag{9.10}$$

The **Power Coefficient** is $\bar{P} = \frac{\bar{B}}{\bar{A}^2}$, and should normally be the minimum possible. Figure 9.4 shows the variation of the design coefficients with radius ratio for a circular step bearing. The optimum ratio for minimum power is when $\bar{R} = 1.89$.

9.3.2. *Stiffness*

Consider the case when a constant flow pump (often a gear pump, or a flow control regulator) is connected to the recess by a large diameter feed hole. This makes the feedhole resistance in Figs. 9.2 and 9.3 negligible, with p_r equalling the pump delivery pressure. Because A and $\bar{A}$ are constants for a given bearing, Eq. (9.5) shows that the load supported does not depend on the viscosity or film thickness but solely on the delivery pressure. Thus,

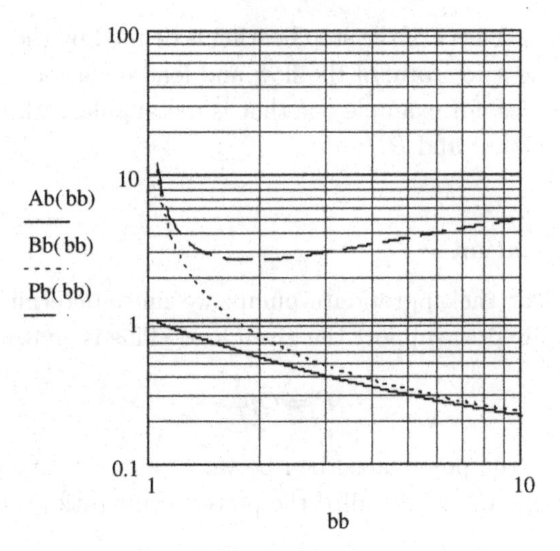

Fig. 9.4. Design coefficients for a circular hydrostatic step bearing. (bb $\equiv \bar{R}$, Ab(bb) $\equiv \bar{A}$, Bb(bb) $\equiv \bar{B}$, Pb(bb) $\equiv \bar{P}$).

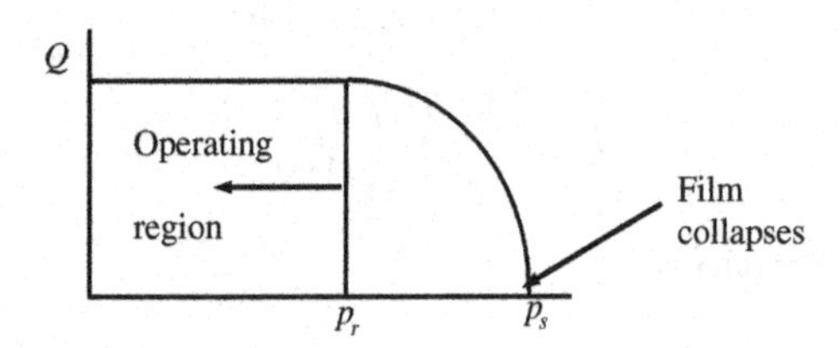

Fig. 9.5. Pump connected directly to a recess.

in practice the bearing runner would either lift off its pad, if p_r was greater than what is needed to support the load, or the film would collapse if it were less, as in Fig. 9.5.

Figure 9.5 demonstrates that the flow is constant in the operating region. For this design, let us find the bearing stiffness. From Eqs. (9.4) and (9.7):

$$W = \frac{\eta Q A}{h^3}\frac{\bar{A}}{\bar{B}}. \tag{9.11}$$

Stiffness is defined by:

$$\lambda_f = -\frac{dW}{dh} = -\frac{3A\bar{A}\eta Q h^{-4}}{\bar{B}}.$$

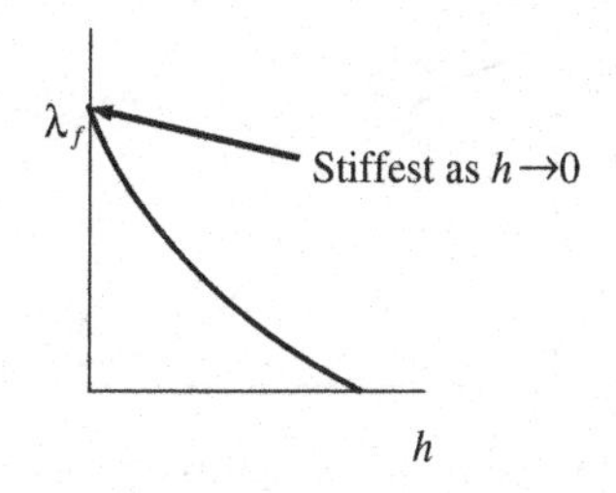

Fig. 9.6. Bearing stiffness using a constant delivery pump.

However from Eq. (9.4) $Q = \frac{p_r h^3}{\eta}\bar{B}$.

Therefore, substituting for Q above and dropping the sign, then:

$$\lambda_f = \frac{3W}{h}. \tag{9.12}$$

Figure 9.6 illustrates the stiffness variation with it approaching a maximum as $h \to 0$. Such a delivery arrangement is satisfactory when a single recess is used. But in the case of a bearing design needing several recesses, like (a), (b) or (c) in Fig. 9.1, each would require a separate supply pump. This is an unsatisfactory arrangement. An alternative design is discussed below.

9.3.3. *Compensators*

The simplest way to control the fluid flow in a hydrostatic bearing is to introduce a resistance in the supply line downstream of a constant supply pressure, p_s, in the pump manifold (A to B in Fig. 9.3). Note that the flow resistance increases as the feedhole diameter decreases compared to its length, it becoming a **capillary tube** when sufficiently long.

Now the pressure drop across a capillary tube of length L_c and diameter d, is $p_s - p_r$ and assuming slow viscous flow, its coefficient is defined by $K_c = \frac{128L_c}{\pi d^4}$, producing a throughput of:

$$Q = \frac{(p_s - p_r)}{K_c\eta}. \tag{9.13}$$

Equating Eqs. (9.4) and (9.13): $\frac{p_r h^3}{\eta}\bar{B} = \frac{(p_s - p_r)}{K_c\eta}$.

Then, substituting W for p_r from Eq. (9.7) we get:

$$W = \frac{p_s A\bar{A}}{h^3 K_c\bar{B} + 1}. \tag{9.14}$$

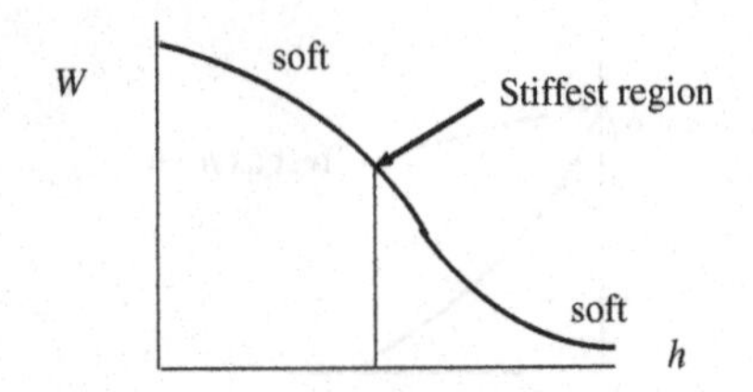

Fig. 9.7. Load variation for a capillary compensated bearing.

We can now sketch Fig. 9.7 to demonstrate how the stiffness varies with load when there is a capillary tube. Where the slope is greatest, there is maximum stiffness.

To obtain the stiffness, differentiate Eq. (9.14) with respect to h.

$$\frac{dW}{dh} = \lambda_c = p_s A\bar{A}\left[\frac{3\bar{B}K_c h^2}{\left(\bar{B}K_c h^3 + 1\right)^2}\right]. \tag{9.15}$$

Substituting p_s in terms of W from Eq. (9.14) we get, after some manipulation:

$$\lambda_c = \frac{3W}{h}\left[\frac{1}{1 + \left(\frac{1}{\bar{B}K_c h^3}\right)}\right]. \tag{9.16}$$

We see from Eq. (9.16) that original Eq. (9.12), for the constant flow stiffness, has been modified by a factor containing h, the geometrical flow factor, and the capillary coefficient. Although this modification makes the bearing less stiff, it becomes a more adjustable design. One big advantage of using capillary compensators is that a single pump having constant pressure manifold can serve several individual bearing pairs. This is illustrated in Fig. 9.8 showing a **multi recess journal bearing**.

If W increases, h reduces at the lower recess, thus increasing the reaction force there (Eq. (9.14)). As h correspondingly increases at the top recess, the reaction force there will be reduced, achieving finally a new equilibrium position. On the other hand, if there is a pulse force (one of short duration) in any direction, the journal will return to its equilibrium position after a time that depends on the film stiffness and its damping properties. There are many different flow control and pressure sensing devices that have been patented to control the behavior of hydrostatic bearings. O'Donoghue and Rowe[1] discuss succinctly the operation of these devices.

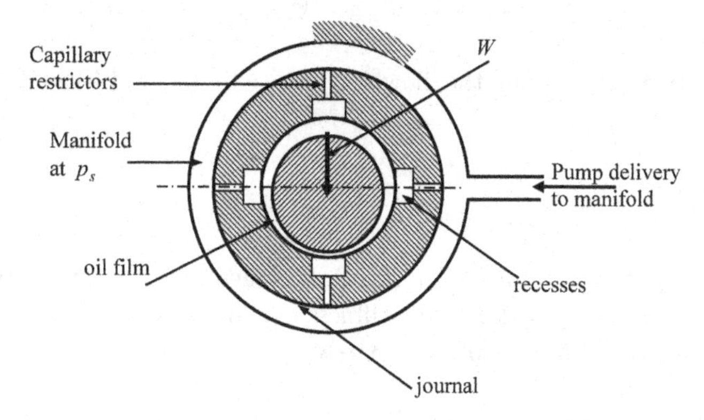

Fig. 9.8. Constant pressure manifold hydrostatic journal bearing.

9.3.4. *Worked Example 1*

A circular hydrostatic central orifice step bearing supports the end of a 0.02 m diameter shaft under a purely axial load. The outside diameter of the bearing is also 0.2 m and its recess diameter is 0.01 m. From a constant pump supply pressure of $10\,\text{MN/m}^2$, an oil of viscosity 25 cP is fed, via a capillary tube, to a pressure in the recess of 5MN/m^2; the bearing film thickness is to be 10^{-4} m.

Determine the load capacity, flow, stiffness and lifting power of the bearing. Choose also suitable capillary tube dimensions if viscous flow is to be maintained within it. (For the tube, assume that $L/d = 100$ is sufficient to maintain purely viscous flow.)

Solution

It's a good idea firstly to calculate the load and flow factors:

$$A = \pi R^2, \quad \bar{R} = \frac{R}{R_0}, \quad \bar{A} = \frac{1}{2}\left(\frac{(1 - 1/\bar{R}^2)}{\ln(\bar{R})}\right),$$

where $R = 0.02\,\text{m}$, $R_0 = 0.01\text{m}$ giving $\bar{A} = 0.541$ and, from Eq. (9.3), $\bar{B} = 0.755$.

Load Capacity

This is the load the bearing is capable of supporting according to the given specification. Use Eq. (9.7) to obtain the load capacity: $W = 3399\,\text{N}$ (**Answer**).

Flow

Use Eq. (9.4) to obtain the flow. It comes to $Q = 1.51 \times 10^{-4}\,\text{m}^3/\text{s}$ (**Answer**).

Stiffness

First K_c must be found from Eq. (9.14). It comes to $K_c = 1.325 \times 10^{12}\,\text{m}^{-3}$ Hence, knowing also the load, the stiffness may be obtained from Eq. (9.16). It comes to $\lambda_c = 5.098 \times 10^7\,\text{Nm}$ (**Answer**).

Power

The power needed to lift the load can be found from Eq. (9.7) to give $P = 755\,\text{Nm/s}$ (**Answer**).

Restrictor Length

Knowing K_c, L_c and d are related by its definition (just above Eq. (10.13)): $L_c = \frac{\pi K_c}{128} d^4$. Also, $L_c/d = 100$. Solving these two equations:

$$D = 1.458\,\text{mm and } L_c = l.458\,\text{mm } (\textbf{Answer}).$$

(The capillary tube need not be supported over its whole length. Often, the top, middle section and outlet part only are secured.)

9.4. Principles of Aerostatic Bearings

9.4.1. *Introduction*

This type of EP bearing deserves a separate section because it comes very close to a frictionless machine, being also capable of excessive rotational surface speeds. These characteristics are sometimes needed in certain applications such as for some drilling machines or small turbo generator sets.

9.4.2. *Flow through the gas film*

Unlike that of hydrostatic bearings, aerostatic bearing theory is quite complicated, the reason being that we must consider in addition the *compressibility* of gas (variable density) in an analysis. We will therefore confine ourselves to the basic theory of a circular aerostatic thrust bearing similar to the hydrostatic example shown in Fig. 9.2.

Another difference between aerostatic and hydrostatic bearings is the role of the recess where, in the latter case, it contributes significantly to the load capacity. The role of the recess in aerostatic bearings is sometimes different, with the recess being omitted completely in certain designs. To stress this difference, we will call it a *pocket.* An additional role for the pocket will be discussed later when we deal with the air supply to the bearing. Figure 9.9 illustrates a central orifice **aerostatic bearing**.

From Eq. (6.12), the film velocity distribution in a **circular thrust bearing** at radius r and thickness h is:

$$u = \frac{1}{2\eta}\frac{dp}{dr}z(z-h). \tag{9.17}$$

Therefore, the mass flow through an annulus of length $2\pi r$ and depth h is:

$$m = 2\pi r\rho \int_0^h u dz.$$

Substituting for u from Eq. (9.17) and integrating across the film:

$$\frac{dp}{dr} = -\frac{6\eta m}{\pi\rho h^3 r}. \tag{9.18}$$

As shown in Fig. 9.7, the negative sign appears because of the negative pressure gradient caused by the falling pressure from the central pocket to the outer rim. For our simplified isothermal solution, we will assume that gas in the film is *incompressible* with its temperature, Θ_M, the same as in the manifold. Incompressible flow requires a constant density. Assume therefore for simplicity that the effective density is at the average pressure

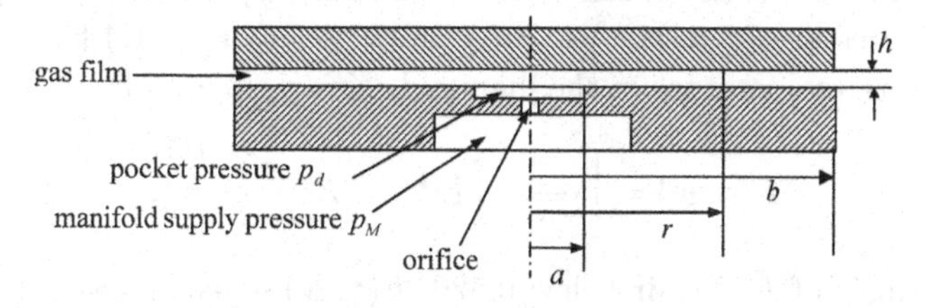

Fig. 9.9. Central orifice pocketed aerostatic thrust bearing.

of the film. Thus:

$$\rho = \frac{p}{\Re\Theta_M} = \frac{(p_d + p_a)}{2\Re\Theta_M}. \tag{9.19}$$

$\Re$ is the gas constant ($\Re = c_p - c_v$), and c_p and c_v are respectively the specific heats of air at constant pressure and volume.

Substituting Eq. (9.19) into (9.18), separating variables and integrating:

$$p_d - p_a = \frac{12\eta m \Re\Theta_M}{\pi h^3}\left(\frac{1}{p_M + p_a}\right)\ln\left(\frac{b}{a}\right).$$

The mass flow is, therefore:

$$m = (p_d - p_a)(p_M + p_a)\frac{\pi h^3}{12\eta\Re\Theta_M \ln(b/a)}. \tag{9.20}$$

In Eq. (9.20) the pocket pressure, p_d, is unknown; so we need another equation. Consider therefore the incoming flow to the film.

9.4.3. *Gas flow through the feed hole*

Just as in the case of the hydrostatic bearing, the incoming flow is supplied to the film through feedholes. In this example, it is a single orifice situated at the bearing center.

The following assumptions are made:

- The pressure at entry to the orifice is the manifold supply pressure, p_M.
- The pressure immediately downstream of the orifice is the pocket pressure, p_d.

For a nozzle of the same diameter as the jet throat, assuming isotropic expansion of the gas, the mass flow is:

$$m = C_D A_t \rho_M (2\Re\Theta_M)^{1/2} F(\gamma, \bar{K}). \tag{9.21}$$

In Eq. (9.21), C_D is the **coefficient of discharge**, ρ_M is the density at the manifold pressure, A_t is the jet throat area, $\gamma = c_p/c_p = 1.4$ for air, Θ_M is the absolute temperature in the manifold, $\bar{K} = p_d/p_M$ and:

$$F(\gamma, K) = \left[\frac{\gamma}{\gamma - 1}\left\{\bar{K}^{2/\gamma} - \bar{K}^{\frac{\gamma+1}{\gamma}}\right\}\right]^{1/2}. \tag{9.22}$$

For all values of $\bar{K}$ for air below 0.526, $F(\gamma, \bar{K})$ equals 0.484. This is the **choked jet condition** we usually try to avoid.[2] As a simplification, for values of $\bar{K}$ exceeding 0.7, assume that changes in density across the nozzle are ignored. Equation (9.22) is therefore modified for incompressible flow of a gas to give:

$$m = C_D A_t \sqrt{\rho_M}[2(p_M - p_d)]^{1/2}. \tag{9.23}$$

As $p_M/\rho_s = \Re\Theta_M$, Eq. (9.23) can be written as:

$$m = C_D A_t \rho_M \sqrt{2\Re\Theta_M}\left(1 - \frac{p_d}{p_M}\right)^{1/2}. \tag{9.24}$$

Thus, $F(\gamma, \bar{K})$ in Eq. (9.21) can be replaced by $\left(1 - \frac{p_d}{p_M}\right)^{1/2}$ for incompressible flow through the nozzle. In our case the nozzle is a sharp edged orifice of diameter, d_o, connected to a pocket, making $A_t = \pi d_o^2/4$. Some designs have the pocket occupying a significant proportion of the bearing area as illustrated in Fig. 9.9. Just as for hydrostatic bearings, this feature increases the load capacity considerably. However, a disadvantage of a large deep pocket containing a gas, is that hammer-blow self–excited oscillations can occur. These can be quite destructive, but a way of preventing them is to make the pocket shallow (about 4–7 times the film thickness). The pocket also has a role in forcing the jet contraction, and subsequent expansion, to occur within it, away from the film commencement. Also, making the pocket diameter too large compared with the bearing outer diameter increases the load capacity but increases considerably the airflow needed because of the reduced film resistance under the sill. A common alternative design is to eliminate the pocket, and have the nozzle connected directly between the supply manifold and the film. It then becomes an annular orifice with $A_t = \pi d_o h$. Annular orifices have the jet contraction in the film itself, thus creating a less stiff bearing but one that is cheaper to manufacture.

9.4.4. *Matching the gas flows*

We can now equate the flows into and out of the film. From Eqs. (9.20) and (9.24):

$$(p_d - p_a)(p_M + p_a)\frac{\pi h^3}{12\eta\Re\Theta_M \ln(b/a)} = C_D A_t \rho_M \sqrt{2\Re\Theta_M}\left(1 - \frac{p_d}{p_M}\right)^{1/2}. \tag{9.25}$$

Our aim is to obtain a gauge pressure ratio, defined as $K_g = \frac{p_d - p_a}{p_M - p_a}$, in terms of the other variables. This was achieved in the following way by Powell[3] who noted that:

$$\left(1 - \frac{p_d}{p_M}\right)^{1/2} = \left[\left(1 - \frac{p_a}{p_M}\right)^{1/2}\left(1 - \frac{p_d - p_a}{p_M - p_a}\right)\right]^{1/2}$$
$$= \left(1 - \frac{p_a}{p_M}\right)^{1/2}(1 - K_g)^{1/2}.$$

Putting this substitution into Eq. (9.25):

$$(p_d - p_a)(p_M + p_a)\frac{\pi h^3}{12\eta\Re\theta_M \ln(b/a)}$$
$$= C_D A_t \rho_M \sqrt{2\Re\Theta_M}\left(1 - \frac{p_a}{p_M}\right)^{1/2}(1 - K_g)^{1/2}.$$

Letting $A_t = \pi d_o^2/4$ and $\rho_M = \frac{p_M}{R\Theta_m}$, after some adjustment we can write:

$$\frac{K_g}{(1 - K_g)^{1/2}} = G_l. \tag{9.26}$$

G_l is a **global design factor** that includes all the sub factors required for a design:

$$G_l = \underbrace{\frac{p_a/p_s}{(1 - (p_a/p_M))^{1/2}(1 + (p_a/p_M))}}_{F_p} \times \underbrace{\frac{24\eta\sqrt{2\Re\Theta_M}C_D d_o^2}{p_a 8h^3}}_{\Lambda_f} \times \underbrace{\ln\left(\frac{b}{a}\right)}_{F_g} \tag{9.27}$$

It composes a pressure factor (F_p), a feeding factor (Λ_f) that groups the orifice, film and lubricant properties, and a shape factor (F_g) that describes the bearing planform geometry. Equation (9.26) can also be written explicitly in terms of G_l as:

$$K_g = \frac{2}{1 + \left(1 + \frac{4}{G_l^2}\right)^{1/2}}. \tag{9.28}$$

Thus, the gauge pressure ratio is given in terms of the bearing properties. We will now find the load capacity and stiffness of the bearing.

9.4.5. *Aerostatic gas bearing load capacity*

By integrating Eq. (9.18) to obtain the absolute pressure distribution and then integrating again, the bearing load capacity can be found. However, there is no need to do these integrations because the load is found from an integration of *gauge* pressures, not absolute pressures, which we have already done for a hydrostatic bearing. It follows that Eq. (9.5) gives the load capacity we need. In terms of absolute pressures it is:

$$\therefore \quad W = \frac{(p_d - p_a)\pi}{2}\left[\frac{b^2 - a^2}{\ln(b/a)}\right]. \tag{9.29}$$

Alternatively, in terms of the gauge pressure ratio and letting $\bar{b} = b/a$ and:

$$\bar{p}_M = \frac{p_m}{p_a},$$

$$W = \frac{K_g(p_M - p_a)\pi b^2\left(1 - \frac{1}{\bar{p}_M}\right)}{2}\left[\frac{1 - \frac{1}{\bar{b}^2}}{\ln\left(\bar{b}\right)}\right]. \tag{9.30}$$

Finally, the mass flow can be found from Eq. (9.20), reproduced below in terms of K_g.

$$m = K_g(p_M^2 - p_a^2)\frac{\pi h^3}{12\eta\Re\Theta_M \ln(\bar{b})}. \tag{9.31}$$

At this stage it is useful to investigate if there is an optimum value of the geometry factor, $\bar{b}$, for minimum power to weight ratio. However, because we have assumed incompressible flow in the air film, Eq. (9.10) again applies where, for minimum power to weight ratio, $\bar{b} = 1.89$, $\bar{A} = 0.565$, $\bar{B} = 0.755$ and $\bar{P} = 2.365$.

9.4.6. *Aerostatic bearing film stiffness*

A bearing that operates at maximum stiffness is a desirable design feature that can be found quite simply for aerostatic thrust bearings. The global design factor, G, in Eq. (9.27), is a function of h and Λ_f. Moreover, for a particular design of bearing, Eq. (9.30) shows us that $W \propto K_g$. Therefore plotting K_g against h for various values of the orifice diameter, d, determines the characteristic shape of the load versus clearance curve. These are shown in Fig. 9.12 for $\bar{p}_s = 5$ and $\bar{b} = 1.89$.

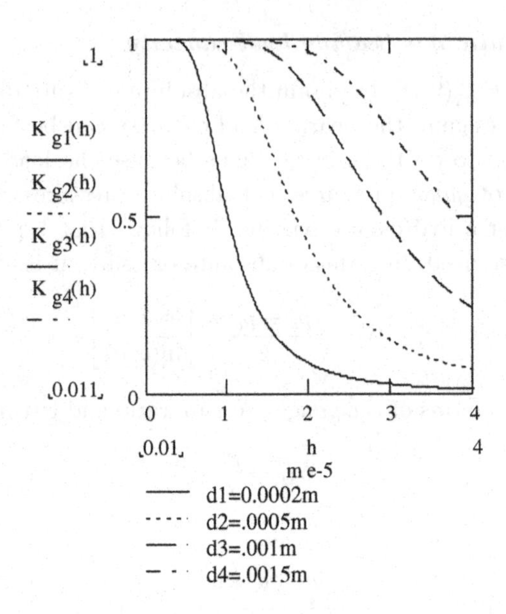

Fig. 9.10. Characteristic shape of gauge pressure against film thickness curve.

We see that the curve shapes in Fig. 9.10 are similar to the stiffness curve for hydrostatic bearings (Fig. 9.7) with the slopes at points on the curve representing the local stiffness. At low values of film thickness the film is *'soft'*, at intermediate values there is the stiff region, becoming soft again at high values. At maximum stiffness for each curve, Grassam and Powell[3] show that $K_g = 0.69$ and therefore, from Eq. (10.26), $G_l = 1.25$. When $K_g = 0.69$, if suffix m means 'associated with maximum stiffness', the measured slope is:

$$\left(\frac{dK_g}{dh}\right)_m = -\frac{0.98}{h_m}. \tag{9.32}$$

Thus, from Eqs. (9.30) and (9.32), the maximum stiffness is:

$$\left(\frac{dW}{dh}\right)_m = K_m = \frac{-0.98p_M(1-(1/\bar{p}_M))\pi b^2(1-(1/\bar{b}^2))}{2h_m \ln(\bar{b})}. \tag{9.33}$$

The corresponding load is given by Eq. (9.30) as:

$$W_m = \frac{0.69p_M(1-(1/\bar{p}_s))\pi b^2(1-(1/\bar{b}^2))}{2\ln(\bar{b})} \tag{9.34}$$

mass flow from Eq. (9.31) becoming:

$$m_m = 0.69(p_M^2 - p_a^2)\frac{\pi h_m^3}{12\eta R\Theta_M \ln(\bar{b})}. \quad (9.35)$$

The above theory enables us to design a central orifice circular aerostatic thrust bearing. The theory of the other type of circular thrust bearing, where there is an annular row of entry orifices surrounding a central hole of radius a, can be found in Ref. 2. A worked example of a central orifice aerostatic thrust bearing design procedure is given below.

9.4.7. *Worked Example 2*

A central orifice aerostatic thrust bearing is required to support a unidirectional maximum vertical load of 2000 N. The available air supply absolute pressure is 5×10^5 Pa. The power required must be the minimum possible for the given load and the bearing film stiffness must also be as high as possible, compatible with this minimum power requirement. Design the bearing.

Let $C_d = 0.65$ and for air, $\Re = 287$ J/Kg°K, $\Theta_M = 293°$ K, $\eta = 1.82 \times 10^{-5}(\text{N/m}^2)s$. We need to find or select the variables a, b, h and d.

We have: $W_s = 2000\,\text{N}$, $p_M = 5 \times 10^5\,\text{N/m}^2$, $\bar{p}_s = 5$.

The procedure is as follows:

(1) For minimum power to lift ratio choose $\bar{b} = b/a = 1.89$

(2) Using Eq. (9.34), we can now calculate the outer radius as $b = 0.064$ m. Hence the inner radius is $a = 0.034$ m.

(3) The maximum stiffness can now be found from Eq. (9.33) as $K_m = 1.12 \times 10^8$ N/m.

(4) To find the orifice diameter (d), we need to use Eq. (9.27) for G_l. First calculate F_p and F_g from their definitions. As $G_l = 1.25$, we can find the feeding factor, $\Lambda_f = \frac{1.25}{F_p F_g} = 10.54$ corresponding to $K_g = 0.69$.

(5) We are now in a position to choose a suitable film thickness in the expression for Λ_f, leaving d as the only unknown. Choosing $h = 2.54 \times 10^{-5}$ m, $d = 0.82 \times 10^{-4}$ m (or the nearest drill size to that value).

(6) Equation (9.31) gives us the mass flow. If it is too large, we can reduce h slightly, though d will increase as a result.

(7) Finally, after determining $\bar{A}$ and $\bar{B}$ and the power to lift ratio (= 0.028), the power itself (P) can be found from Eq. (9.9). If any of these are outside the specification, or if b needs to be made smaller from

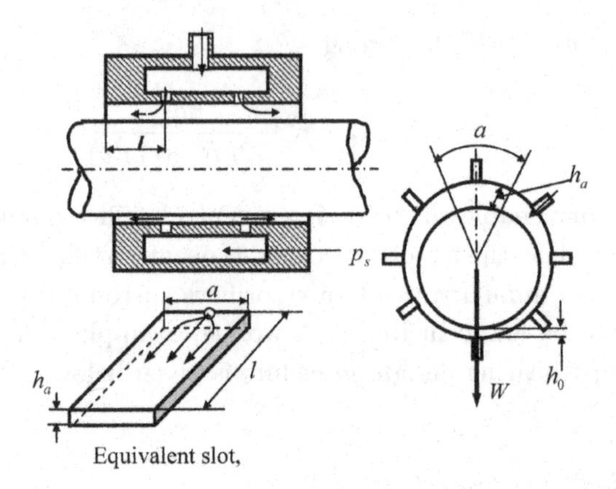

Fig. 9.11. Aerostatic journal bearing.

space considerations, then we can try adjusting $\bar{b}$, the low power to lift ratio allowing us to do this, though K_s and d will also change. The advantage of doing all the calculations with a package like MathCAD, is rapid, because any alterations to the design are seen immediately.

9.4.8. *Aerostatic journal bearings*

Unlike hydrostatic journal bearings, **aerostatic journal bearings** can support only moderate radial loads, but are capable of excessive speeds (a 0.025 m diameter journal can rotate at 100,000 rev/min). The same approximate design methods, as were used for the central orifice thrust bearing, apply to the journal bearing with additional lifting force coming from the hydrodynamic action caused by the rotational speed (a hybrid bearing). Figure 9.11 illustrates the way gas is fed into the journal clearance space.

The journal bearing in Fig. 9.11 has two rows of supply orifices and is without pockets. If the journal is not rotating and is sufficiently long, compared with the circumferential spacing of the orifices, most of the air will leak axially from the film. In order to obtain a simple mathematical model, the bearing is divided into equivalent discrete angular parallel slots across each orifice, of circumferential width, a, of length, l and of center height h_a, varying round the bearing because of its eccentricity. The theory we used for the thrust bearing is then applied to each slot to obtain a local pressure distribution (assumed circumferentially uniform there). The local radial load is found by integrating the pressure distribution over each slot. The vertical components of these reaction loads are then summed up to equal the applied load, W. Correction factors are used to account for the jet axial flow dispersion from the orifices, and the assumption of no circumferential flow between the slots. Grassam and Powell,[3] and Powell[4] explain in detail the design procedure.

9.4.9. *Aerostatic journal bearing instability*

Aerostatic journal bearings need to be designed carefully because of self-excited vibrations that occur when operating as hybrids at certain rotational speeds. We have already discussed this type of vibration in Chapter 8 (see Sec. 8.9.3). There, they were explained when considering a vertical rigid rotor. If the bearing is again vertical and the journal geometric center is made to whirl at some speed, ω, about the fixed bearing center, while the journal itself is spinning at about its own geometric center at speed Ω in the same sense, then if $\Omega = 2\omega$, it is equivalent to the hydrodynamic film boundaries having zero surface speed, thus causing the oil film to collapse. This condition is called the **critical speed** of the rotor. In the case of an aerostatic bearing supporting the rotor, ω occurs if the rotor center is given a small perturbation. Because an externally pressurized gas film surrounds the journal, it possesses dominantly aerostatic elastic stiffness. When disturbed, the rotor center will rotate about the bearing center at the system natural frequency of $\omega = \omega_n$. Self-excited vibrations will therefore occur when $\Omega = 2\omega_n$ (called the **half speed whirl** condition). In principle, Ω should not exceed this value.

For a loaded *horizontal* rigid rotor, running at steady state eccentricity, ε, Powell[4] shows that approximately

$$\Omega = \left(\frac{2+\varepsilon^2}{1-\varepsilon^2}\right)\omega_n. \tag{9.36}$$

Thus, the critical speed will be lowest when the rotor is vertical and $\varepsilon = 0$. If h_0 is the least film thickness (at the bottom of the journal when not rotating, see Fig. 9.12) and g the acceleration due to gravity, Powell gives the natural frequency as:

$$\omega_n = \sqrt{\frac{2g}{h_0}} \cdot s^{-1} \tag{9.37}$$

Equation (9.37) is designed for $\varepsilon = 0.5$ but applies also to $0 < \varepsilon < 0.5$.

There is further discussion of aerostatic journal bearing design and instability in Refs. 5, 6 and 7.

9.5. Closure

Chapter 9 has introduced you to externally pressurized bearing design. There, we concentrated on the simplest type of bearing, but for the readers who wish to know more about other designs, please refer to the references we have supplied. In the next chapter, we will deal with concentrated contact fluid film lubrication. The oil films found there will be ten times thinner than those discussed so far, relying heavily on the elastic distortion of the bounding surfaces and the increase in viscosity resulting from the high pressures generated.

References

1. O'Donoghue J. P. and Rowe, W. O., Hydrostatic bearing design, *Tribology International* (February 1969).
2. Powell, J. E., Moye, M. H. and Dwight, P. R., Fundamental theory and experiments on hydrostatic air bearings, *I. Mech. E. Lubrication and Wear Convention* (1963).
3. Grassam, N. S. and Powell, J. W., *Gas Lubricated Bearings*, Butterworths, London (1964).
4. Powell, J. E. *The Design of Aerostatic Bearings*, The Machinery Publishing Company (1970).
5. Uneeb, M. and Gohar, R., Three dimensional incompresseble flow of a gas through feed holes into an annular space of finite length, *Lubrication Science* **9** (1997) 142–160.
6. Uneeb, M. and Gohar, R. Sleeve dampers for externally pressurized gas journal bearings, *Lubrication Science* **9** (1997) 409–434.
7. Cazan, A. Gohar, R. and Safa, M. M. A., Externally pressurized gas bearings in a mixed configuration, *J. Multi-Body Dynamics Proc. I Mech. Engrs.* **216** (2002) 181–189.

CHAPTER 10

ELASTOHYDRODYNAMIC LUBRICATION (EHL)

10.1. Introduction

Referring back to the Stribeck Chart (Fig. 1.3), we have yet to discuss the fluid film zone that falls between points C and D. The behavior of bearings in this zone is in a regime called **Elastohydrodynamic Lubrication (EHL)**. Compared with the 25 μm films found in hydrodynamic bearings, films of generally less than 2 μm characterize EHL. Their geometry has been classified by parts (b), (c) and (d) of Fig. 3.6, indicating that the touching line or point contact between the elements is counterformal or slightly conformal. Such thin films can make surface roughness an important part of bearing design in this zone. We will, therefore, assume in the elementary theory below, that the surfaces are smooth enough to ignore their roughness features. Examples, where EHL films are produced, include rolling element bearings, gears, cam-followers and thin shell journal bearings.

10.2. Principles of EHL

Consider firstly a *rigid*, long smooth cylindrical roller under a moderate load per unit length of $P' = 10^5$ N/m, rolling at constant entrainment speed $U = 2$ m/s against a smooth plane or another similar roller, as in Fig. 6.8. Let the reduced radius of the equivalent roller be $R = 0.01$ m. Assume firstly that a hydrodynamic **isoviscous** oil film of viscosity $\eta_0 = 0.1$ Pas is present between their surfaces. What is the least film thickness, h_0, under these conditions? As a rough estimate we can use Eq. (6.29) for line contacts: $h_0 = 4\frac{\eta_0 UR}{P'}$. Making the above substitutions we get $h_0 = 1.6 \times 10^{-8}$ m, such a value being less than the CLA of a super-finished rough surface. (Table 2.2).

The corresponding maximum pressure can be found from Eq. (6.28) by differentiating it with respect to x and equating to zero, or by plotting p against x as shown in Fig. 10.1. A maximum pressure of 25.7 GPa is as impossibly high as the minimum film thickness is low! These characteristics

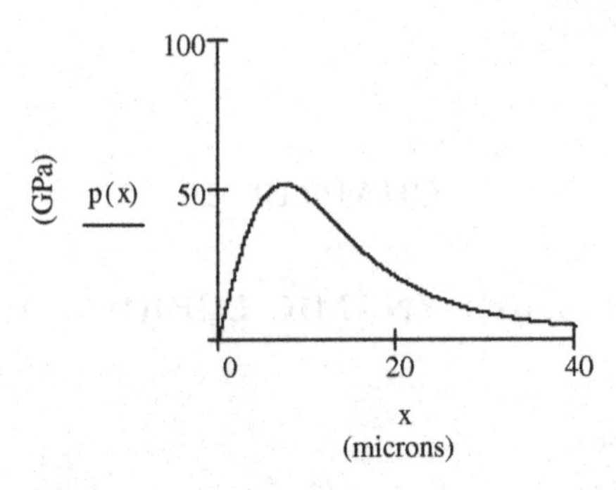

Fig. 10.1. Pressure distribution under a rigid counterformal contact (inlet is on the right).

do not accord with what is found in a practical situation. It is well known that both rolling element bearings and involute gears operate with much thicker films without wear or contact stress failure problems under such conditions, so there must be other factors that have been neglected in our analysis.

The answer is to be found partly in this Chapter's title letters '**EHL**'. We have seen in Chapter 3 that, according to static contact theory, if the equivalent roller were of steel, under the same load intensity as in the above example, the elastic distortion would create a flat rectangular footprint. Using the relevant formulae found in Table Appendix 3.1, $a = 0.564\,\text{mm}$ and $p_0 = 0.59\,\text{GPa}$. Thus, the footprint width far exceeds the region of significant pressures found in Fig. 10.1, and the corresponding maximum static pressure is far less.

Finally, supposing this footprint shape could be maintained under EHL conditions, so creating a parallel film conjunction. In addition, from Eq. (5.7), let the viscosity of the oil increase exponentially under pressures close to 0.59 GPa. With all these ingredients we are now dealing with an **elastic-piezoviscous** problem of EHL. The theory behind this approach, outlined below, is found in pioneering paper attributed to Grubin.[1]

Assuming the pressure inlet is far out (called a **fully flooded inlet** or **drowned inlet**) and using the Swift-Steiber exit condition (Fig. 6.9), together with distorted geometry, and piezoviscous oil behavior, the more accurate numerical solutions shown in Fig. 10.2 were obtained by Dowson and Higginson in another pioneering paper.[2] They demonstrate that the resulting EHL film thickness and pressure distribution shapes, for steady conditions, contain characteristics that appear in both the above extreme examples quoted.

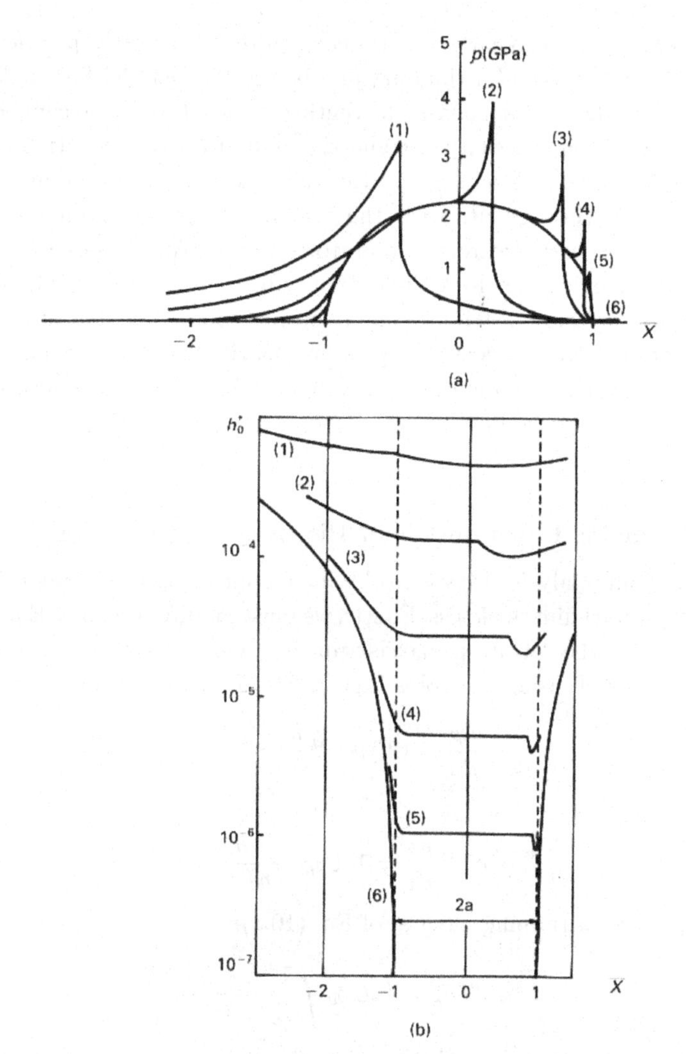

Fig. 10.2. EHL steady state line contact (a) pressure distributions (b) film shapes. Reproduced by permission of the Institution of Mechanical Engineers.

The independent variable for the various shapes seen is the **entrainment speed**, U ($U = (U_1 + U_2)/2$), defined also in Chapter 6, with the load kept constant. At very low speeds (Condition 6) the film appears to have a parallel conjunction along most of its width under a mainly elliptically shaped (Hertzian) pressure distribution, apart from a hardly discernable pressure discontinuity close to the exit, accompanied by a localized film reduction there. At intermediate speeds there is more clearly seen a hydrodynamic inlet wedge prior to a nearly parallel film, followed by extensive film thinning just before the parallel film end, thus creating a wedge (analogous to a Venturi-meter throat). Accompanying this film reduction is a sharp rise and then fall of pressure to atmospheric. At excessive speeds (Condition 1) the characteristics are becoming more hydrodynamic throughout, with the trailing wedge extending to make the film shape tend towards its undistorted form. At the same time, the original trailing edge pressure discontinuity has also extended, eliminating the elliptical pressure distribution and becoming more like that of Fig. 10.1 Clearly, at such high speeds, the pressure distribution has become spread out with a realistic maximum value. Note that the film thickness (log scale employed) distorts the inlet wedge shapes.

10.3. Reynolds Equation Under Piezoviscous Conditions

The aim of our analysis below is to obtain an approximate expression for the assumed parallel film thickness. Firstly, we must modify Reynolds Equation, to include for the oil, its **piezoviscous** behavior. Taking our equivalent roller as a model, using Reynolds Eqs. (6.23) and (5.7) we now have:

$$\frac{dp}{dx} = 12U\eta_0 e^{\alpha p}\frac{h - h_c}{h^3},$$

or

$$e^{-\alpha p}\frac{dp}{dx} = 12U\eta_0\frac{h - h_c}{h^3}. \tag{10.1}$$

First we create a running integral of Eq. (10.1):

$$\int_0^p e^{-\alpha p}dp = 12U\eta_0\int_{-\infty}^{-x}\frac{h - h_c}{h^3}dx.$$

Integrating by parts for the *LHS*:

$$-\frac{1}{\alpha}\left(e^{-\alpha p} - 1\right) = RHS.$$

Define a **'reduced pressure'** as:

$$p_r = \frac{1}{\alpha}\left(1 - e^{-\alpha p}\right). \tag{10.2}$$

Note that in Eq. (10.2), as $p \to 0, p_r \to p^*$ and as $p \to \text{large}, p_r \to 1/\alpha$.

*Expand Eq. (10.2) as a series and neglect higher order terms.

Differentiate Eq. (10.2) with respect to x and substitute into Eq. (10.1) to give:

$$\frac{dp_r}{dx} = 12U\eta_0 \frac{h - h_c}{h^3}. \tag{10.3}$$

Equation (10.3) is the '**psuedo isoviscous**' version of Eq. (10.1): From observation of Fig. 10.2 at low speeds, assume that, because p is quite high at the start of the nearly parallel region that $p_r = 1/\alpha$ there, making $dp_r/dx = 0$ there. Because the reduced pressure gradient has become zero there, the film must be parallel under a constant reduced pressure of $1/\alpha$. Thus, in terms of *actual* pressures, the presence of a flat parallel oil film means that the pressure distribution over the footprint is elliptical, as under static (Hertzian) conditions (Chapter 3). Furthermore, because of these conditions, the equivalent distorted roller profile must have deflected by amount $w(x)$ in the inlet region prior to the parallel film. Its shape may be found from elastic contact theory. Figure 10.3 illustrates the EHL film shape we have so far obtained.

In order to solve Eq. (10.3) we need to find the deformed inlet film shape, defined by h_s, prior to the start of the conjunction (the parallel part). A solution is found by solving Eq. (3.13) applied to the equivalent roller surface of reduced radius, R. If $\bar{x} = x/a$ and p_0 is the maximum Hertzian pressure, then from Ref. 3, up to $\bar{x} = -1$:

$$h_s = \frac{2P'}{\pi E^*}\left[-\bar{x}\sqrt{\bar{x}^2 - 1} - \ln(\bar{x} + \sqrt{\bar{x}^2 - 1})\right].$$

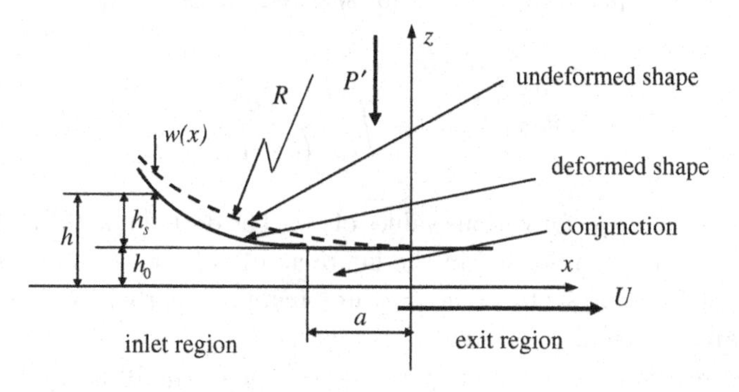

Fig. 10.3. Film shape in an elastic line contact.

Therefore, the film thickness is $h = h_0 + h_s$, up to the parallel film start, or:

$$h = h_0 + \frac{P'}{\pi E^*} 2\left[-\bar{x}\sqrt{\bar{x}^2 - 1} - \ln(\bar{x} + \sqrt{\bar{x}^2 - 1})\right] \tag{10.4}$$

Letting $\bar{\delta} = 2\left[-\bar{x}\sqrt{\bar{x}^2-1} - \ln(\bar{x} + \sqrt{\bar{x}^2-1})\right]$, $H = \frac{\pi E^* h}{P'}$ and $H_0 = \frac{\pi E^* h_0}{P'}$, Eq. (10.3) becomes

$$\frac{dp_r}{dx} = \frac{(12U\eta/\pi E^*)(H - H_0)}{(P'H/\pi E^*)}. \tag{10.5}$$

As $H = H_0 + \bar{\delta}$, after some manipulation, Eq. (10.5) becomes $\frac{dp_r}{dx} = \frac{12U\eta}{(P'/\pi E^*)^2}\left[\frac{\bar{\delta}}{(H_0+\bar{\delta})^3}\right]$. Substituting $\bar{x} = x/a$ and letting $\bar{p}_r = (P'/\pi E^*)^2(p_r/12U\eta a)$, we can write a dimensionless equation:

$$\frac{d\bar{p}_r}{d\bar{x}} = \frac{\bar{\delta}}{\left(H_0 + \bar{\delta}\right)^3}. \tag{10.6}$$

Remember, the variables with an over-bar are internal being functions of $\bar{x}$. H_0 is an external variable containing h_0. All are defined above. We are seeking the values of $\bar{p}_r$ at the beginning of the parallel film region $\bar{x} = -1$. In order to do this, we must chose a series of values of H_0 and for each, compute Eq. (10.6) numerically, say from $\bar{x} = -2$, where $\bar{p}_r \approx 0$, to $\bar{x} = -1$, where $p_r = 1/\alpha$ is assumed to occur, making, $\bar{p}_r = (P'/\pi E^*)^2(1/12\alpha U\eta a)$ there. The only other expression we now need is the semi footprint width, a in the expression for $\bar{p}_r$. Appendix Table 3.1 gives: $a = \left(\frac{4P'R}{\pi E^*}\right)^{1/2}$. Using this final substitution enables Eq. (10.6) to be integrated numerically to obtain values of $\bar{p}_r$, up to $\bar{x} = -1$, for selected values of H_0.
Thus:

$$\bar{p}_r\left(\bar{x} = -1\right) = \int_{-2}^{-1} \frac{\bar{\delta} d\bar{x}}{\left(\bar{\delta} + H_0\right)^3}. \tag{10.7}$$

Figure 10.4 shows $\bar{p}_r$ for various values of H_0. The dotted line (b) is the inlet pressure distribution defined in the figure as $p_0^* = \bar{p}_r$. Just as in Fig. 10.2, the dotted line represents a true pressure, merging approximately with the **Hertzian elliptical pressure**.

The relationship between $\bar{p}_r(\bar{x} = -1)$ and H_0 is given by $\bar{p}_r = 0.0986 H_0^{-11/8}$. If the definitions of H_0, $\bar{p}_r(\bar{x} = -1)$ and a above are inserted

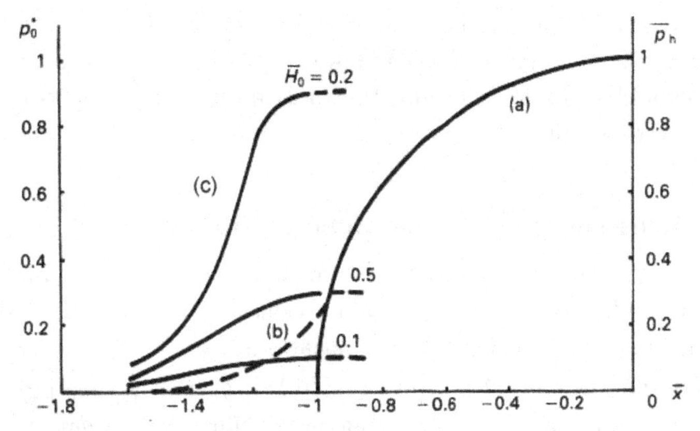

Fig. 10.4. Approximate pressure distribution for a line contact. (a) Elliptical pressure distribution over the parallel film; (b) EHL inlet pressure distribution; (c) reduced pressures p_0^* at inlet.

into Eq. (10.7), we finally obtain:

$$\frac{h_0}{R} = 2.076\left(\frac{\alpha\eta_0 U}{R}\right)^{8/11}\left(\frac{E^* R}{P'}\right)^{1/11} \tag{10.8}$$

Equation (10.8) predicts approximately the parallel film thickness in terms of load, rolling speed, geometry and piezoviscous properties of the lubricant. Equation (10.8) has some important characteristics:

- The film thickness depends strongly on the product of lubricant viscosity and rolling speed, but is very insensitive to load. This is because, unlike in hydrodynamic bearings, the parallel shape of the EHL film, widens as the load increases.
- From Fig. 10.2, the *minimum* film thickness occurs at the exit end of the film. Dowson and Higginson[2] suggest that this thickness is about 75% of h_0.

Equation (10.8) is quite accurate at high loads and/or low speeds. For example, let us find the film thickness under a roller in a heavily loaded roller bearing. Let:

$$R = 0.00635\,\text{m},\quad U = 2\,\text{ms}^{-1},\quad P' = 3.94\times10^5\,\text{Nm}^{-1},$$
$$E^* = 110\times10^9\,\text{Nm}^{-2},\quad \eta_0 = 0.01\,\text{Nm}^{-2}\,\text{s},\quad \alpha = 20\times10^{-9}\,\text{N}^{-1}\,\text{m}^2\,\text{s}.$$

Substituting these values into Eq. (10.8), $h_0 = 2\times10^{-7}\,m$. Additionally, using the formula for a, in Appendix Table 3.1, $a = 1.7\times10^{-4}\,\text{m}$, so that the cross section aspect ratio of the whole film is about 1700. This emphasizes the assumption we made in Chapter 6, that in order to justify using Reynolds equation the film thickness must be small in comparison with its width (x direction).

10.4. Discussion on EHL Line Contact Film Shape

Equation (10.8) does not predict the film thinning seen in Fig. 10.2, which occurs near the exit end of the film. Consider now the flow through the complete film. Ignoring density variations, as: $U = (U_1 + U_2)/2$, the volumetric flow of the oil is, from Eq. (6.18), $q_x = -\frac{h^3}{12\eta_0 e^{\alpha p}}\left[\frac{dp}{dx}\right] + Uh$. In order to maintain a constant q_x along the film width, there is, for any given value of U, an interaction between the sign and value of the pressure gradient, the pressure dependent viscosity and the film thickness. This interaction creates an optimum pressure distribution and a film thickness that is not quite parallel in the region where it was assumed to be, when deriving Eq. (10.8). In particular in some cases after the local maximum (Hertzian) pressure has been reached at $x = 0$, and the pressure gradient has changed sign, in order to maintain flow continuity in this region, there is another sign reversal with a steep rise and fall in pressure. This is accompanied by considerable film thinning followed by an increase to exit. Kostreva[4] has analyzed the conditions determining where this pressure rise and fall occurs.

10.5. Circular Footprint EHL Contacts

Just as in Chapter 3, the two extreme categories of EHL are line contacts yielding a long band footprint, and point contacts yielding a circular footprint. We have derived Eq. (10.8) for an EHL line contact; but because of the point contact geometry, we should account for the side leakage in both the rigid-isoviscous and EHL solutions. Reference 3 gives a solution for rigid spherical bodies that is the equivalent of the rigid line contact example in Sec. 10.2. No accurate analytical EHL solution exists for circular or elliptical footprints, so we can take the opportunity below only to outline the numerical solution method.

10.6. Numerical Predictions

In order to obtain a numerical solution to the general EHL problem, the following governing equations must be solved simultaneously. This can be

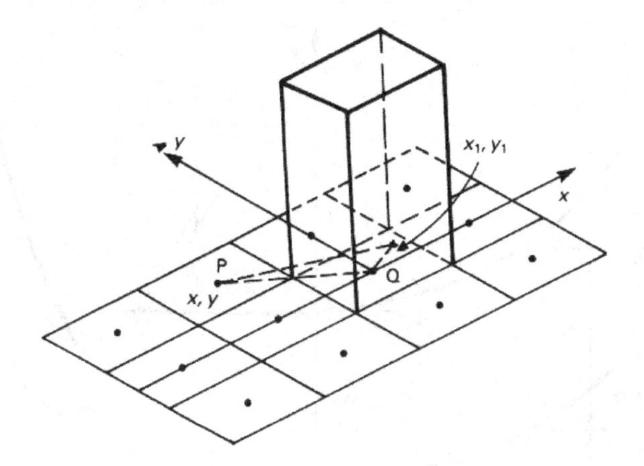

Fig. 10.5. Flat roofed elements.

achieved most simply by forming a rectangular grid, as in Fig. 10.5, that covers the expected region of pressure with flat roofed rectangular elements of heights that will eventually define the wanted pressure distribution.

In order to obtain a numerical solution, we must satisfy the governing equations. These are discussed below.

(1) *Elastic deflection*

Referring to Fig. 10.5, Eq. (3.13), reproduced below:

$$w(x,y) = \frac{1}{\pi E^*} \iint_A \frac{p dx_1 dy_1}{[(x-x_1)^2 + (y-y_1)^2]^{1/2}} \tag{10.13}$$

can be used to find the deflection at a point P (x, y), defining a grid element center, due to unit pressure at point (x_1, y_1) within another element's base, centered at point Q. Solving analytically the double integral over the element base area produces the deflection at x, y due to a single rectangular pressure distribution of unit height.[3]

Equation (3.13) gives the deflection on the flat surface of an elastic half space caused by the pressure distribution. In our case the surface is *curved*, which brings us to the next governing equation we will need.

(2) *Film shape*

We only needed to consider the simplified film shape outside the main pressure region, when deriving Eq. (10.8). In this more general case we must

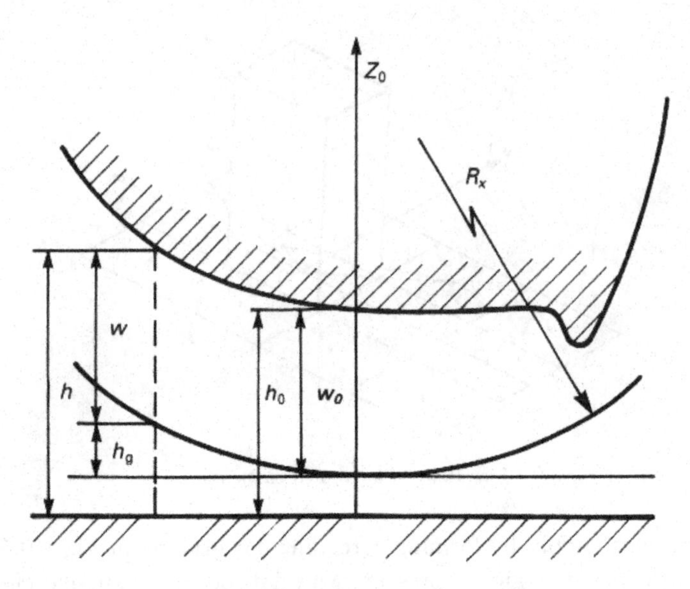

Fig. 10.6. EHL film geometry.

cover the whole region. Figure 10.6 depicts the undistorted and distorted film shapes in the $x - z$ plane, the undistorted surface being approximated by a parabola, as first discussed in Chapter 3.

From Fig. 10.6, the film thickness is given by:

$$h = h_0 + (w - w_0) + h_g, \tag{10.9}$$

where the undistorted gap height in the xz plane is $h_g = x^2/2R_x$ and $w - w_0$ is found from Eq. (3.13). We next consider the lubricant properties.

(3) *Equations of state of the lubricant*

As the numerical solution is comprehensive, with as few assumptions as possible, we need no longer assume that the oil density is constant, so Eq. (5.13) is used. We have seen in Sec. 10.2 that the oil viscosity variation with pressure is very important for EHL. Therefore, in a numerical solution the more accurate Eq. (5.8) is appropriate.

(4) *Reynolds equation*

For a numerical solution, **Reynolds Equation**, reproduced below is used in a form that is appropriate to the specification. Assuming it is steady state, in the x direction only, with U invariable along the film, Eq. (6.20) becomes:

$$\frac{\partial}{\partial x}\left[\frac{\rho h^3}{\eta}\frac{\partial p}{\partial x}\right]+\frac{\partial}{\partial y}\left[\frac{\rho h^3}{\eta}\frac{\partial p}{\partial y}\right]=12U\left\{\frac{\partial}{\partial x}(\rho h)\right\}. \tag{10.10}$$

(5) *Load*

The load is the integration of the pressure distribution. If A the area over which the pressure acts:

$$W=\iint_A p\,dx_1 dy_1. \tag{10.11}$$

This completes our review of the seven governing equations, but in a solution, boundary conditions also must be considered. The simplest inlet boundary condition along the left boundary of Fig. 10.5, is $p = 0$ at where the boundary is far out enough not to affect the results (called a **flooded inlet**). Generally, five times a is sufficient distance upstream from the y axis, and similarly about $1.5a$ from the x axis along each side boundary. The exit boundary condition is usually the Reynolds one (Swift-Steiber) discussed in Sec. 6.5.3. When solving numerically, the exit boundary is allowed to float by using the conditional statement that $p = dp/dx = 0$ somewhere before the grid exit boundary. The final position and shape are determined after convergence of the numerical procedure.

10.7. Computer Solution

There is insufficient space in this elementary textbook to go into the details of numerical procedures, of which there are many.[3] The deflection, Reynolds and load equations are put into finite difference form and the resulting sets of linear equations are solved iteratively in the manner of the flowchart shown in Fig. 10.7.

Before showing some numerical solutions, it is instructive if these numerical results are compared with experiments. Because of EHL film thickness (usually less than 1μm) the main experimental technique has been to map the film shape using optical interferometry. This process will be outlined in the next section.

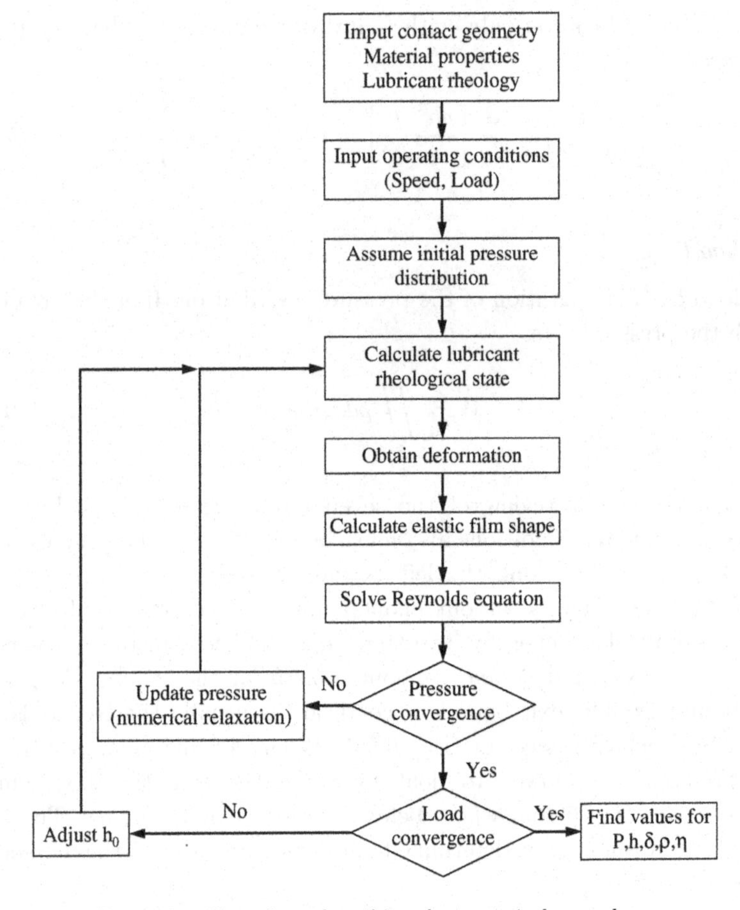

Fig. 10.7. Flow chart describing the numerical procedure.

10.8. Experimental Methods

The most important experimental method associated with EHL is **optical interferometry**, because it enabled researchers to map the oil film.[5]

10.8.1. *Optical interferometry*

The principles of optical interferomery are found in numerous text books, see for example Tolanski.[6] Referring to Fig. 10.8, consider a wafer thin glass plate of thickness t, with its refractive index differing from that of the surrounding medium. If a collimated beam of monochromatic light falls on

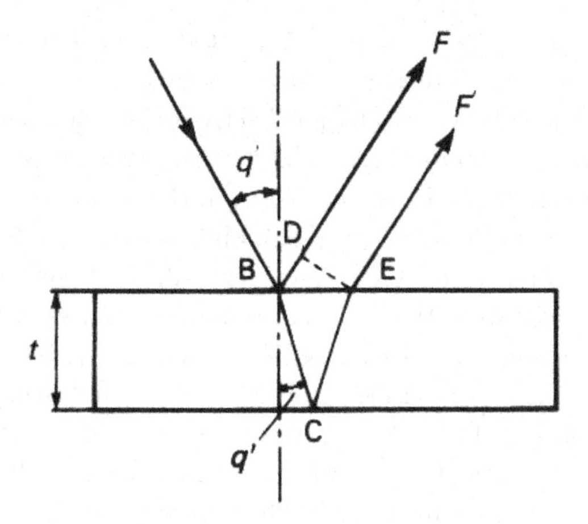

Fig. 10.8. Two beam interference fringes.

it at nearly normal incidence q, it splits into partial reflections at the plate top and bottom surfaces.

Two rays emerge, BF off the top surface and DF off the bottom surface. If λ is the wavelength of light in the film medium (the glass wafer) and assuming that q' is small, it can be shown that the difference between the distances traveled by the two rays (the *path difference*) is[5]:

$$S = 2t + \lambda/2.$$

Whenever $S = 0$, the emerging rays reinforce so that the eye will see a light area called a **bright interference fringe**.

If N is the fringe order, $N = 0, 1, 2, 3, \ldots$ the general condition for a bright fringe is:

$$t = (N + 1/2)\,\lambda/2.$$

For a **dark interference fringe**, the condition is:

$$t = N\lambda/2.$$

Clearly, as to whether the fringes are light or dark depends on the glass wafer thickness. This is called **two beam interferometry** with the fringe intensity variation following a $\cos^2$ law. Now consider the plate at a small wedge angle. It will now be crossed by alternating dark and light fringes that are lines of constant $2t/\lambda$ (contours), with the vertical distance between a pair of adjacent light and dark fringes of $\lambda/4$. The wavelength of the monochromatic light (λ) in the plate is related to the wavelength in air by $\lambda = \lambda_a/n$, where n is the refractive index of glass.

If the model now becomes a high refractive index glass flat plate loading a steel ball, with a drop of oil trapped between them, replacing the glass wafer with a wavelength of light in it of λ, rather faint concentric circular fringes will be seen because the glass and oil have similar refractive indices. In particular, around the contact point there will be a grayish circular area, indicating approximately the Hertzian diameter with zero film thickness there. If the ball is now rotated, an EHL oil film $h(x, y)$ builds up, causing the originally dark central circle to become white, indicating a rise in the film thickness of $\lambda/4$. Thus, with increasing speeds the film thickness may be measured by counting the black to white changes, as well as the general shape of the oil film at any speed. The sharpness of the fringes can be dramatically improved if a 200 Å chromium layer is deposited on the plate's lower surfaces so as to raise its reflectivity close to that of the ball. This method, called **multiple beam interferometry** is commonly used in EHL experiments to map the oil film. Some examples will be given later.

A general purpose interferometry test rig for measuring the oil film is shown in Fig. 10.9. The light used here is **duochromatic**, meaning light

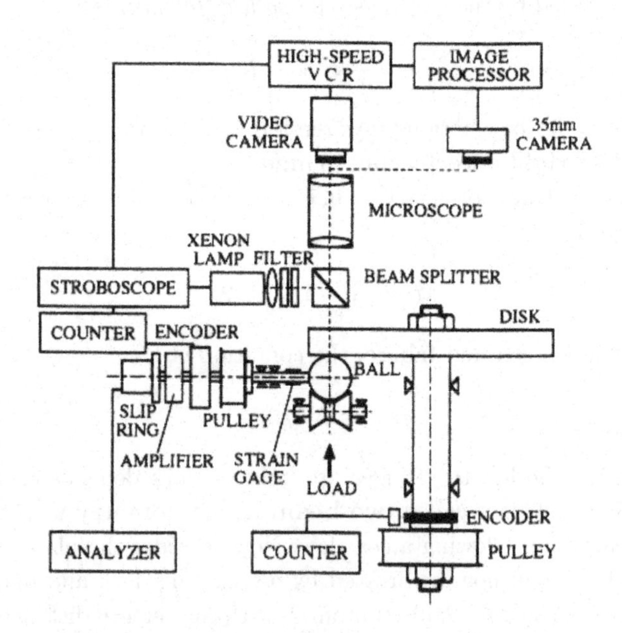

Fig. 10.9. Interferometry film thickness apparatus.

containing two wavelengths instead of one. This adds to the fringe density and hence to the accuracy.

If the filters are removed, leaving only the white light source, colored fringes of the spectrum of several orders are produced, thus improving the accuracy of measurement. This method was developed further for mapping very thin films, see for example Cann *et al.*[7]

Using a sodium light source, Figs. 10.10(a) and (b) show some interference fringe pictures of an **EHL point contact** with (c) and (d) numerical solutions to (b).

The first points to note are the similarities with the line contact solution. At the slower speed, (a), there is clearly a plane area (black), over which the bulk of the pressure distribution occurs. Towards the exit there is a film constriction (now a white crescent) about $\lambda/4$ deeper than the plane area (0.147 μm). In extending sideways the constriction has also reduced the side leakage. Just as in line contacts, the film is seeking its optimum shape. With a higher speed, the picture in (b) shows a relaxation of the contact surface just as it did for a line contact in Fig. 10.2. Referring to the pocket step bearing in Fig. 7.10, a similar film shape is precisely what Kettleborough was seeking in his thrust bearing design. The corresponding contours of pressure in (d), have similarities to the line contact Fig. 10.2. There is a local maximum pressure peak prior to a futher rise in pressure towards the exit, followed by a sharp reduction towards the exit. We also show in Fig. 10.11 an interference fringe picture of the *end* part of a **finite line contact.**[10]

In practice, all line contacts must have ends, with a similar film behavior there to the point contacts shown in Fig. 10.10. In this example the film thinning close to the roller end is severe because the roller used was fairly sharp ended. As anticipated in Sec. 10.5, the thinning continues to the left along the roller length (y direction), becoming less severe and eventually

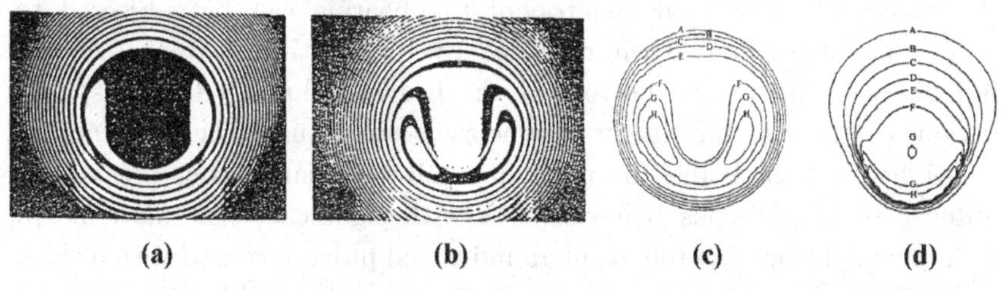

Fig. 10.10. Interference fringe pictures[8] (a) slow speed, (b) higher speed, (c) numerical solution contours for (b),[9] (d) numerical solution isobars for (b).[9]

Fig. 10.11. Interference fringe picture of the end of a finite line contact (exit at the bottom).

constant far from the ends. The rising pressure, accompanying the film thinning (not shown here) is again a maximum at the ends, reducing somewhat in the y direction, eventually following the behavior of Fig. 10.2. Reference 11 shows several examples of the rising EHL end pressures encountered in sharp ended rollers.To reduce these high pressures, the designer should always round off the roller ends, a good practice in many engineering applications (called a **dub-off**). Another interesting feature of Fig. 10.11, is the clearly seen exit wake after the pressure disribution ends (Reynolds condition). The oil has broken up into streamers mixed with air while attempting to fill the enlarging gap it is encountering there.

10.9. Results from Numerical Solutions

10.9.1. *Regimes of concentrated contact lubrication*

Hitherto, we have assumed that the contacting bodies are elastic and the oil has piezoviscous behavior. However, this need not always be the case. Apart from isoviscous-rigid behavior we have already studied in Chapter 6, there are intermediate conditions that can occur. Take for example a gravity loaded angular contact ball bearing. If the preloading is insufficient, the precessing balls near the top of the bearing can be subjected to very light loads. Thus, their **regimes of lubrication** may change fom full piezo-viscous at the bottom of the bearing, through to rigid-piezo-viscous conditions, that finally become rigid-isoviscous at the top. Another possibility is if the bodies are made of a soft material, causing a complete absence of piezoviscous behavior, but with significant deformation. An example might be steel rollers, in an industrial process, coated with a thick layer of soft elastic material. The design chart in Fig. 10.12 covers all these regimes for a long line contact.

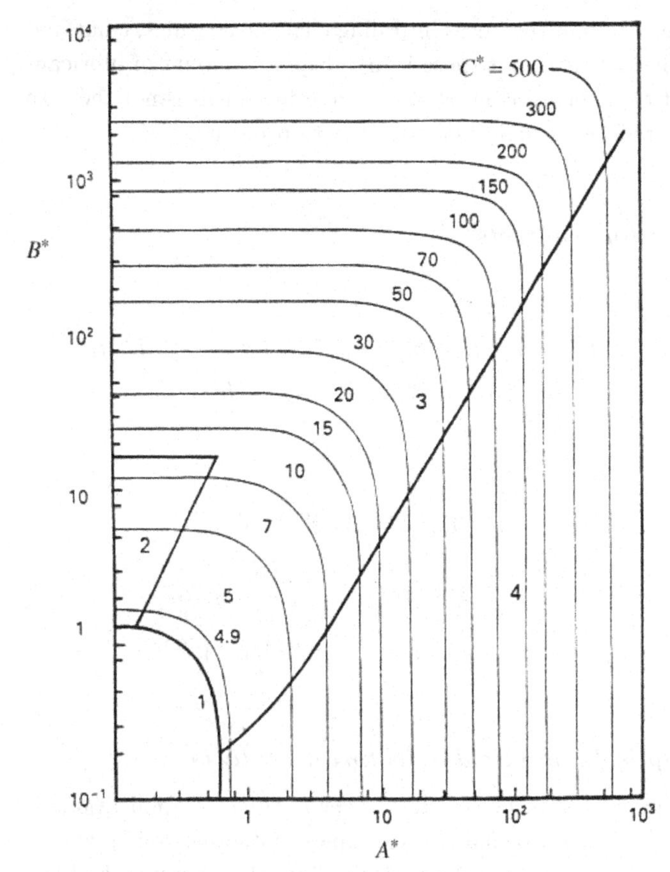

Fig. 10.12. Regimes of lubrication chart for a long line contact. h_m = minimum film thickness (reproduced by permission of ESDU Ltd and I. Mech. E, London) 1:IR, 2:PR, 3:PE, 4:IE.

$$A^* = \left(\frac{P'^2}{\eta_0 UE^*R}\right)^{1/2} \quad B^* = \left(\frac{\alpha^2 P'^3}{\eta_0 UR^2}\right)^{1/2} \quad C^* = P'h_m/\eta_0 UR.$$

10.9.2. *Concentrated contact numerical solution relationships*

The relationships between the variables in piezoviscous EHL can be conveniently put into regression expressions. Starting with the long line contact, expressions for the parallel part of the film (h_0) and the least film thickness (h_m) are found from results like those in Fig. 10.2. All the piezoviscous EHL regression equations, given below, can be used for predicting the film thickness in rolling element bearings, cams and involute gears. They assume that there is an adequate amount of lubricant present, sufficient to flood the contact area. Sometimes, this is not the case, making the film thickness predictions somewhat optimistic.

(a) *Long line contacts*

Let:

$$h_m^* = h_m/R, \quad h_0^* = h_0/R, \quad U^* = \eta_0 U/E^* R,$$
$$W^* = P'/E^* R, \quad G^* = E^* \alpha.$$

Then[12,13]:

$$h_0^* = 3(U^*)^{0.69}(G^*)^{0.56}(W^*)^{-0.1}, \tag{10.12}$$

and

$$h_m^* = 2.58(U^*)^{0.7}(G^*)^{0.54}(W^*)^{-0.13}. \tag{10.13}$$

(b) *Elliptical and circular footprint contacts*

EHL elliptical footprint contacts cover a range of applications where the geometry is counterformal (for example, contacting rollers with their spin axes crossed) or slightly conformal contacts (for example, ball bearing ball to race contacts). Various geometry categories were given in Figure 3.6 and there is the worked example1 in Sec. 3.4.3.

The film thickness expressions below for such contacts[14,15] are general insofar as they can apply to circular footprints as well. However, an elliptical footprint aspect ratio cannot be extended to produce what is nearly a line contact, because this geometry is outside the computation range of the original numerical procedure. Referring to Fig. 10.13, if $\vec{V_e}$ defines the total entrainment velocity vector and V_e its magnitude, having components U and V respectively along the x and y axes, then: $V_e = \sqrt{U^2 + V^2}$.

Also, if R_e and R_s are orthogonal radii of curva ture of the bodies along and normal to V_e, then:

$$\frac{1}{R_e} = \frac{\cos^2\theta}{R_x} + \frac{\sin^2\theta}{R_y} \quad \text{and} \quad \frac{1}{R_s} = \frac{\sin^2\theta}{R_x} + \frac{\cos^2\theta}{R_y}, \tag{10.14}$$

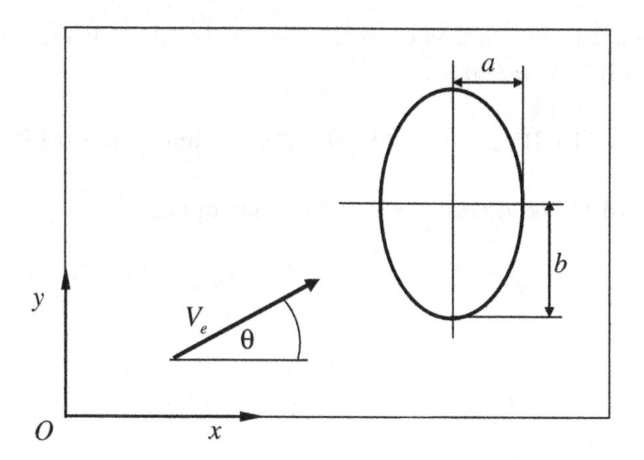

Fig. 10.13. Elliptical contact footprint in the region of integration.

where R_x and R_y are respectively the reduced radii of curvature in the xz and yz planes, as discussed in Example 3.4.3.

If $V_e^* = \eta_0 V_e / E^* R_e, W_e^* = W/E^* R_e^2, G_e^* = E^* \alpha, H_m^* = h_m/R_e$, and $H_0^* = h_0/R_e$.

Minimum film thickness

$$H_m^* = 3.39(V_e^*)^{0.68}(W_e^*)^{-0.073}(G_e^*)^{0.49}\{1 - \exp(-0.67(R_e/R_s)^{0.67})\}. \tag{10.15}$$

Centerline film thickness

$$H_0^* = 3.92(V_e^*)^{0.68}(W_e^*)^{-0.073}(G_e^*)^{0.49}\{1 - \exp(-1.23(R_e/R_s{}^{0.67}))\}. \tag{10.16}$$

Worked Example 10.1

To apply Eqs. (10.15) and (10.16) to an **elliptical footprint contact**, with entrainment only along the x axis, let $\theta = 0$, making $R_e = R_x$ and $R_s = R_y$. Using the data from Example 3.4.3:

$$R_x = 0.0072\,\text{m}, \quad R_x/R_y = 9.712.$$

In addition, let the total entraining velocity be U = 10 m/s, (in the x direction only), $W = 500\,\text{N}$,

$$E^* = 110\,\text{GPa}, \quad \alpha = 20 \times 10^{-9}\,\text{Pas}^{-1} \quad \text{and} \quad \eta_0 = 0.1\,\text{Pas}.$$

Equation (10.15) becomes for x direction entrainment:

$$H_m^* = 3.39\,(U^*)^{0.68}\,(W^*)^{-0.073}\,(G^*)^{0.49}\left\{1 - \exp\left(-0.67\,(R_x/R_y)^{0.67}\right)\right\}, \tag{10.17}$$

where $U^* = \eta_0 U/E^* R_x, W^* = W/E^* R_x^2, G^* = E^*\alpha$.

With these substitutions in Eq. (10.17), $H_m^* = 5.162 \times 10^{-5}$. As $h_m = H_m^* R_x$, $h_m = 3.72 \times 10^{-7}\,\text{m}$ (**Answer**).

And from Eq. (10.16), similarly modified for entrainment in the x direction, $h_0 = 4.49 \times 10^{-7}\,\text{m}$ (**Answer**).

The above example is typical example of the contact area film thickness found under piezoviscous EHL conditions. Finally, Fig. 10.14 shows experimental fringe contours for elliptical EHL contact shape along and angled to the x direction, together with the numerical solutions.[3]

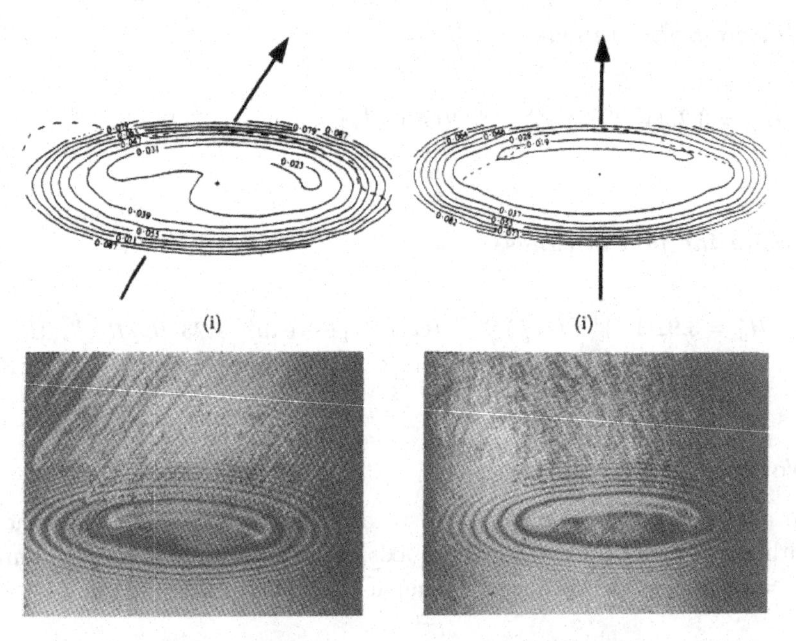

Fig. 10.14. Oil flow in elliptical footprint EHL contacts.

Observe that in the case of the angled flow, the least film thickness occurs on the right side of the contact area. In both fringe pictures, the exiting cavitated flow (top part) clearly follows the entrainment direction. In the section below, the principles of friction in EHL contacts will be discussed.

10.10. Friction Forces in EHL Contacts

As with the other categories of fluid film bearing we have so far covered, friction is an important design factor. In the case of rolling element bearings and gears the total friction should be the minimum possible, but for friction drives an appropriate synthetic lubricant should be selected to maximize the friction.

Let us now discuss friction in relation to EHL. As with other types of bearing, the dominant component of friction is due to sliding.[3] Therefore, assume here that the rolling friction component is negligible. So far, we have employed a *Newtonian* treatment of the lubricant behavior, when dealing with friction (Chapters 7 and 8). However, the results obtained for piezoviscous EHL are not realistic, especially at high loads, mainly on account of the exponential dependence on pressure of the oil viscosity. This gives excessive friction forces that do not generally agree with experimental results.[3] Various researchers have, therefore, considered the *rheological* properties of the oil. The *constitutive equations* formed, model the relationship between force and deformation rate in the oil at the pressures and temperatures existing within the conjunction of the EHL contact.

The aim of this section is basically to familiarize the reader with the principles of the lubricant film rheological properties.

10.10.1. *A viscoelastic model of the oil behavior*

Johnson and Tevaarwerk[15] and Hirst and Moore[16] have proposed a simple constitutive equation for an EHL film in shear, which includes non-linear stress-strain behavior within the conjunction. The total **shear strain rate** for an element passing through the conjunction, under Couette flow only, is (and see Sec. 5.2):

$$\dot{\gamma} = du/dz = \Delta U/h. \tag{10.18}$$

If the lubricant is *linear viscous*, as we have normally assumed, the shear stress is:

$$\tau = \eta \dot{\gamma}_v \tag{10.19}$$

where viscosity, η, describes Newtonian behavior. If the lubricant is considered *linear elastic* under the high pressures it may encounter, by definition of **elastic shear strain**:

$$\gamma_e = \tau / G_o, \tag{10.20}$$

where G_0 is called the **elastic shear modulus** of the lubricant. The **elastic shear strain rate** is, therefore:

$$\dot{\gamma}_e = \dot{\tau} / G_o. \tag{10.21}$$

Consider the lubricant now to be **viscoelastic** when subjected to unidirectional shearing. The total shear strain rate is then:

$$\dot{\gamma}_t = \dot{\gamma}_v + \dot{\gamma}_e.$$

If instead we consider the viscous part of the total strain to be *non-linear viscous*, then:

$$\dot{\gamma}_t = \dot{\tau}/G_0 + F(\tau). \tag{10.22}$$

If τ_0 is a reference or **Eyring stress**, one definition of $F(\tau)$, called the **Rees-Eyring model**, is[15]:

$$F(\tau) = (\tau_0/\eta)\sinh(\tau/\tau_0). \tag{10.23}$$

Thus, dropping the suffix t, Eq. (10.22) becomes:

$$\dot{\gamma} = \dot{\tau}/G_0 + (\tau_0/\eta)\sinh(\tau/\tau_0), \tag{10.24}$$

$\dot{\gamma}$ is the imposed shear rate defined by Eq. (10.18), τ and η are the shear stress and viscosity at a point in the film conjunction being considered. The properties G_o, η and τ_0 all vary with p and film temperature, θ_f. Equation (10.24) includes the nonlinear Newtonian effects observed in disc machine experiments. These are the nonlinear relationship between τ and $\dot{\gamma}$ and the viscoelastic behavior arising from the transient nature of an EHL film under shear. If $\tau = \tau_0, \sinh(\tau/\tau_0) \approx (\tau/\tau_0)$, making the lubricant elastic linear viscous (a **Maxwell liquid**). As written, Eq. (10.24) represents a

nonlinear Maxwell liquid. Provided $\tau < \tau_0, \sinh(\tau/\tau_0)$ differs from by less than 15%, making τ_0 effectively the limit of linearity, separating the Newtonian from the **non-Newtonian** behavior. The *reference stress* τ_0 is found from disc machine experiments by plotting τ against $\dot{\gamma}$ and noting where the linearity ceases. The elastic term, defined by G_o, is also found from separate experiments.

To show more clearly the relative significance of the terms, Eq. (10.24) should be written non-dimensionally. Let x' be measured from the start of the EHL conjunction of length $2a$. Then $D = \eta U/2G_o a$, $\tau^* = \tau/\tau_0$ and $x'^* = x'/2a$. Equation (10.24) then becomes:

$$\dot{\gamma}\eta = \tau_0 \left[D\frac{d\tau^*}{dx'^*} + \sinh(\tau^*) \right], \tag{10.25}$$

where for an element distance x' from the conjunction start at time t, $t = x'/U = (2a/U)x'^*$.

D is called the **Deborah Number**. It is the ratio of the relaxation time of the lubricant ($\eta U/2aG_0$) to the time of passage of an element through the EHL conjunction ($2a/U$). When D is large, viscoelastic behavior results, making the elastic term in Eq. (10.25) become significant. This occurs under high pressures, making η excessive, G_o however, only increasing linearly with pressure.[17] When $D = 1$, because η and/or U are low, $\sinh \tau^* \to \tau^*$, making Eq. (10.25) $\dot{\gamma} = \tau/\eta$, which is Eq. (10.19) for a Newtonian lubricant.

10.10.2. *Limiting shear stress*

We have so far considered the lubricant properties τ_0, G_o and η. What will happen if the pressures become really excessive? In this case the lubricant appears to become fully elastic, resembling an amorphous glassy solid with its property governed by the transient shear modulus, G_o. Just as with the elastic/plastic solids we discussed in Chapter 3 (see Fig. 3.1(b), the lubricant flows at a constant stress, τ_c, like an elastic-plastic solid behaves under direct stress. In this case,

$$\tau = \tau_c, \quad \gamma = \tau_c/G_o.$$

Equation (10.24) can be fitted to disc machine experimental results.[17] The contacting rotating discs, used in these machines, can vary their slide-roll ratio and the frictional force on each surface can be measured with high accuracy. The discs are also fitted with thermocouples to monitor their surface temperatures allowing them to be controlled. Equation (10.24) can be fitted to the experimental results by assuming that all the variables are based on *average* values in the film conjunction. Thus, $\bar{\tau} = F/A$ where F is the measured friction or traction force, $p_m = P'/2a$, where P' is the applied load intensity, h_0 is found from Eq. (10.12) and assuming rolling friction is negligible, $\dot{\gamma} = \Delta U/h_0$.

Results of the disc machine experiments determine the average values of the lubricant properties: $\bar{\tau}_0, \bar{\tau}_c, G_o$ (the latter found independently), $\bar{\theta}_f$ (estimated mean film temperature) and $\bar{\eta}$ (a function of the mean temperature and pressure). The form these results take is illustrated in Fig. 10.15 taken from Ref. 17.

Referring to Fig. 10.1, and Eq. (10.25) we can now construct Table 10.1 to describe idealized EHL traction behavior.

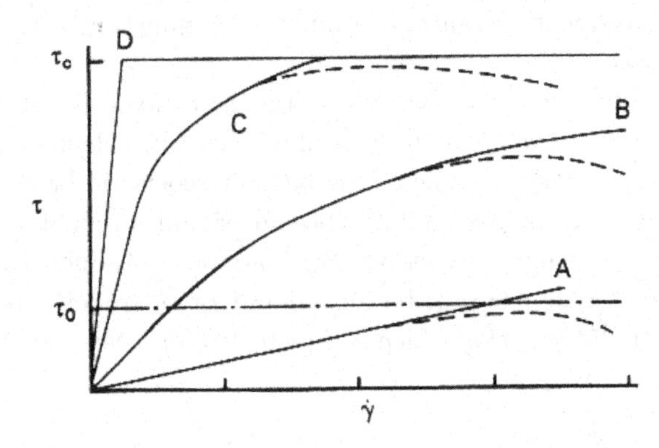

Fig. 10.15. Illustration of traction curves describing the rheological model.

Table 10.1. Traction behavior regimes.

Curve	Conditions	Behavior
A	$D \ll 1,\ \bar{\tau}^* < 1$	Elastic term negligible Straight line until, approaching τ_0, it diminishes because of thermal effects (dotted).
B	$D \ll 1,\ \bar{\tau}^* > 1$	Elastic term negligible. Straight line until graph becomes influenced by hyperbolic sine term (Eyring) after τ_0 is exceeded, eventually diminishing from thermal effects (dotted).
C	$D \gg 1$	Entirely elastic, until approaching τ_c, thermal effects eventually make it diminish (dotted).
D		Viscous region eliminated completely, becomes elastic-plastic solid when τ_c is reached.

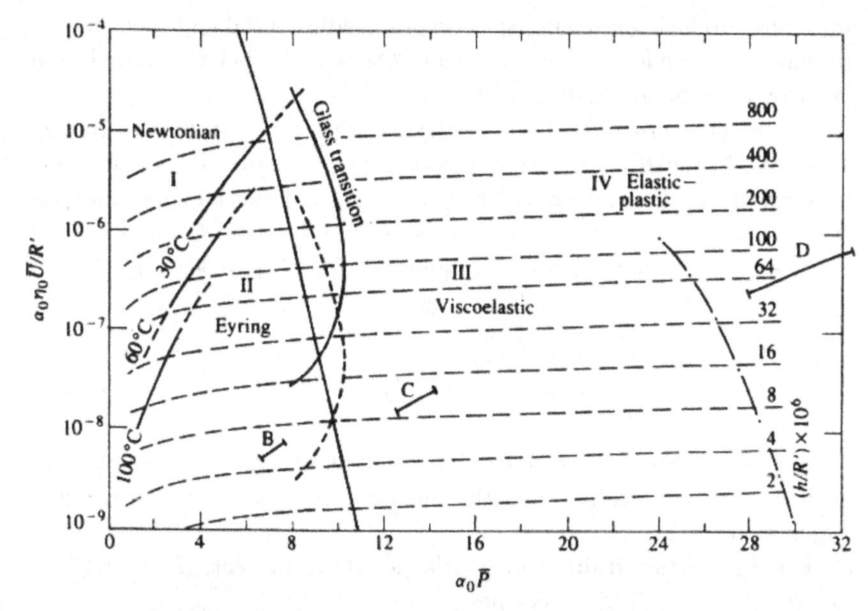

Fig. 10.16. Traction map for a mineral oil (Ref. 18); reproduced by permission of the Institution of Mechanical Engineers.

10.10.3. *Regimes of traction*

Figure 10.16 shows a map of the various **traction** regimes described in Table 10.1. Knowing a given set of conditions for the production of an EHL film, of conjunction thickness h_0, we can calculate from Eq. (10.12), the corresponding position on the map.

The estimates of the friction coefficient based on Eq. (10.24) are modified to suit the appropriate regime. The various oil properties under shear are found from disc machine experiments using averaged variables as discussed above. Curve B in Fig. 10.15, where the Eyring term dominates. Region II in Fig. 10.16 is the most common in many practical situations.

10.10.4. *Determination of the friction coefficient*

So far we have assumed that the frictional properties of an Eyring fluid, under EHL conditions, are isothermal. More realistically, the heat generated due to sliding between the surfaces will influence the nonlinear friction force by reducing further the effective viscosity. These thermal effects are shown dotted in Fig. 10.15. The aim of this section is to give a simplified numerical procedure producing a relationship between sliding speed and coefficient of friction that includes thermal behavior. The analysis below is based on a paper by Johnson and Greenwood.[20]

As temperature is involved in the analysis, so must be the Energy Eq. (6.38). In Sec. 6.7.1 we also discussed heat dissipation along and across lubricant films, where in particular we demonstrated that for very thin films (EHL) there was mainly heat loss *across* the film by conduction. As most of this loss will occur in the parallel conjunction, from Eq. (6.38) in its general form (see footnote there) it follows that:

$$q = \tau \frac{\partial U}{\partial z} = -k_t \frac{\partial^2 \theta}{\partial z^2}, \tag{10.26}$$

where q is the rate of heat generation per unit volume, k_t is the thermal conductivity of the oil and θ is the temperature at a point z in the film conjunction.

For an **Eyring fluid**, and no elastic effects present, if $\tau^* = \eta\dot{\gamma} = \eta\partial u/\partial z$. Equation (10.24) becomes:

$$\tau^* = \tau_0 \sinh(\tau/\tau_0) \tag{10.27}$$

As $\tau^* = \eta\partial u/\partial z$ Eq. (10.26) can be written as:

$$\frac{\tau\tau^*}{\eta} = -k_t \frac{\partial^2 \theta}{\partial z^2}. \tag{10.28}$$

Integration of Eq. (10.28) with respect to z gives†:

$$\left(\frac{k_t d\theta}{dz}\right)^2 = \tau\tau^* \int_\theta^{\theta_c} \frac{2k_t d\theta}{\eta}, \tag{10.29}$$

where τ and τ^* are not functions of z and $d\theta/dz = 0$ at mid plane in the film.

From Eq. (10.29) two expressions can be obtained:

(a) Another integration with respect to z, gives an expression for the conjunction film thickness where it is assumed that there are equal

†To perform this integration it helps to use the substitution $\bar{y} = d\theta/dz$.

heat flows to the film surfaces, each at temperature θ_s

$$h\,(\tau\tau*)^{1/2} = 2\int_{\theta_s}^{\theta_c} \frac{2k_t d\theta}{\left\{\int_{\theta}^{\theta_c} 2k_t/\eta\, d\theta\right\}^{1/2}}. \tag{10.30}$$

(b) The total heat, produced, per unit area is $Q_h = \tau\Delta U = k_t(\partial\theta/\partial z)$, which results in another expression, obtained by using Eq. (10.29) again, to give:

$$\tau\Delta U = \left(\tau\tau^* \int_{\theta}^{\theta_c} \frac{2k_t d\theta}{\eta}\right)^{1/2}. \tag{10.31}$$

Further simplifications of Eqs. (10.30) and (10.31) are:

(a) Assume for simplicity that the temperature, θ_0, at the film pressure commencement, is the same as that of the boundary solids, which are perfectly conducting.
(b) All the internal variables are represented by their *average values*, as discussed briefly in Sec. 10.10.2. For example, $\bar{\eta}_s = \eta_0 \exp(\alpha\bar{p})^{\ddagger}$ $\bar{\eta} = \bar{\eta}_s \exp\left\{-\beta_0\Delta\bar{\theta}\right\}$ where $\Delta\bar{\theta} = \bar{\theta}_f - \theta_0$ is the average temperature rise of the film above θ_0

Equations (10.30) and (10.31) are then integrated to produce design expressions linking traction, temperature and sliding speed:

Let

$$X = \{\exp(\beta_0\Delta\bar{\theta}) - 1\}^{1/2}, \quad X^* = \frac{\sinh^{-1} X}{(1+X^2)^{1/2}},$$

$$\Delta U^* = \frac{\Delta U\bar{\eta}_s}{\bar{\tau}_0 h_0} \quad \text{and} \quad A = \frac{1}{\bar{\tau}_0 h_0}\left(\frac{2k_t\bar{\eta}_s}{\beta_0}\right)^{1/2},$$

Then

$$\frac{\bar{\tau}}{\bar{\tau}_0} = \frac{4A^2X^*X}{\Delta U^*}, \tag{10.32}$$

$$\sinh\left[\frac{4A^2X^*X}{\Delta U^*}\right] = \Delta U^*\left(\frac{X^*}{X}\right). \tag{10.33}$$

[‡]α_0 should be based on the more accurate Roelands expressions found in Sec. 5.9, especially if the average pressure exceeds 1 GPa.

In addition, the mean temperature rise of the film above that at pressure commencement is obtained from the definition of X above. To solve Eqs. (10.32) and (10.33), for a line contact having *constant* rolling velocity but varying sliding velocity, proceed as follows:

(1) From the specification enter: α_0, β_0, $\bar{\tau}_0$, η_0, $k_t\theta_0$, P' and the rolling speed for the film thickness $U = (U_1 + U_2)2$ h Determine h_0 from Eq. (10.8) and $\bar{p} = p_m$ from Table 3.1 Hence determine A. Note that A is constant being independent of ΔU.
(2) Guess a value of ΔU. and hence of ΔU^*.
(3) Start a numerical procedure by *choosing* a value of $\Delta\theta$.
(4) Hence determine X and X^* from this value.
(5) Let $\bar{Y} = \frac{4A^2X^*X}{\Delta U^*}$. Then $\bar{Y}\sinh(\bar{Y}) = 4A^2X^{*2}$ must be satisfied.
(6) Determine $\bar{Y}$ from this relationship numerically. A program like Mathcad 2001i, finds it quickly if a realistic first guess of ΔU^* from step 2 is made.
(7) Knowing $\bar{Y}$, from Eq. (10.7), as $\bar{Y} = \bar{\tau}/\bar{\tau}_0$, $\bar{\tau}$ can be determined.
(8) The coefficient of friction, μ, is found from $\mu = \bar{\tau}/p_m$.
(9) To find the correct value of ΔU^*, we get this from the expression in step 5 above with the *solution* value of $\bar{Y}$ used.

The procedure is repeated until sufficient points for a plot are obtained that includes the specification value of ΔU. The above analysis also applies approximately to EHL point contacts with the relevant expressions for $\bar{p}$ from Chapter 3, and h_0 from Eq. (10.16) with appropriate simplifications. (An alternative approach, especially when both rolling and sliding speeds are specified, is to use Eq. (10.33) directly to find the temperature rise. This situation arises for example when there is pure sliding, in which case $\Delta U/U = 2$ always.)

Figure 10.17 shows results for a constant rolling velocity disc machine, based on the above procedure.

The Eyring *isothermal* solution, shown in Fig. 10.16, can be obtained from Eq. (10.24) by omitting the elastic term. We have $\dot{\gamma} = (\tau_0/\eta)\sinh(\tau/\tau_0)$. Moreover, as we are assuming rolling friction is insignificant, $\dot{\gamma} = \Delta U/h_0$. Using this substitution, $\sinh(\tau/\tau_0) = \eta\Delta U/\tau_0 h_0 = \Delta U^*$. Now $\mu = \bar{\tau}/p_m$, so that for an **Eyring fluid** under isothermal conditions

$$\mu = \tau_0/p_m \sinh^{-1}(\Delta U^*) \tag{10.34}$$

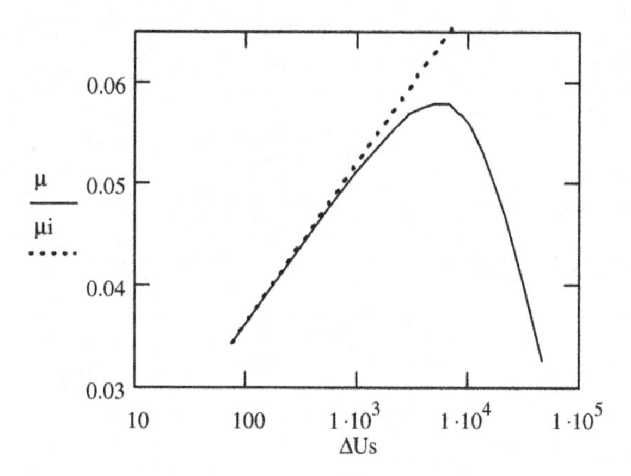

Fig. 10.17. EHL sliding friction coefficient variation. The dotted line, μI, is the isothermal solution.

Values of μ between 0.03 and 0.06 are typical of EHL contacts. Note that thermal conditions at higher speeds and/or poorly conducting bounding solids, can cause μ to reduce.

10.11. Rough Surface EHL

We mentioned in the introduction to this chapter that surfaces in practice are rough. Therefore, this fact should be considered in an EHL analysis. There is no reason why low slope features should not react on a micro scale to the high EHL pressures they encounter. Certainly in a dry contact they do, as we saw in Chapters 3 (Sec. 3.6) and 4, where some asperity summits were flattened and even worn off under sliding conditions (Sec. 4.5.2).

The numerical simulation shown in Fig. 10.18, from Venner and ten Nepal,[21] is of a roller in sliding EHL line contact with a roller bearing race that has simulated roughness features. It illustrates how roughness modifies the smooth surface assumptions we have so far made. In this case the features selected all have wavelengths exceeding 1 μm, of low slope such as a lapped surface. The EHL pressures have deformed them elastically, from their original undeformed state, as depicted by Fig. 10.18(a), to their deformed shape in (b). The corresponding pressure distribution is shown in (c).

Table 10.2. The characteristics of the film profiles in Fig. 10.18.

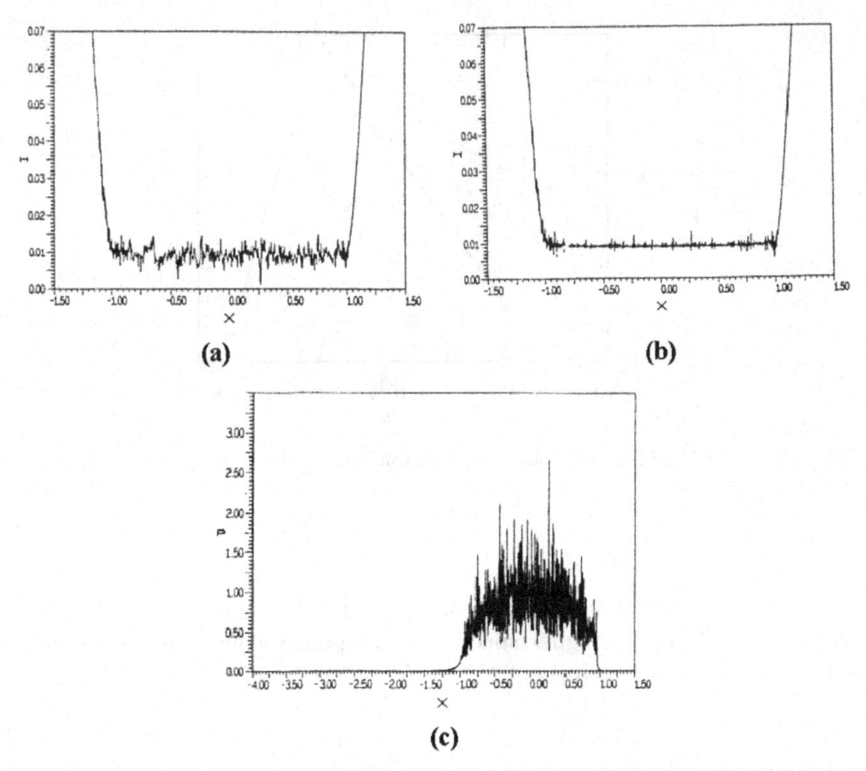

Fig. 10.18. EHL line contact conjunction film thickness and pressure distribution for rough surfaces. (a) Smooth film profile with the undeformed roughness superimposed on it; (b) actual distorted film profile; (c) calculated pressure distribution for the rough surface. Reproduced by permission of ASME.

Table 10.2. Roughness distortion in a line contact, maximum Hertz pressure 2 GPa, sliding speed 1 m/s.

	Average conjunction film thickness μm	R_a μm (CLA)	σ μm (RMS)
H(case (a))	0.167	0.0286	0.0359
h (case (b))	0.165	0.01	0.0151

The elastically deformed features in (b) have caused severe pressure rippling in (c) with the smooth surface pressure distribution running approximately along their mean values. The average film thickness in the conjunction has reduced only slightly. The pressure rippling seen is also associated with **micro EHL** where local films are formed from the wedges between the longer wave surface roughness features.[3] Even though some of the pressure ripples are very high, they do not affect much the **bulk shear stress** distribution within the deeper bounding body hinterlands. They do however influence the *local* stress distribution just under the roughness features. In Fig. 11.7 the local surface roughness orthogonal stress distribution, caused by such pressure ripples, illustrates this. Their effect on bearing fatigue life is discussed in that chapter.

A point to note from Table 10.2, is that we can find the Stribeck number, defined by $\lambda_s = h_0/\sigma$ (Chapter 1). Using case (a), $\lambda_s = 4.65$ and using case (b), $\lambda_s = 10.93$, there being a dramatic improvement in the value. Nevertheless, it is better to introduce a factor of safety when designing and assume that the surface features remain undistorted. Another reason for a conservative estimate is the presence of debris in the oil generated from the surface running in process or from atmospheric pollution. A solution is to avoid using re-cycled oil or to employ a magnetic filter.

10.12. Closure

Chapter 10 has covered the important regime of steady state EHL between smooth contacting surfaces. A point to note is that these piezoviscous EHL films are generally an order of magnitude thinner than those of other fluid film bearings. Such films are formed even though the initial geometry and oil viscosity appear unfavourable for their formation. The fact that these two properties alter to suit the imposed conditions is the wonderful part of EHL. In subsequent chapters, we will deal with unsteady EHL films as well as the isoviscous EHL of thin shell engine bearings. Finally, in Chapter 13, even thinner films will be discussed when we deal with the recent advances in Nano Technology, where the gap between the surfaces is of molecular thickness, being separated by surface and intermolecular forces that mostly replace those caused by hydrodynamic action.

References

1. Grubin, A. N., Contact stresses in toothed gears and worm gears, *Book 30 CSRI for Technology and Mechanical Engineering,* Moscow (1949) DSRI Trans, No. 337.
2. Dowson, D. and Higginson, G. R., *Journal of Mechanical Engineering Science* **4**(2), (1962) 121–126.
3. Gohar, R., *Elastohydrodynamics*, 2nd Ed., Imperial College Press London (2001).

4. Kostreva, M. M., Pressure spikes and stability considerations in EHD models, *Trans ASME J. Lub. Tech.* **106** (1984) 386–395.
5. Cameron, A. and Gohar, R., Theoretical and experimental studies of the oil film in lubricated point contact, *Proc. Roy. A* **291** (1966) 520–536.
6. Tolanski, S., *An introduction to Interferometry*, 4th Ed., Longmans London (1966).
7. Cann, P. H., Hutchinson, J. and Spikes, H. A., The development of a spacer layer imaging method (SLIM) for EHL contacts, *Tibol. Trans.* **39** (1996) 915–921.
8. Gohar, R., Oil film thickness and friction in EHD point contacts, *Trans ASME J. Lub. Tech.* **93** (1971).
9. Jalali-Vahid, D., Rahnejat, H. and Gohar, R., *J. Phys. D: Appl. Phys.* **31** (1998) 2725–2732.
10. Bahadoran, H. and Gohar, R., End closure in EHL line contacts, *J. Mech. Eng. Sci.* **16** (1974) 276–278.
11. Kushwaha, M., Rahnejat, H. and Gohar, R., Aligned and misaligned contacts of rollers in EHL finite line conjunctions, *Proc. I. Mech. E. Part C J. of Mech. Eng. Sci.* **216** (2002) 1051–1070.
12. Dowson, D. and Toyoda, A., A central film thickness formula for EHD line contacts *Leeds-Lyon Symposium* (1978) Leeds (1979) 60–65.
13. Dowson, D. and Higginson, G. R., *Proc I Mech. E.* **182**, 3A (1968) 151–167.
14. Chittenden, R. J., Dowson, D., Dunn J. F. and Taylor, C. M., A theoretical analysis of EHL concentrated contacts Parts I and II, *Proc Roy. Soc. Lond. A* **387** (1985) 245–269 and 271–295.
15. Johnson, K. L. and Tevaarwerk, J. L., Shear behavior of EHD films, *Proc Roy. Soc. Lond A* **356** (1977) 215–236.
16. Hirst, W. and Moore, A. G., The effect of temperature and traction in EHL, *Phil. Trans. Roy. Soc. A* **298** (1980) 215–236.
17. Evans, C. R. and Johnson, K. L., The rheological properties of EHD lubricants, *Proc. I. Mech. E. Part C* **200** (1986) 303–312.
18. Evans, C. R. and Johnson, K. L., Regimes of traction in EHD lubrication, *Proc. I. Mech. E. Part C* **200** (1986) 313–324.
19. Crook, A. W., The lubrication of rollers III A, *Theoretical Discussion of Friction.*
20. Johnson, K. L. and Greenwood, J. A., Thermal analysis of an eyring fluid in elastohydrodynamic traction, *Wear* **61** (1980) 353–374.
21. Venner, C. H. and ten Napel, W. E., Surface roughness effects in EHL line contact, *Trans ASME J. Tribol.* **114** (1992) 612–622.

CHAPTER 11

FATIGUE LIFE OF ROLLING ELEMENT BEARINGS

11.1. Introduction

In Chapter 4 we briefly discussed the various types of fatigue suffered by contacting surfaces in relative motion. This chapter will give an introduction to the *prediction* of their **fatigue life** with particular reference to rolling element bearings. Frequently, the failure of a bearing disables also the machinery it supports. Thus, for safety and commercial reasons it is important to have some idea of the predicted lifespan of the bearing. There are, of course, other reasons for rolling bearing failure, some of these being:

- Wear aggravated by severely starved metal-to-metal contacts (Chapter 4)
- Localized plastic deformation caused by wear debris being drawn into the conjunction
- Overloading caused by bulk plastic deformation of the contacting bodies (Chapters 3 and 4)
- Overheating from an outside source or by flash temperatures from gross relative sliding of the surfaces (Chapter 4)
- Faulty assembly of the bearing, for example causing edge loading of cylindrical bearing rolling elements.

These deficiencies can be considerably reduced by careful design of the bearings, regular servicing and ensuring that coherent EHL films are present between their contacting surfaces Assuming that all these causes of failure have been removed, eventually, the bearings must fail because, like us, they have a finite lifespan.

As we pointed out in Secs. 4.4.3–4.4.6, the cause of eventual fatigue failure of rolling element bearings is from periodic stressing of their elements. In the case of ball bearings, they are generally radially preloaded by a negative tolerance to give an interference fit. Despite the overall steady preload throughout the bearing, individual points on the races and the balls still suffer from periodic loading. Taking a single ball first, a point on one

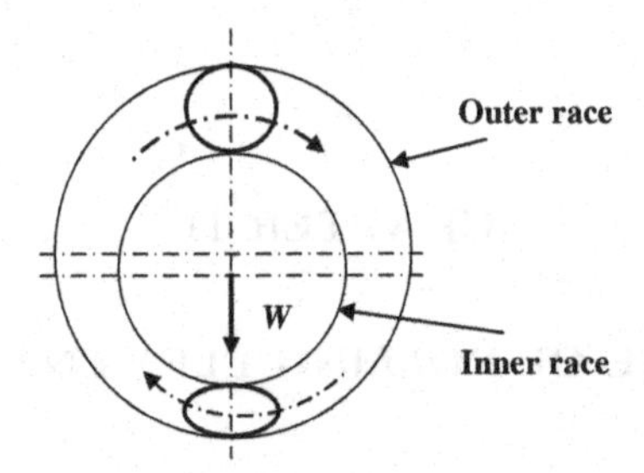

Fig. 11.1. Periodic loading of a rolling element race.

of its surfaces is *only* loaded when that point contacts the races during its rolling motion. The same applies to a point on a race as the balls pass over it. The greater the number of balls, the higher the loading frequency, but the lower the individual load the point sustains. When a steady external load is applied in addition to the preload, the total load distribution is altered. This situation is explained by Fig. 11.1, showing two positions of one of the orbiting balls between the races.

In the top position, the outer race sees a total force less than the preload, while in the bottom position it is greater than the preload. Clearly, the load distribution on the bearing also affects its fatigue life.

There are also other important factors to consider that can affect the fatigue life of a bearing. One is that its material is by no means homogeneous. For example, the manufacturing process may cause residual stresses, or leave defects such as voids and cracks, of the same size order as the roughness of the contacting surfaces and a similar distance below them (a few micrometers). Under the periodic forces endured by the races, and the balls that orbit round them, what is known as **microstressing** causes these defects to enlarge and eventually break out onto the race surfaces, appearing as small fracture craters known as **microspalls**. By themselves, they might not necessarily cause failure, but they can act as initiators of the process. A coherent EHL film reduces microstress damage considerably.

At a greater depth below these near surface layers, there exists a relatively low stress part of the Hertzian zone (see Fig. 3.7) until eventually the main **macrostress** region is reached, that is at a depth of about 0.5 of the contact radius. For the same reasons as happens close to the surfaces, existing cracks can enlarge there under the applied periodic stresses, sometimes migrating to the material surface. The resulting fatigue damage is a pit, called a **macrospall**, typically about $25\,\mu$m deep. The time a crack takes to expand onto a race surface, before appearing as a microspall or macrospall, is a measure of a bearing's fatigue life.

11.2. Failure Stress Hypotheses

From the description of the fatigue process, the stress fields below a Hertzian contact must be known in order to predict the bearing lifespan. From what we have said in the introduction to this chapter, there are two identifiable stress fields below the contact, one being just below the asperities and the other within the subsurface stress field, where the bulk stresses are. We must first discuss the microstress field that results from the bearing load.

11.2.1. *Asperity stress field*

Just as we discussed in Chapters 3 and 4, when dealing with the force between contacting asperities, the maximum shear stress is chosen as the criterion for a representative spherical asperity. In this case it is defined by Eq. (3.26) as:

$$\tau_{\max} = \frac{1}{2}(|\sigma_1 - \sigma_2|). \tag{11.1}$$

It is half the difference between the maximum and minimum principal stresses within the asperity. The contours of $\tau_{\max}$ are plotted in Fig. 3.7 for a two dimensional elastic line contact. For a circular contact, the position of $\tau_{\max}$ is on the asperity z axis about $0.5a$ below the asperity summit contact footprint.

Ioannides[1] points out that the fatigue damage from this near surface stress field has become more significant in modern precision bearings, because their manufacture has considerably improved the quality of the material in the deeper subsurface stress field.

11.2.2. *Subsurface stress field*

The simple theory of fatigue failure is still based on the subsurface stress field, where the bulk stresses occur. The choice of hypothesis in this case is usually the **alternating orthogonal shear stress** field. The stresses sought are the maximum orthogonal shear stresses, τ_{xz} and τ_{zx}, which occur in pairs at 90° to each other in auxiliary planes normal to the xz plane below the x axis.[2,3]

In a ball bearing, the footprint major axis (2b) is normal to the ball rolling direction. The maximum shear stress, $\tau_{yz}(\max)$, is in an auxiliary xy plane 0.49 b below the y axis at $y = \pm 0.9\,b$ (nearly under each end of the major axis of the footprint ellipse). However, the local maximum amplitude of $\tau_0/p_0 = 0.25$, is less than the amplitude of $\tau_{yz}(\max)$, but occurs twice

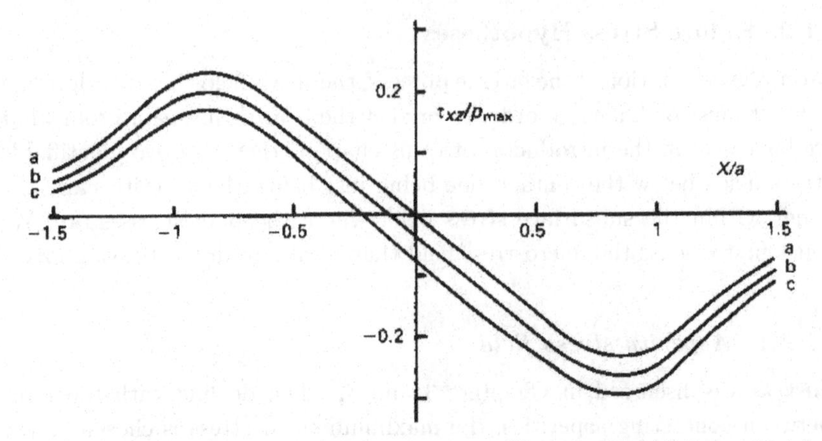

Fig. 11.2. Variation of orthogonal shear stress in an elliptical rolling/sliding contact for different friction levels (Reproduced by permission of the American Society of Mechanical Engineers).[3]

per revolution of the ball. Figure 11.2 shows the variation of the orthogonal shear stress in pure rolling (a) and also with additional sliding friction {(b) and (c)}. The double amplitude of 0.5 p_0($p_{\max}$ in the figure) is believed to be the cause of fatigue life to failure in the **subsurface stress field**. The behavior of the material at the critical depth, where the orthogonal shear stresses exists, is illustrated in Fig. 11.3.

Consider case (a) in Fig. 11.2 (pure rolling). As shown in Fig. 11.3, during one pass of a ball over a point on the outer race, the material is first sheared to the right and then to the left. Thus, there are two **stress reversals** per pass. With additional surface traction, (b) and (c) in Fig. 11.2, the double amplitude stress ratio remains approximately at 0.5.

11.3. Fatigue Life Dispersion

If the *random* weaknesses described above for the subsurface stress field are present in the bearing material hinterland of a given batch of otherwise

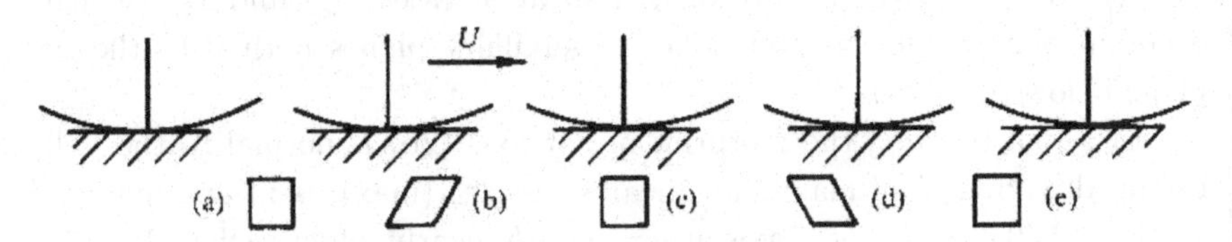

Fig. 11.3. Element subjected to reversing orthogonal shear stress.

identical bearings operating under identical conditions, their fatigue lives will vary. The timing of these fatigue failures will vary according to some dispersion that can be forecast using statistical methods. By assuming dry Hertzian conditions applied, Lundberg and Palmgren (**LP**)[4] were able to define the number of cycles that the inner race in a rolling element bearing can endure before the first macrospall appeared on one of the race surfaces. They then employed what is known as the Weibull distribution[5] to determine the fatigue life dispersion.

Firstly, there are some important parameters that need defining below:

τ_0 = Maximum specified value of the pulsating orthogonal shear stress
Z_0 = Depth below the load carrying track where τ_0 occurs
S = Probability of survival (or reliability) expressed as a fraction or a percentage
a = Half width of the elliptical contact normal to the rolling (y) direction ($a > b$)
u = Number of stress cycles per revolution (here $u = 2$)
L = Fatigue life in millions of revolutions of the inner race
r_r = Inner race radius at groove bottom
l = Rolling element path length, so that $l = 2\pi r_r$
L_s = Fatigue life in millions of revolutions of the inner race corresponding to a probability of survival of $(1\text{–}S)\%$ of a group of bearings
L_{10} = Fatigue life in millions of revolutions of the inner race corresponding to the probability that 90% of a group of bearings survive
N = Number of stress cycles endured
n = Number of bearings in a batch
n_s = Number of bearings in a batch that have successfully endured L_S revolutions in tests.

It follows that: $N = uL$, $S = n_s/n$. Figure 11.4 illustrates a 'rule of thumb' relationship between L and S for a batch of typical bearings, which is approximately linear down to $S = 0.5$

Figure 11.4 shows, for example, that approximately after two million revolutions, 10% of the bearings have failed by fatigue, i.e. after L_{10} millions of revolutions, the probability of *survival* of the remaining ones is $S = 0.9$. Alternatively, the probability of failure is $F = 1 - S$, F replacing S as the abscissa.

The fundamental law of **Weibull theory** is that the probability of **fatigue failure**, F, due to some distribution of stress, σ, over volume V_o

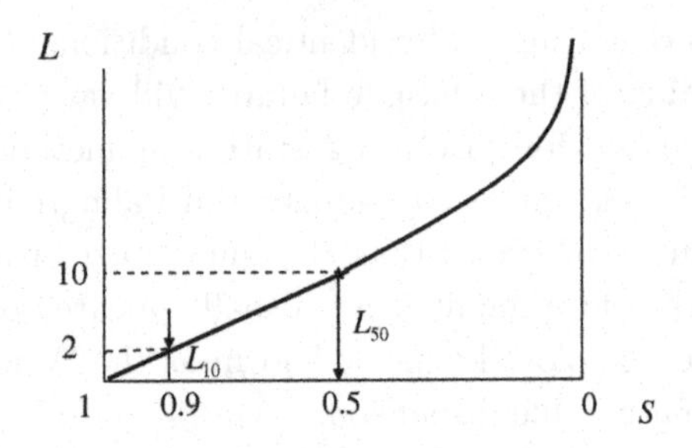

Fig. 11.4. Approximate fatigue life distribution (not to sale).

of a race material, having a characteristic, $n(\sigma)$, is:

$$\ln(1 - F) = -\int_{V_o} n(\sigma)dV_o. \tag{11.2}$$

The above theory assumes that an initial crack inevitably leads to a spall. Starting with Eq. (11.2) **LP** showed that the following approximate expression for bearings with perfect surfaces subjected to periodic stress is given by:

$$\ln \frac{1}{S} \approx \frac{\tau_0^c V_o L^e u^e}{Z_0^h}. \tag{11.3}$$

As illustrated in Fig. 11.5, if $l = 2\pi r_r$ for the inner race, the half volume of the stressed material below the annulus is $V_o \approx Z_0 al$.

Alternatively, if the constant for any bearing type under a specified load and rotational speed is defined by $A = \tau_0^c Z_0^{1-h} alu^e$, Eq. (11.4) states that for a fatigue life L_s, the probability of survival of a group of bearings

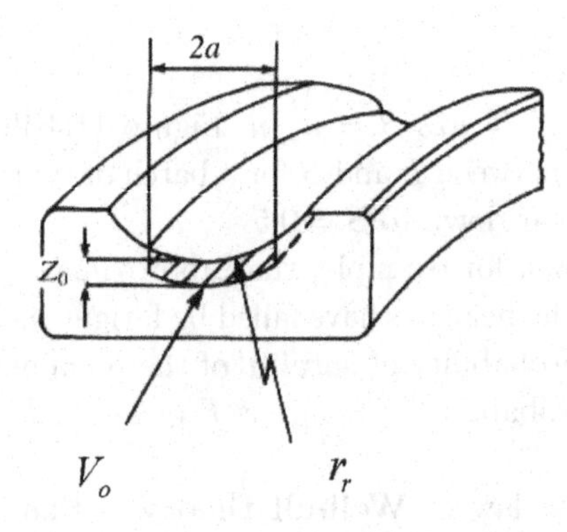

Fig. 11.5. Volume at fatigue risk in an inner race.

can be written as:

$$\ln\frac{1}{S} = AL_S^e. \tag{11.4}$$

Taking logs again:

$$\ln\ln\frac{1}{S} = e\ln L_S + \ln A. \tag{11.5}$$

If plotted on especially designed Weibull paper, the experimentally measured value of e defines the **Weibull Slope** or **shape dispersion parameter**. It shows the fatigue life dispersion for a group of bearings and fits the data quite accurately. If, therefore, a group of bearings is tested under a prescribed value of A, on account of their random weaknesses some will fail earlier than others, allowing their behavior to be found when Weibull paper is used to plot S against L. If we choose any two points L_1, S_1 and L_2, S_2 on the best straight line, e can be found from Eq. (11.5) as:

$$e = \frac{\ln\frac{\ln(1/S_1)}{\ln(1/S_2)}}{(L_1/L_2)}. \tag{11.6}$$

After many tests, **LP** found $e = 10/9$ for ball bearings and $e = 9/8$ for roller bearings. The L_{10} and L_{50} lives, marked in Fig. 11.4, are frequently used as an indicator of bearing performance. At $L = L_{10}, S = 0.9$, allowing us to find A from Eq. (11.5) as:

$$A = \frac{\ln(1/0.9)}{L_{10}}.$$

For the bearings tested by **LP**, Eq. (11.5) can now be written as:

$$\ln\frac{1}{S} = 0.1053\left(\frac{L_s}{L_{10}}\right)^{10/9}. \tag{11.7}$$

Harris[2] makes Eq. (11.7) valid between $0.4 < S < 0.93$.

Hence, knowing L_{10} and the measured value of e, L_S can be found for any chosen reliability.

Worked Example 1

A particular type of radial ball bearing is tested and is found to have a fatigue life of 150 million revolutions with $S = 0.9$. What fatigue life would there be if S were 0.95?

Solution

Here, $L_{10} = 150$ million revolutions. If $S = 0.95$, we must find L_5. Using Eq. (11.7):

$$\ln \frac{1}{0.95} = 0.1053 \left(\frac{L_5}{150 \times 10^6} \right)^{10/9}.$$

Therefore, $L_5 = 82.9$ million revolutions (**Answer**), (Alternatively, using the 'rule of thumb' relationship from Fig. 11.4, we can write: $\frac{L_{10}}{1-0.9} \approx L_5/1 - 0.95$, giving: $L_5 = 75$ million revolutions.)

Worked Example 2

A group of 100 ball bearings of a given type are being tested. Up until now, 20 have failed. Estimate the expected L_{10} life of the remaining bearings.

After 20% have failed, the probability of survival of the rest is $S_a = 0.8$. Again using Eq. (11.7), the corresponding consumed life, L_a, may be obtained from:

$$\ln \frac{1}{0.8} = 0.1053 \left(\frac{L_a}{L_{10}} \right)^{10/9}.$$

Therefore,

$$L_a = 1.485 L_{10}. \tag{a}$$

After an additional L_{10} of the surviving 80 bearings has occurred, the number of bearings remaining is $0.9 \times 80 = 72$ making $S_b = 0.72$.

Hence, the consumed life L_b is obtained from:

$$\ln \frac{1}{S_b} = 0.1053 \left(\frac{L_b}{L_{10}} \right)^{10/9},$$

$$\therefore \quad \ln \frac{1}{0.72} = 0.1053 \left(\frac{L_b}{L_{10}} \right)^{10/9}.$$

Thus,

$$L_b = 2.766 L_{10}. \tag{b}$$

Therefore, the *additional* L_{10} life of the surviving bearings is: $\bar{L}_{10} = L_b - L_a = 2.778 L_{10} - 1.486 L_{10} = 1.498 L_{10}$ (**Answer**).

11.4. Effect of Load on Fatigue Life

As pointed out in the introduction to this chapter the effect of bearing load is an important factor in its fatigue life span. The right hand side of Eq. (11.2) includes the bearing stressed volume, maximum shear stress and fatigue life for a single contact. From elasticity theory in Chapter 3, these factors can be related to the load the bearing can endure for a given L_{10} life. **LP** also extended this theory to cover both the inner and outer races of a complete bearing. The **dynamic load capacity**, W_c(dynamic load rating), is defined as the load a bearing can carry to give a life of a million inner race revolutions with 90% probability of survival. For any other applied load, W, the fatigue life of a complete ball bearing, with 90% probability of survival, can be calculated from the approximate expression:

$$L_{10} = \left(\frac{W_c}{W}\right)^3. \tag{11.8}$$

Furthermore, if there are multiple bearings in a machine, each with a life of $L_1, L_2, \ldots$, the life of the complete assembly is given approximately by:

$$L = [(L_1)^{-3/2} + (L_2)^{-3/2} + \cdots]^{3/2}. \tag{11.9}$$

11.5. Effect of EHL on Fatigue Life

So far we have assumed that the overall conditions at the bearing rolling contacts are dry Hertzian between smooth surfaces and with no lubricant film present. The nature and dry contact behavior of these surfaces was discussed in Chapters 2 and 3. Chapter 10 then dealt extensively with EHL contacts, both for smooth and rough surfaces. Until recently, the effect of roughness on EHL films was based on a surface roughness factor, $\lambda_s = h_m/\sigma$ (Chapter 1) where h_m is the minimum film thickness for smooth surfaces (Chapter 10) and σ is the root mean square of the combined roughness of the surfaces (Chapter 2). If $\lambda_s = 1$, then there is no EHL film, but if $\lambda_s > 3$ then a coherent EHL film exists with the rough surfaces having little influence on the film thickness.[6] Nevertheless, more recent research[6] indicates that the asperities do indeed deform elastically under the EHL pressure, just as they do in a dry contact (Chapter 3). The new research demonstrates that asperity deformation increases the rough surface film thickness and hence the value of λ_s. A coherent EHL film thickness should increase the fatigue life, because the overall pressure distribution is more spread out, especially

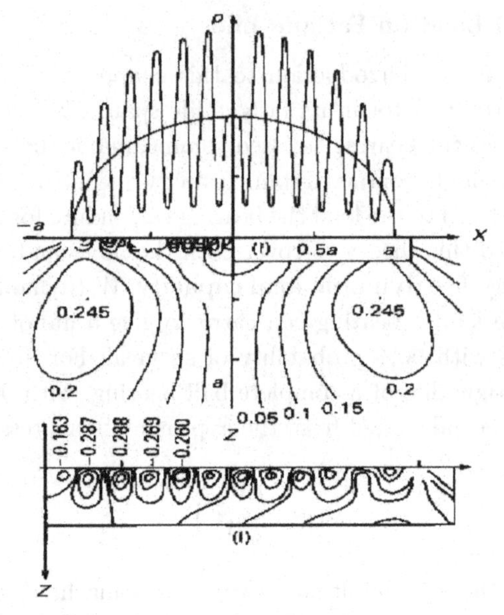

Fig. 11.6. Orthogonal shear stress distribution under a rough surface (Reproduced in part by permission of Wear).

in the inlet region, than by assuming the dry Hertzian contact conditions. The presence of a pressure discontinuity in the outlet region of the EHD film (Fig. 10.2) has not been found to influence much the fatigue life of the bearings.[6] However, the downside to the distorted roughness theory is that there are local pressure increases needed to distort the roughness features. The net effect on fatigue behavior of rolling element bearings is again to produce two distinctly different zones in the interior stress field. One is the orthogonal shear stress, τ_0 field located well below the contact surfaces, while the other zone, just below the surfaces, is associated with the pressure rippling, caused by these two distinct stress regions, which occur either under a dry or micro EHL contacts. Observe that some localized pressure peaks, emanating from the roughness features exceed the maximum bulk contact stress.

11.6. A Unified Life Model

A more general development from the initial theory derived by **LP**, has recently been incorporated into a unified life model for rolling element bearing fatigue failure by Ioannides and Harris[7] and, more recently, a review by Ioannides[1] embraces all the factors discussed above, together with simulations of:

- The nature of the surface topography itself, in particular the influence on the EHL film of the roughness feature average slope.
- The bearing material strength and its resistance to crack propogation
- Factors defining the stress raising properties of inclusions within the material matrix
- The number of defects per unit area found on the surfaces and per unit volume within the stressed matrix
- The shape of the circumferential load distribution round the bearing
- Effect of surface friction, in particular in relation to crack propagation
- Consideration of environmental factors on bearing life, such as atmospheric contamination
- Incorporation into the model simulations of debris damage to the race surfaces
- Effect on bearing fatigue life of dynamic contact and bending stresses, causing flexing of the shafts supported by the bearings as well as their housings
- Influence of the complete surrounding structure on the bearing fatigue life, for instance of an automobile. The computational difficulties here are the vastly different physical scales involved, varying from 0.5 μm at the bearing surface roughness to 5 m defining the complete structure (such as in a truck).[1]

11.7. Closure

We have introduced here the principles of rolling bearing fatigue life prediction. The model described has been based on the original one devised by Lundberg and Palmgren because it forms the basis of all the subsequent developments. The chapter was concluded by an introduction to a unified theory of rolling element bearing fatigue failure that embraces, not only the bearings themselves, but also the influence of the surrounding support structure.

References

1. Ioannides, E., Tribology and rolling element bearings: The partnership continues, *D. J. Groen Prize Lecture*, *Inst. Mech. Engrs.* (2004).

2. Harris, T. A., *Rolling Bearing Analysis*, J. Wiley and Sons NY (1990).
3. Kannel, J. W. and Trevaarwerk, J. l., Sub surface stress evaluations under rolling/sliding contacts. *Trans ASME J. Tribology* **106** (1984) 96–103.
4. Lundberg, G. and Palmgren, A., Dynamic capacity of rolling bearings, *Acta Polytech. Mech. Eng. Ser.* **1**, R.S.A.E.E., No. 3, 7 (1947).
5. Weibull, W., A Statistical representation of fatigue failure in solids, *Acta Polytech. Mech. Eng. Ser.* **1**, R.S.A.E.E., No. 9, 49 (1949).
6. Gohar, R., *Elastohydrodynamics*, Imperial College Press (2001).
7. Ioannides, E. and Harris, T. A., A new fatigue life model for rolling bearings, *ASME Journ. of Tribology* **107**(3) (1985) 367–378.
8. Michau, B., Berth, D. and Godet, N., Influence of pressure modulation in line Hertzian contact on the internal stress field, *Wear* **28** (1974) 187–195.

CHAPTER 12

TRANSIENT ELASTOHYDRODYNAMIC LUBRICATION

12.1. Introduction

The aim of this chapter is to familiarize the reader with some typical industrial problems associated with EHL and how using simplified solutions can produce adequate results. The approach used is to discuss the problem and give the governing equations that need to be solved numerically.

In Chapter 10 we described the mechanism of **elastohydrodynamic lubrication** (EHL) under *steady state* conditions (Eq. (10.10)). This meant that we assumed steady entrainment of the lubricant into the conjunction at speed U, while ignoring the effect of lubricant being squeezed by approach of contacting solids (in the z direction). Equation (10.10) is, therefore, a special case of (6.20), where the **squeeze film velocity**: $2\frac{d}{dt}(\rho h)$ was ignored. In practice, however, conditions experienced in lubricated contacts are often unsteady. For example, bearings when subjected to cyclic or shock loads, there bounding surfaces may approach or separate, giving rise to a squeeze effect. This squeezing action increases the contact pressure distribution during its application, thus enhancing the load carrying capacity. Consequently, for a given contact load a higher film thickness would be expected than that supplied by extrapolated regression equations such as (10.12) and (10.13), which are based on a series of steady state numerical solutions.

One approach used to include the effect of squeeze in such equations has been to assume a range of values for squeeze film velocity: $w_s = \frac{d}{dt}(\rho h)$, prior to the regression of the numerical results at a given oil film location. Mostofi and Gohar[1] have provided an expression for the minimum film thickness for **elliptical point contact**, modified by Rahnejat[2] to include the effect of squeeze as:

$$G^{*2} h_m^{*\$} = 14.04 (U^* G^{*4})^n (W^* G^{*3})^m (1 - 0.683 e^{-0.669 e_p^*})(1 - 0.75 e^{-132 w_s^*}), \tag{12.1}$$

where, n and m are coefficients, dependent on the direction of lubricant entrainment into the contact, shown by θ in Fig. 10.13, as:

$$n = 0.649 - 0.0875\cos^3\theta \quad \text{and} \quad m = 0.0865\cos^3\theta - 0.045.$$

When the flow is along the minor axis of the footprint, then: $\theta = \pi/2$.

Rahnejat[2] has also provided a similar expression for a **finite line contact** (contact of a roller against a flat) for the central oil film thickness as:

$$h^* = 1.67G^{*^{0.421}} U^{*^{0.541}} W^{*^{0.059}} e^{-96.775w_s^*}. \qquad (12.2)$$

The non-dimensional groups, indicated by an asterisk superscript are defined in Chapter 10.

Note that in both these expressions when: $w_s^* = \frac{1}{U}\frac{dh}{dt} = 0$, pure rolling steady state conditions are achieved. The squeeze film operates when: $w_s^* < 0$ (approaching surfaces), and the bodies are separating when: $w_s^* > 0$, where squeeze effect does not operate.

We can now use these expressions to study *unsteady* (transient) behavior of the lubricant film, providing we know the variation in load, W^*, which must be obtained by considering the dynamics of the system. This means that dynamics of the system and EHL of the contact must be solved in an iterative manner, as shown for example by Rahnejat and Gohar[3] for a ball bearing. Note that in a step-by-step time solution method with a time step Δt, a first order approximation for squeeze film effect can be obtained as: $\frac{dh}{dt} = \frac{h_i - h_{i-1}}{\Delta t}$, where subscript i denotes the time step number.

Whilst the predictions made using this approach are more realistic than the steady state conditions, this solution is still not really unsteady, because we have obtained the contact load prior to the estimation of film thickness. A more appropriate method would have been to obtain the contact load as generated by the lubricant pressure distribution. Therefore, the approach highlighted above does not fully integrate dynamics and lubrication and so may be regarded as *quasi-static*. The distinction becomes more apparent when we describe some case studies.

12.2. A Valve Train System

Figure 12.1 shows a single overhead cam (OHC) **valve train** arrangement. The **cam** rotates at an angular velocity, ω as the valve reciprocates with

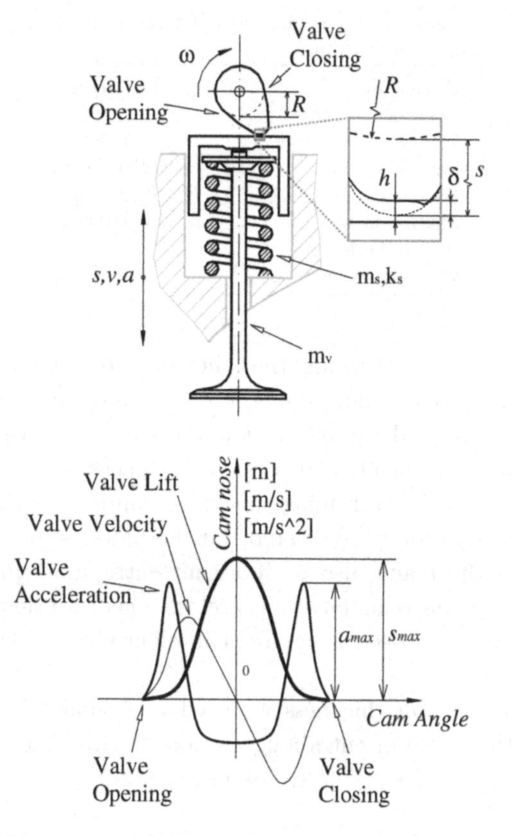

Fig. 12.1. An overhead cam and its kinematic characteristics.

displacement, s (being the cam lift), velocity, v and acceleration, a. The cam profile consists of a base circle of radius R, and a maximum lift of s. The important parameters for this valve train are listed in Table 12.1. This is a typical polynomial cam used in automotive engines.[4] As the cam rotates, the valve opens at the beginning and closes at the end of the event as shown in Fig. 12.1. While the contact is on the base circle, the valve remains closed. The **valve lift** against cam angle is also shown in the lower figure, with the maximum lift clearly taking place when the contact is on the cam nose tip. This position is considered to represent the cam angle of 0°.

The valve velocity, $v = \frac{ds}{dt}$ changes sign at the cam nose tip. Also, when the entraining velocity reverses at its maxima (on the either sides of the cam nose in this case) **inlet reversals** occur in lubricant entrainment. This will become clearer in (b) below.

Table 12.1. Typical data for an overhead cam valve train.

Engine speed:	$n = 1000$ [rpm]
Mass of the valve:	$m_v = 0.1$ [kg]
Mass of the spring:	$m_{sp} = 0.04$ [kg]
Spring constant:	$k = 14000$ [N/m]
Base circle radius:	$R = 0.025$ [m]
Maximum cam lift:	$s_{\max} = 0.009$ [m]
Atmospheric dynamic viscosity of the oil:	$\eta_0 = 0.007$ [Pa s]

An inlet flow reversal means that the speed of the entraining motion calculated as the mean velocity of the contacting surfaces, U (cam and the **tappet**) changes direction. It follows that using the extrapolated Eqs. (10.2) or (12.2), when $U = 0$, a zero film thickness should be predicted. In reality the speed of entraining motion is quite high (here $\approx$2.6 m/s), so that the film is rapidly re-created. Furthermore, some film is retained by the squeeze effect and also by lubricant entrapment in the valleys of rough surfaces of the boundaries. Therefore, there are few signs of wear at such locations, as indeed is apparent in older engines, even after many revolutions.

To predict the film thickness we need to evaluate both the contact load, W and the speed of entraining motion, U. Additionally, we need to determine the value of $w_s = \frac{dh}{dt}$ to use in Eq. (12.2).

(a) *Contact load*

The contact load is obtained by considering the dynamics of the system, in this case, the force balance on the valve at any instant of time, as:

$$W = F_e + F_p - F_i. \tag{12.3}$$

The elastic restoring force is supplied by the **valve spring** due to its net displacement (see Fig. 12.1), which is given by:

$$z = s - \delta + h, \tag{12.4}$$

being taken from the position of the central oil film thickness, h in the **cam-tappet contact**. Note that the lift of the cam is a polynomial function as already mentioned. Thus, $s = f(\theta)$ and since $\theta = \omega t$, then z is a function of time.

Now if we assume the valve spring to have a constant stiffness, k (and this is not usually the case in varied-pitch thick valve spring coils), then:

$$F_e = kz, \tag{12.5}$$

F_p is the valve spring preload, which is introduced in order to reduce any sudden oscillations, caused by loss of contact between cam and tappet or impact of the valve against the valve seat. Such behavior can have serious adverse effects and is referred to as **valve spring surge.**[5] Thus:

$$F_p = kz_p, \tag{12.6}$$

where z_p represents the amount of preload (compression of the spring from its unloaded free length).

The inertial force is given by Newton's second law of motion as: $F_i = m_{eq}a$, where: m_{eq} is the equivalent mass of all parts in translational motion in the z direction. The equivalent mass, therefore, is the mass of the valve plus 1/3 of the valve spring mass, which cannot be ignored (see any standard text on vibrations). Thus:

$$F_i = \left(m_v + \frac{1}{3}m_s\right)a. \tag{12.7}$$

Finally, as $s = f(\theta)$, then: $a = \frac{d^2z}{dt^2} \propto \frac{d^2}{dt^2}f(\theta) = j_\theta$, referred to as the **geometrical acceleration** in cam design parlance. Therefore, we are able to find the contact load, W at any instant of time.

The dynamics of the system represented as an instantaneous force balance can best be appreciated by noting the way $F_e + F_p$ and F_i vary with cam angle in Fig. 12.2. At the instances of **valve opening and closure** the net force is the chosen preload, F_p. It can be appreciated that with insufficient preload and at higher values of ω, the inertial force rise rate, being faster than the generated spring restoring force (see Fig. 12.2) would cause untoward motions of the valve. As the valve opens and the cam contact proceeds up to the cam nose, the inertial force follows the characteristics of valve acceleration, shown in Fig. 12.1. At the cam nose with a fully open valve, the inertial force levels off, and the elastic restoring force in the valve spring attains its maximum value. From here onwards the cam rotation relieves the compression of the valve spring and the inertial force increases. Finally, the inertial force decreases as the valve fully closes and the contact returns to the cam base circle.

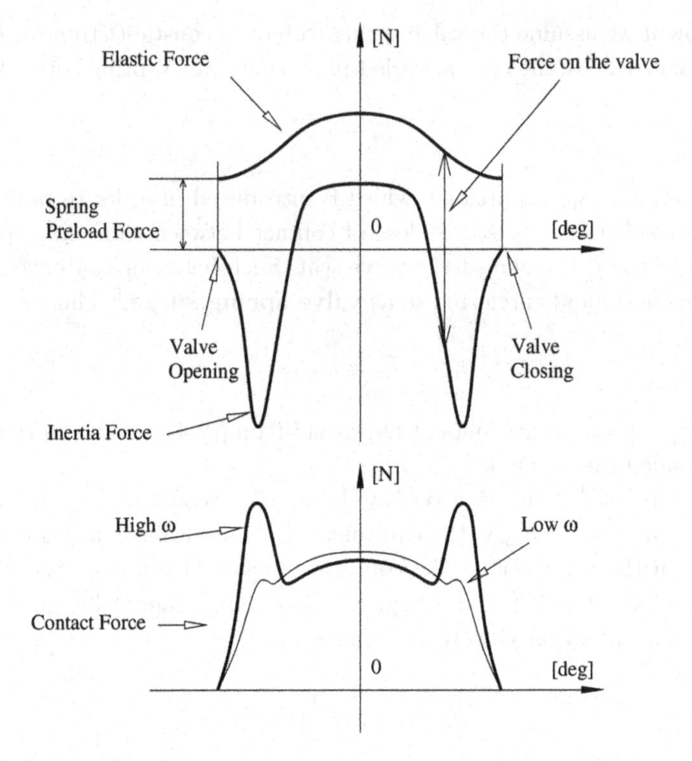

Fig. 12.2. Variation of forces during a valve cycle.

The peak inertial forces experienced by the valve are as the result of the aforementioned geometrical acceleration induced by the valve lift function and the cam rotational speed:

$$j_\theta = \frac{d^2 s}{d\theta^2} = \frac{d^2 s}{d\left(\omega t\right)^2} = \frac{1}{\omega^2}\frac{d^2 s}{dt^2}. \tag{12.8}$$

Therefore, $a = \frac{d^2 s}{dt^2} = \omega^2 j_\theta$, and we can see that as ω increases the inertial force increases, resulting in a larger contact force, as shown in Fig. 12.2.

(b) *Speed of entraining motion*

Figure 12.3 shows an instantaneous point of contact, P, between the cam and the tappet. The velocity of the tappet at this point is:

$$v_T = \omega j_\theta. \tag{12.9}$$

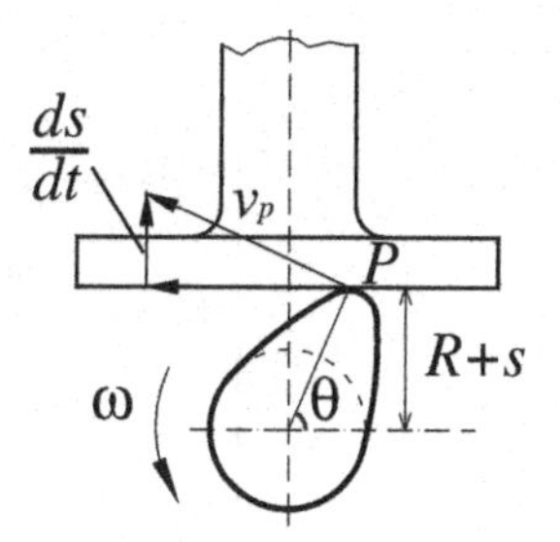

Fig. 12.3. Cam-tappet contact.

That of the cam surface is given as:

$$v_C = \omega r, \tag{12.10}$$

where r is the instantaneous radius of the cam at P, given by:

$$r = R + s + j_\theta. \tag{12.11}$$

The speed of lubricant entrainment is obtained as the average value of these surface velocities, as defined in Chapter 10. Thus:

$$U = \frac{1}{2}\omega(R + s + 2j_\theta). \tag{12.12}$$

From Eqs. (12.8) and (12.12) we can see that with $\omega \neq 0$, $U = 0$ when $\frac{d^2s}{dt^2} = -\frac{1}{2}\omega^2(R+s)$ or $\frac{ds}{dt} = \max$. These occur when the horizontal component of v_p (see Fig. 12.3) becomes zero. These positions are on the either sides of the cam nose tip (Fig. 12.4). It is clear that their locations are a function of the cam-lift design (i.e. s). They usually occur at the two positions already mentioned (one prior to, and the other after the cam nose-to-tappet contact). However, there may be two such points on either side of the nose in succession, as in a **cycloidal cam**, described later.

Since U and W have been determined, and material and lubricant properties are known for a given cam-tappet pair, the lubricant film thickness can be evaluated using the aforementioned extrapolated oil film thickness formulae at any instant of time. Figure 12.5 shows the variation of central film thickness using these formulae as alternatives. In both cases the inlet reversal positions ($U = 0$) forecast zero film thickness, as described above. Using Eq. (12.2), the full line shows a greater film thickness because of the higher load carrying capacity obtained from the inclusion of the **squeeze film effect**.

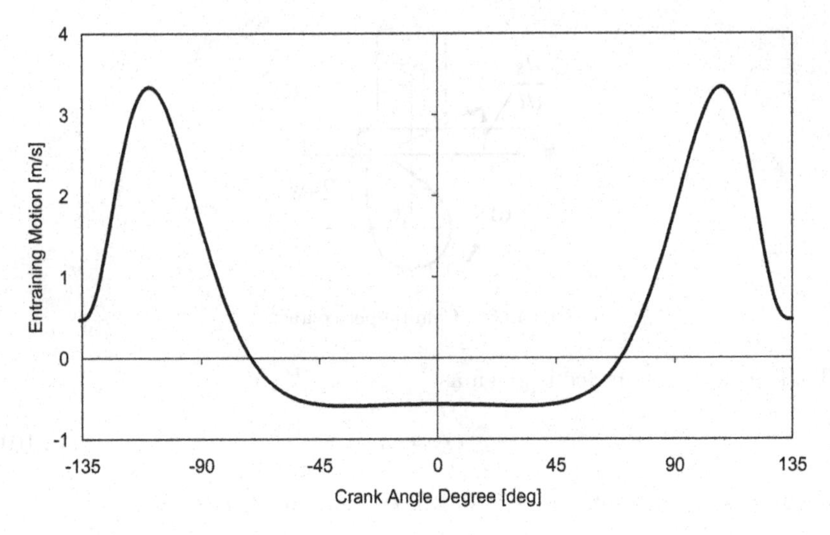

Fig. 12.4. Variation of entraining motion in a cam cycle.

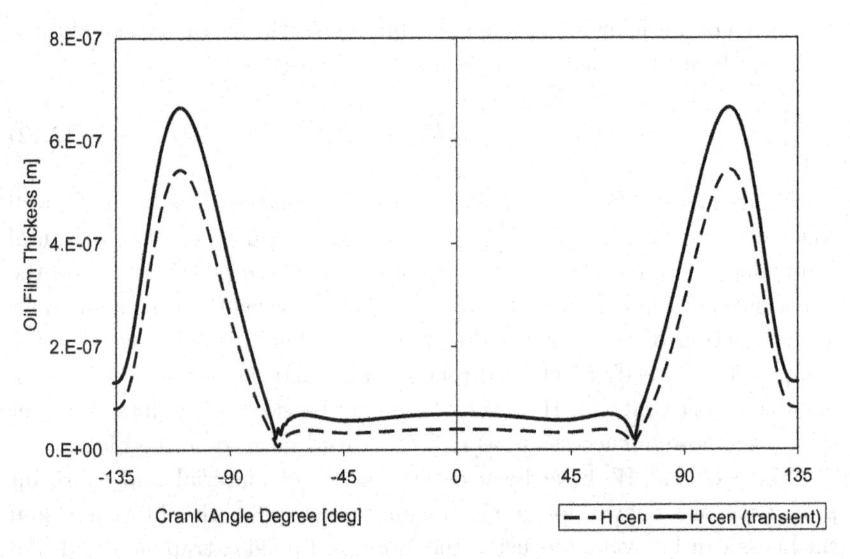

Fig. 12.5. Variation in film thickness in a cam cycle with and without squeeze.

Note that the film thickness, which is in tenths of micrometer on the flanks, nearly vanishes at the inlet reversal points and is less than a tenth of micrometer in the region of the **cam nose**. We can better visualize how it varies during the cycle in Fig. 12.6, which uses polar co-ordinates.

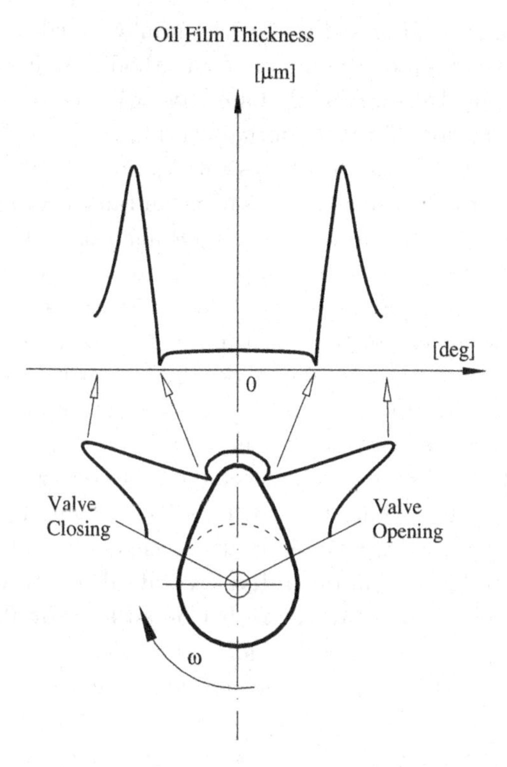

Fig. 12.6. Film thickness variation in polar coordinates.

The above analysis is *quasi-static*, for two reasons. Firstly, the extrapolated equations were initially obtained through the assumption of a range of values for the problem non-dimensional groups and, in the case of Eq. (12.2), for the squeeze velocity without regarding the existing inertial dynamics of the system. Secondly, in the calculation of z the film thickness obtained at a time step is included in the next one without taking into account the load carrying capacity of the film. In other words, the Reynolds equation is *not* solved to find pressure distribution, and then the contact load by integration.

In a transient analysis we follow the same computation procedure as that shown in the flow chart of Fig. 10.7, except that the contact load is *not* an input but an outcome due to the elastic film of the force balance used in Eq. (12.3). The unknown to be determined here is $\ddot{z}$ (acceleration of the valve). A step-by-step integration method is then used to calculate $\dot{z}$ and subsequently z. The convergence criterion, replacing the load balance in Fig. 10.7 is usually based on displacement, velocity or acceleration, depending on the method of solution or the time-based integration algorithm used. Later, we will see a simple example demonstrating this for thin shell journal bearings.

It is, therefore, clear that the transient solution for valve trains, taking into account the tribological conditions in a cam-tappet conjunction is quite complex, compared to the quasi-static simple analysis described above. So, what are the repercussions of using approximate quasi-static solutions? It was shown by Kushwaha and Rahnejat[6] that under transient conditions the continuity in the history of the film thickness is maintained. This means that the film thickness at any instant depends on its pre-existing value at a prior instant (unlike the quasi-static analysis reported above). Also, the film thickness is affected by the dynamics of the system and the resulting kinematic conditions in the contact, including the squeeze film effect. This means that the predicted films under transient conditions are in fact thicker than those using the extrapolated oil film thickness formulae. Figure 12.7 shows these findings for a **modified cycloidal cam**, where 4 **inlet reversals** take place, owing to the rather broad flat cam flanks on either side of the cam nose. They account for less incidence of wear in practice than those predicted by quasi-static solutions. Numerical predictions of

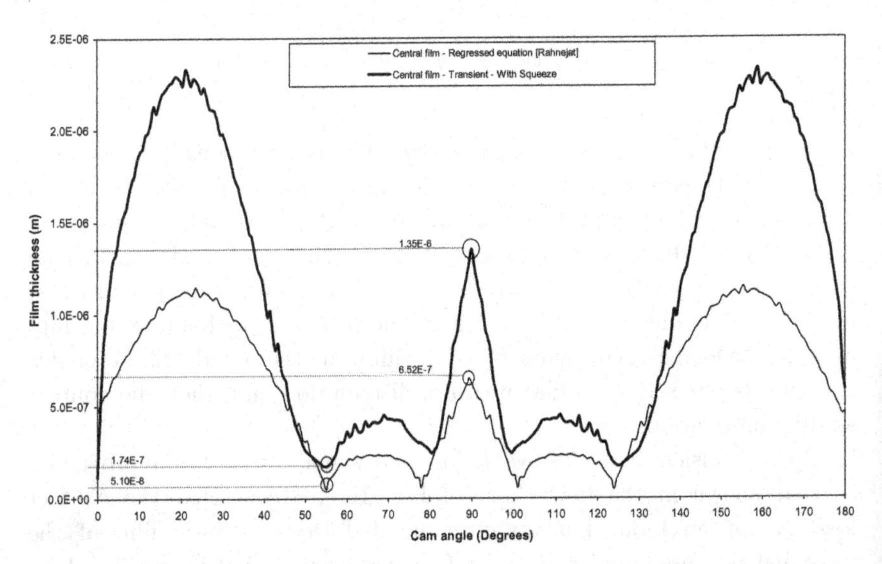

Fig. 12.7. Comparison of transient and quasi-steady lubricant film variation for a modified cycloidal cam (Kushwaha and Rahnejat[6]) with the kind permission of the Institution of Mechanical Engineers.

transient lubricated contacts also show that the pressure distribution is more spread out with lower peak values than when using quasi-static assumptions. This means that simplified quasi-static solutions predict onset of fatigue prematurely. The conclusions from transient studies indicate more favorable conditions than those from simplified quasi-steady analysis. This is fortuitous, since the more accessible quasi-steady methods considerably reduce computation times, thus being more suitable to practical use within typical industrial time-scales.

12.2.1. *An example*

To illustrate the simpler highlighted quasi-steady method, we take the case of a polynomial cam, which is widely used in automotive engines. Obviously the type of cam used determines the lift s, velocity v and acceleration a. For a **polynomial cam**, the lift is typically given as:

$$s = s_{\max} + C_2\bar{\theta}^2 + C_m\bar{\theta}^m + C_p\bar{\theta}^p + C_q\bar{\theta}^q + C_r\bar{\theta}^r; \quad \bar{\theta} = \theta/\hat{\theta}, \tag{12.13}$$

where $s_{\max}$ is the maximum amount of lift at the cam nose tip, m, p, q, r are polynomial orders chosen, and C_2, C_m, C_p, C_q, C_r are constants. $\acute{\theta}$ is the angle from the beginning of the cam event (polynomial law) until the maximum lift is reached at the cam nose tip (see Fig. 12.8).

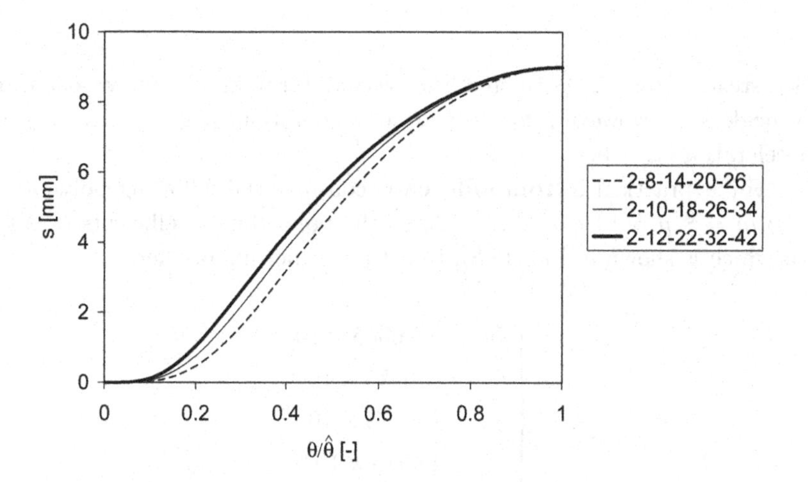

Fig. 12.8. Valve lift for 3 different polynomial laws.

The above constants are given as:

$$\begin{cases} C_2 = -s_{\max} \cdot m \cdot p \cdot q \cdot r/[(m-2)(p-2)(q-2)(r-2)] \\ C_m = -s_{\max} \cdot 2 \cdot p \cdot q \cdot r/[(2-m)(p-m)(q-m)(r-m)] \\ C_p = -s_{\max} \cdot 2 \cdot m \cdot q \cdot r/[(2-p)(m-p)(q-p)(r-p)] \\ C_q = -s_{\max} \cdot 2 \cdot m \cdot p \cdot r/[(2-q)(m-q)(p-q)(r-q)] \\ C_r = -s_{\max} \cdot 2 \cdot m \cdot p \cdot q/[(2-r)(m-r)(p-r)(q-r)]\,. \end{cases}$$

To show how the lift, s is affected by a chosen polynomial law, we have chosen a number of alternatives, given by the law 2-m-p-q-r (see Fig. 12.8). Note that the figure shows half the symmetric cam event profile up to the nose tip.

We can also express $\dot{s}$ and $\ddot{s}$ as:

$$\ddot{s} = [2C_2\bar{\theta} + mC_m\bar{\theta}^{m-1} + pC_p\bar{\theta}^{p-1} + qC_q\bar{\theta}^{q-1} + rC_r\bar{\theta}^{r-1}]\omega/\hat{\theta},$$

$$\text{where: } \dot{\bar{\theta}} = \dot{\theta}/\hat{\theta} = \omega/\hat{\theta}$$

$$\begin{aligned} \ddot{s} = {} & [2C_2 + m(m-1)C_m\bar{\alpha}^{m-2} + p(p-1)C_p\bar{\alpha}^{p-2} \\ & + q(q-1)C_q\bar{\alpha}^{q-2} + r(r-1)C_r\bar{\alpha}^{r-2}][\omega/\alpha_t]^2 \\ & + [2C_2\bar{\theta} + mC_m\bar{\theta}^{m-1} + pC_p\bar{\theta}^{p-1} + qC_q\bar{\theta}^{q-1} + rC_r\bar{\theta}^{r-1}]\dot{\omega}/\hat{\theta}. \end{aligned} \tag{12.14}$$

For steady state $\dot{\omega} = 0$ and the second term in the above equation is neglected. However, for a transient condition (e.g. accelerating or decelerating) $\dot{\omega} \neq 0$.

For a practical **automobile cam** consider the following polynomial law: $m = 8, p = 14, q = 20, r = 26$ as the polynomial coefficients (the lift for which is shown in Fig. 12.8), thus the coefficients become:

$$\begin{cases} C_2 = -1.0355 \times 10^{-2} \\ C_m = 1.0355 \times 10^{-2} \\ C_p = -8.875 \times 10^{-3} \\ C_q = 4.1418 \times 10^{-3} \\ C_r = -7.965 \times 10^{-4}. \end{cases}$$

Now we can also use the following data from an engine:

$$\hat{\theta} = 134.5° = 2.34\,\text{rad}, \quad m_v = 0.100\,\text{kg (mass of the valve)},$$
$$m_{sp} = 0.04\,\text{kg (mass of the valve spring)},$$

Thus: $m = m_v + m_{sp}/3 = 100 + 40/3 = 0.113\,\text{kg}$. If the engine is running at the crankshaft speed of 1400 rpm, then camshaft speed is 700 rpm, and thus the rotational speed of the cam is: $\omega = 2\pi N = 2\pi \frac{700}{60} = 73.3\,\text{rad/s}$.

We also need the following data:

Lubricant dynamic viscosity at atmospheric pressure, $\eta_0 = 0.007$ Pa s
Piezo-viscosity index, $\alpha = 1.45 \times 10^{-8}\ \text{Pa}^{-1}$
Modulus of elasticity of cam and tappet (both made of steel):
$E_1 = E_2 = 210 \times 10^9\,\text{N/m}^2$
Poisson's ratios: $\nu_1 = \nu_2 = 0.3$
Valve spring stiffness, $k = 14{,}000$ N/mm
preload, $z_p = 13$ mm
maximum lift, $s_{\max} = 9$ mm
Cam base circle radius, $R = 18$ mm
Cam width, $L = 10$ mm

Now as the time advances in a simulation study, the cam contact with the tappet is given by $\theta = \omega t$. We can obtain the conditions, such as contact kinematics, load and film thickness at any location. For instance, let us look at cam nose tip-to-flat tappet contact condition. At this location: $\theta = \hat{\theta} \Rightarrow \bar{\theta} = 1$. Now we can use Eq. (12.14) to obtain the value of valve acceleration: $a = \ddot{s}$ as:

$$a = \ddot{s} = 0.0079 \left[\omega\hat{\theta}\right]^2, \quad \text{thus:} \quad a = \ddot{s} = 0.0079 \left[\frac{73.3}{2.34}\right]^2 = 7.75\ \text{m/s}^2.$$

And Eq. (12.7) to obtain the inertial force:

$$F_i = m\ddot{s} = 113.3 \times 10^{-3} \times 7.75 = 878.075 \times 10^{-3}\,\text{N}$$

At the maximum lift: $z = s_{\max} + z_p = 22$ mm, and the total elastic force is given as:

$$F_e = k(s_{\max} + z_p) = 14000 \times 22 \times 10^{-3} = 14 \times 22 = 308\,\text{N}$$

Thus, the contact force is obtained as: $W = F_e - F_i = 308.45\,\text{N}$.

To find the speed of entraining motion, we first use Eq. (12.11) to find the instantaneous radius as:

$$r = R + s + \frac{\ddot{s}}{\omega^2} = 18 \times 10^{-3} + 9 \times 10^{-3} + \frac{7.75}{73.3 \times 73.3}$$
$$= 18 \times 10^{-3} + 9 \times 10^{-3} + 1.44 \times 10^{-3}$$
$$= 28.44 \times 10^{-3}\,\text{m}.$$

Then, using Eq. (12.12):

$$U = \left(r + \tfrac{\ddot{s}}{\omega^2}\right)\tfrac{\omega}{2} = \left(28.44 \times 10^{-3} + \tfrac{7.75}{73.3 \times 73.3}\right) \times \tfrac{52.36}{2} = 0.78\,\text{m/s}.$$

Also,

$$\tfrac{1}{E^*} = \tfrac{1}{2}\left\{\tfrac{1-\nu^2}{E_1} + \tfrac{1-\nu^2}{E_2}\right\} = \tfrac{1-0.09}{210 \times 10^9} = \tfrac{1}{230 \times 10^9}, \quad E^* = 230\,\text{GPa}.$$

We can now calculate all the non-dimensional groups required for the extrapolated oil film thickness formulae, such as those given above:

$$U^* = \frac{\eta_0 U}{E^* r} = \frac{0.007 \times 0.78}{230 \times 10^9 \times 28.4 \times 10^{-3}} = \frac{0.007 \times 0.78}{230 \times 28.4 \times 10^6} = 8.38 \times 10^{-13},$$

$$W^* = \frac{W}{E^* r L} = \frac{308.45}{230 \times 10^9 \times 28.4 \times 10^{-3} \times 10^{-2}} = \frac{308.45}{230 \times 28.4 \times 10^4}$$
$$= 4.7 \times 10^{-6},$$

$$G^* = \alpha E^* = 1.45 \times 10^{-8} \times 230 \times 10^9 = 3335.$$

At the cam nose tip: $w_s^* = \frac{1}{U}\frac{\partial h}{\partial t} = 0$, since: $\frac{\partial h}{\partial t} \approx 0$, thus (using Eq. (12.2)):

$$h^* = 1.67 G^{*^{0.421}} U^{*^{0.541}} W^{*^{0.059}} e^{-96.775 w_s^*}$$
$$= 1.67 \times 3335^{0.421} \times (8.38 \times 10^{-13})^{0.541} \times (4.7 \times 10^{-6})^{0.059}$$
$$= 2100.4 \times 10^{-8},$$

which gives a film thickness of:

$$h = h^*r = 2100.4 \times 10^{-8} \times 28.44 \times 10^{-3} = 5.97 \times 10^{-7}\,\text{m} \approx 0.6\mu\text{m}.$$

The same procedure is repeated as the cam contact point with the tappet alters in time.

As an exercise you can calculate the oil film thickness, which arises near the inlet reversal positions. These are positions, where the speed of entraining motion becomes zero (the film will be zero at these positions according to the extrapolated oil film equation). For the polynomial cam used in our example, these positions are shown in Fig. 12.4. The remaining data that you need are given in our example above. You will find that the film thickness near these inlet reversal positions would be much smaller than that calculated in our example, because the film thickness is mostly maintained by the squeeze film effect.

12.3. EHL of Thin Shell Journal Bearings

In Chapter 8 we dealt with hydrodynamic journal bearings. In these cases, the **bearing bush or shell** is usually of sufficient thickness and made of a material of high elastic modulus such as steel. Together, with these attributes, the large contact area between the journal and the bush (leading to relatively low pressures) inhibits the deformation of surfaces, thus encouraging a hydrodynamic lubrication regime. These bearings are used in many applications such as in large power generation plants. In other applications where there are relatively high loads and small journal to bushing clearances thin films may result, which can lead to instances of metal-to-metal contact. An attempt to discourage this is made by using materials of *lower* elastic modulus for the bush or the journal or both, as well as by reducing the thickness of the shell. At sufficiently high loads the shell will be subject to both *global* and *local* deformation. Global deformation refers to the tri-axial state of stress in any hollow shell subjected to load, which can suffer direct and shear stresses causing it to bend, twist and buckle. Local deformation results from the generated pressure in a contact, for example between the journal and the bush. Therefore, in many applications the introduction of thin shells, made of materials of lower elastic modulus can induce elastic deformation, which creates an additional gap for the lubricant film to occupy. This idea has been used successfully in IC engines in the piston to cylinder liner contact and between the journal and bush in crankshaft support journal bearings. The geometry between the journal and the bush in an IC engine is conformal, but not closely conformal (see Fig. 3.6(b)). The bearing bush, fitted into the cylinder block, is a thin shell of 2–5 mm thickness, made of **babbitt**, a tin-based alloy of modulus of elasticity 55–62 GPa. For example in high torque 1.8 liter diesel engines, with peak combustion pressures of 100–150 bar, the bearing loads can reach 10–15 KN, which lead to elastic deformation of the shell, with maximum contact pressures of the order of 50 MPa. Other journal bearings may have brass bushing to encourage similar conditions at high centrifugal forces due to small eccentricities, but at very high rotational speeds, such as in turbochargers running in excess of 100,000 rpm. The high applied loads and generated pressures are sufficient to cause *global* and *local* deformations, but not high enough to alter the viscosity of the fluid by any significant amount (see Chapter 5). Thus, we can refer to such conditions as **isoviscous elastic**.

Now referring back to Eq. (8.2), we can see that when shell deformation takes place, the rigid hydrodynamic film (**isoviscous rigid** condition) will have an additional gap due to the local deformation, δ. Thus, the film thickness at any location, given by θ, is now obtained as:

$$h = c(1 + \varepsilon \cos\theta) + \delta. \tag{12.15}$$

This film thickness, comprising the rigid hydrodynamic film and the local deformation, can now be referred to as the **elastic film**.

If the journal bearing under investigation can be considered as of short length, then to obtain a full solution we have two Eqs. (8.4) and (12.15). However, note that in this case an additional unknown, δ, exists, for which we need to find an expression.

The deflection at any point in the contact arises from the effect of all the pressure elements in the overall pressure distribution, see Eq. (3.13) and Sec. 10.6. However, when the boundary solids have low elastic moduli, the effect of the local pressure directly above a position has far more significance on the local deformation there than from other pressure elements remote from it. This simplifies the calculation process and leads to an analytical expression. The approach is referred to as the **column method**, because we simply ignore the effect of other pressure columns not acting directly above the location, where the value of δ is being determined.

In using this approach we need to be aware of the extent of its applicability. In short, it is a good approximation for conforming contacts of materials of low elastic modulus, such as the thin shell journal bearings described above.

A typical example is a piston skirt-to-cylinder liner contact where the liner thickness is 2–5 mm. Sometimes the skirt or the liner or both are made of an aluminium-based alloy, as in motor sport applications. For these cases the error using this approach is 5–10% of a full numerical solution of the type described in Chapter 10.

Another important point to note is that most thin shell engine bearings have a finite length, in which case Eq. (8.4) does not apply. Hence, a numerical solution of Reynolds equation is necessary. Nevertheless, a reasonable approximate solution can be obtained using the column method approach. Such a solution will be favored in industry due to tight time-scales allowed for engine development programmes.

For a thin shell, the tri-axial state of stress-strain can be represented as:

$$\{\sigma\} = [D](\{\varepsilon_s\} - \{\varepsilon_{s0}\}) + \{\sigma_0\}, \tag{12.16}$$

where σ and ε_s are stress and strain, and the subscript 0 refers to any pre-stressing or pre-straining of the material. Here $[D]$ is the influence coefficient matrix, with its elements providing relationships between the stress and strain components in terms of mechanical properties of the contacting solids of modulus of elasticity, E, shear modulus, G and Poisson's ratio, υ (see Chapter 3). Using the stress-strain relationships:

$$\begin{cases} \varepsilon_x = \dfrac{1}{E}[\sigma_x - \upsilon(\sigma_y + \sigma_z)] \\ \varepsilon_y = \dfrac{1}{E}[\sigma_y - \upsilon(\sigma_z + \sigma_x)] \\ \varepsilon_z = \dfrac{1}{E}[\sigma_z - \upsilon(\sigma_x + \sigma_y)], \end{cases} \tag{12.17}$$

$$\gamma_{xy} = \left\{\frac{2(1+\upsilon)}{E}\right\}\tau_{xy}, \tag{12.18}$$

$$\gamma_{xz} = \frac{1}{G}\tau_{xz}, \tag{12.19}$$

$$\gamma_{yz} = \frac{1}{G}\tau_{yz}. \tag{12.20}$$

Thus, the influence coefficient matrix is obtained as:

$$D = \frac{E(1-v)}{(1+v)(1-2v)} \begin{bmatrix} 1 & \frac{v}{1-v} & \frac{v}{1-v} & 0 & 0 & 0 \\ \frac{v}{1-v} & 1 & \frac{v}{1-v} & 0 & 0 & 0 \\ \frac{v}{1-v} & \frac{v}{1-v} & 1 & 0 & 0 & 0 \\ 0 & 0 & 0 & \frac{1-2v}{2(1-v)} & 0 & 0 \\ 0 & 0 & 0 & 0 & \frac{1-2v}{2(1-v)} & 0 \\ 0 & 0 & 0 & 0 & 0 & \frac{1-2v}{2(1-v)} \end{bmatrix}. \tag{12.21}$$

Note that: $G = \frac{E}{2(1+v)}$.

Equation (12.16), with the inclusion of Eq. (12.21), provides the general state of stress in a solid. Referring to Chapter 3, note that in contact mechanics problems we are interested in the *local*deformation of elastic solids in contact, and consider that the generated pressures are due to the compression of the contiguous surfaces. Thus: $\|p\| = \|\sigma_z\|$, when $\sigma_z < 0$ in the contact, and $p = 0$, when $\sigma_z > 0$ outside the contact area. This considerably simplifies the problem of evaluating δ. Thus, referring to (12.21), we are only interested in the element (3,3) of the matrix:

$$p = \frac{E(1-v)}{(1+v)(1-2v)} \varepsilon_z. \tag{12.22}$$

The strain is $\varepsilon_z = \frac{\delta}{d_s}$, where d_s is the shell thickness.

Thus, using the above relations:

$$\delta = \frac{(1-2v)(1+v)d_s}{E(1-v)} p. \tag{12.23}$$

We can now see that Eqs. (12.15), (8.4) and (12.23) can be used to find h, δ and p for a thin shell low elastic modulus journal bearing. An iterative procedure is required for this, in which we can initially assume $\delta = 0$ and find h from (12.15). Note that: $\frac{dh}{d\theta} = -(c\varepsilon \sin\theta)\, R$ and we can find p from Eq. (8.4). Then, p can be used to find δ from Eq. (12.23), the procedure being repeated until a specified convergence criterion is satisfied, such as

$$\sum_i \left| \frac{(p_i^{old} - p_i^{new})}{p_i^{old}} \right| > 10^{-4}. \tag{12.24}$$

This procedure is shown by the inner loop in the flowchart of Fig. 12.9. Note that in the quasi-static analysis we assume an eccentricity ratio, ε. For a realistic solution, ε is determined by solution of inertial dynamics of the system. For a shaft supported by a narrow journal bearing, 3 degrees of freedom motion may be considered. These are translational motions along the line of centers, x direction in Fig. 8.3, normal to the line of centers (z direction) and rotational motion about the y axis (normal to the zx

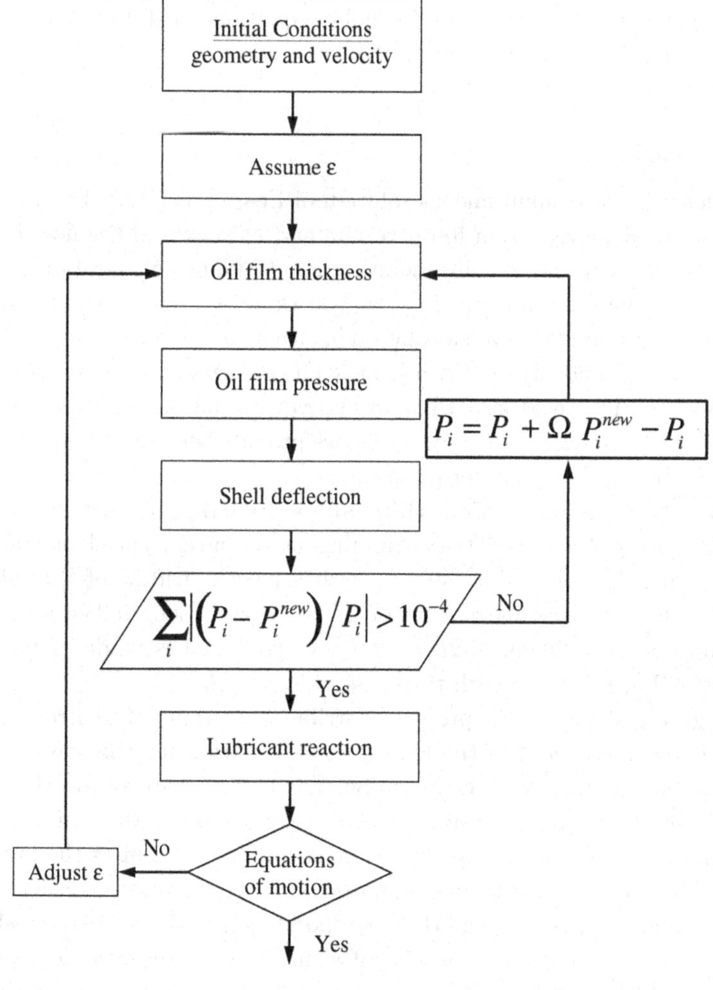

Fig. 12.9. Quasi-static and transient solutions for thin shell journal bearings.

plane). Any applied force, F and the gravity load W must be resolved into the x and z directions, resisted by the lubricant reactions: W_x and W_z (as functions of ε, see Chapter 8). The rotational degree of freedom is a function of both shaft inertial torque and friction torque (also a function of ε, see Eq. (8.16)) multiplied by the radius of the shaft/journal). The solution of the equations of motion yield the eccentricity ratio, ε, and the attitude angle, ψ (see Fig. 8.3) through an iterative procedure. This approach forms a transient analysis of **thin shell journal bearing**, and includes both the inner and outer loops of the flowchart in Fig. 12.9.

Finally, note that if a bearing has a finite length, the use of Eq. (8.4) is not permitted, and the full Reynolds equation must be used instead, together with Eqs. (12.15) and (12.23).

An example

It is clear that the simultaneous solution of Eqs. (8.4), (12.15) and (12.23) can only be achieved in an iterative manner, as shown in the flowchart of Fig. 12.8. This would usually require a good number of iterations. Thus, a hand calculation is not practicable. You would therefore need to write a computer program to obtain a solution in either case: a quasi-static solution (value for ε is known) or a transient solution (value of ε also subject to change). To give an idea of what results you would expect to obtain and the physical behavior of such bearings, we provide here an example already obtained by such a computer program.

We take the case of a **thin shell bearing** subject to a load variation from 1200 to 5000 N, running at a journal speed of 500 rpm. The journal diameter, $D = 40$ mm, with a length (width) of $L = 20$ mm (thus a diameter-to-length ratio of 2, i.e. a short-length bearing). The clearance is, $c = 20\,\mu$m ($c/R = 1/1000$). The shell is made of **babbitt** ($E = 65$ GPa, $\upsilon = 0.33$) with thickness, $d_s = 5$ mm.

Figure 12.10 shows the pressure distributions (a) and the corresponding oil film thickness profiles (b) for the load variation for this case. Notice that as the load increases the hydrodynamic (rigid) shape of the film alters as deformation becomes visible at the outlet region of the contact. This deformation of the shell helps in retaining a lubricant film as the load has more than trebled and the maximum pressure has increased considerably.

Pressure distribution and the corresponding film shape are often shown in three dimensional plots for ease of visualization. An example is shown is Fig. 12.11 for the case of a load of 5000 N. Note that the film shape is

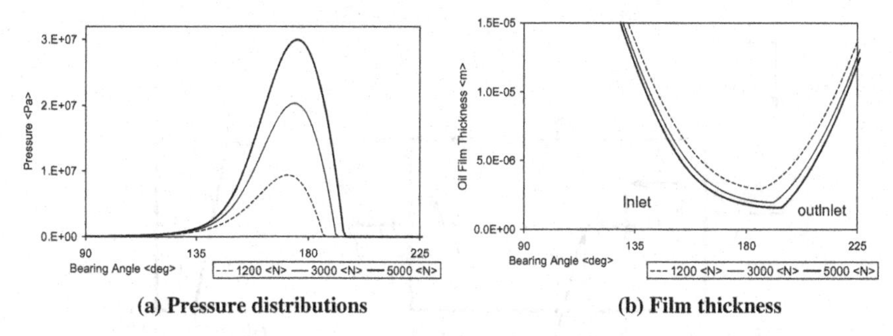

(a) Pressure distributions **(b) Film thickness**

Fig. 12.10. Pressure distribution and film shapes for a thin shell bearing under different loads.

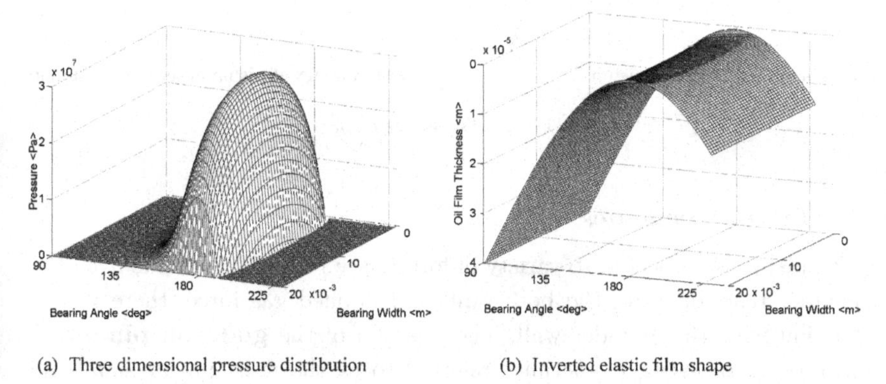

(a) Three dimensional pressure distribution (b) Inverted elastic film shape

Fig. 12.11. Three-dimensional pressure distribution and film shape for the case of 5000 N load.

intentionally inverted in order to observe the shell local deformation at the exit, which contributes to an increasing gap, and thus the film thickness.

12.4. Transient EHL of Piston-to-Cylinder Liner

There are a number of lubricated conjunctions between the piston and cylinder bore. These include contacts between the **piston ring-pack** and the bore, and the **piston skirt** and the bore. Here we confine ourselves to the analysis of the latter.

During the combustion cycle the piston crown is subjected to a variation in the gas force, resulting in corresponding changes in the contact load. The speed of the piston also varies during the cycle, as well as it being subject to reversals at the top and bottom dead centers.

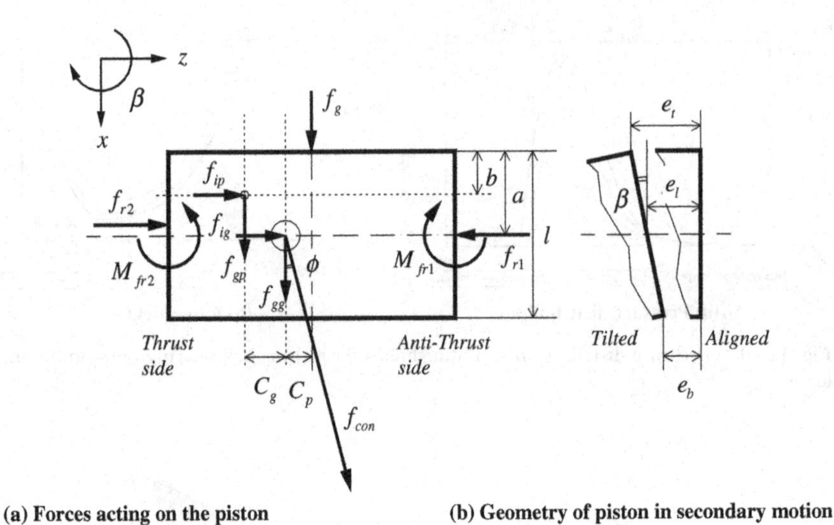

Fig. 12.12. Dynamics of piston.

(a) *Inertial dynamics*

The piston is subjected to variety of forces (see Fig. 12.12). Aside from the contact reaction from the bore wall, and applied gas force, there is also friction with the cylinder wall. The position of the **gudgeon pin** or the **pin bore bearing** (sometimes referred to as the wrist-pin or small-end bearing at the smaller end of the connecting rod) is offset from the axis of symmetry of the piston. This is an attempt to adhere the piston to one side of the bore, and reduce its lateral oscillations within the confine of its clearance with the bore. This offset, therefore, reduces the chance of impact between of the piston and the cylinder wall.

Note that the piston undergoes a 3-degrees-of-freedom motion. These are its axial component (primary motion, along the x direction in Fig. 12.12), lateral (in z direction) and tilting motion (rotation about y axis, denoted by β). The lateral and tilting motions are referred to as **secondary motions** of the piston. The primary motion of the piston is its ideal function, whilst the lateral motions take place partly due to the asymmetrical positioning of the **gudgeon pin** with the respect to the cylinder bore.

The secondary motions of the piston; in tilting: β, and lateral excursion: e_ℓ can be substituted by its deviations from the cylinder surface at its top, e_t and at its bottom, e_b from the center-line of the cylinder bore. Thus, referring to Fig. 12.11(b) for small tilts, $\tan\beta \approx \beta$, giving:

$$e_t = e_l + a\beta, \tag{12.25}$$

and,

$$e_b = e_l - (l - a)\beta, \tag{12.26}$$

where, l is the length of the piston skirt, and a is the axial distance from the gudgeon pin to the top of the piston skirt.

These motions occur as the result of forces applied to the piston at any instant. Therefore, we need to consider the inertial balance of the piston. The net force, made up of the contact forces between the piston skirt and the cylinder on the thrust and anti-thrust sides, is:

$$\sum f_z = f_{r1} - f_{r2} = f_{ip} + f_{ig} - f_{con}\sin\phi, \tag{12.27}$$

where f_{r1} and f_{r2} are the contact forces, f_{con} is the elastic resisting force in the connecting rod, and f_{ip} and f_{ig} are the lateral inertial forces of the pin and the piston respectively:

$$f_{ip} = m_p\left(\ddot{e}_t + \frac{b}{l}(\ddot{e}_t - \ddot{e}_b)\right), \tag{12.28}$$

and,

$$f_{ig} = m_g\left(\ddot{e}_t + \frac{a}{l}(\ddot{e}_b - \ddot{e}_t)\right), \tag{12.29}$$

where m_p and m_g are the masses of the piston and the gudgeon pin respectively.

The connecting rod force, f_s is given by:

$$f_s = \tan\phi(f_{gp} + f_{gg} + f_g), \tag{12.30}$$

where f_g is the instantaneous value of the combustion gas force applied to the piston crown, f_{gg} is the inertial force of the pin, and f_{gp} is the inertial force of the piston in its primary motion. This is given by: $f_{gp} = m_p\ddot{x}$. The combustion process determines the primary motion. Therefore, $\ddot{x}$ is known at any instant of time. Rahnejat[7] gives this as:

$$\ddot{x} = r\omega^2\left(\cos\omega t + \frac{r}{L}\cos 2\omega t + \cdots\right). \tag{12.31}$$

Note that the terms up to second **engine order** (multiples of crankshaft angular velocity, ω) have been only included. In practice, the significance of higher order terms depends on the ratio of crank radius to connecting rod length (r/L). In the example highlighted later, higher engine orders up to the 8ω have been included in the analysis.

The piston also tilts due to a net moment about the y axis at the position of its pin:

$$\sum M_y = M_{fr1} + M_{fr2} = -f_{ip}(a-b) - f_g C_p + f_{gp} C_g - M_c, \qquad (12.32)$$

where, $M_c = \frac{I_p}{l}(\ddot{e}_t - \ddot{e}_b)$, and M_{fr1} and M_{fr2} are the moments, caused by the contact forces.

Using the above equations, we can now represent the inertial dynamics of the piston in matrix form, in terms of its excursions of the top and bottom eccentricities: e_t and e_b as:

$$\begin{bmatrix} m_g\left(1-\frac{a}{L}\right)+m_p\left(1-\frac{b}{L}\right) & m_g\frac{a}{L}+m_p\frac{b}{L} \\ \frac{I_p}{L}+m_p(a-b)\left(1-\frac{b}{L}\right) & m_p(a-b)\frac{b}{L}-\frac{I_p}{L} \end{bmatrix} \begin{bmatrix} \ddot{e}_t \\ \ddot{e}_b \end{bmatrix}$$

$$= \begin{bmatrix} f_{r1}+f_{r2}+f_{con} \\ m_{fr1}+m_{fr2}+m_{con} \end{bmatrix}. \qquad (12.33)$$

From this pair of coupled equations, we can obtain $\ddot{e}_t$ and $\ddot{e}_b$ at any given time, t. Then, using a step-by-step integration algorithm (such as the linear acceleration method, developed by Newmark,[8] see also Timoshenko *et al.*[9]), the variables $\dot{e}_t, e_t, \dot{e}_b$ and e_b can be obtained at each time step Δt in terms of the past time history of the system (previous time steps) in an iterative manner. Having obtained these quantities, all other forces and moments for the next time step can be evaluated, using the equations given above.

Within each time step, we need to iterate, having commenced with an initial condition from a previous step. Various convergence criteria may be used for this purpose, for example:

$$\begin{aligned} \left|\frac{\ddot{e}_{t,n}-\ddot{e}_{t,n-1}}{\ddot{e}_{t,n}}\times 100\right| &< 1 \\ \left|\frac{\ddot{e}_{b,n}-\ddot{e}_{b,n-1}}{\ddot{e}_{b,n}}\times 100\right| &< 1 \end{aligned} . \qquad (12.34)$$

We can thus obtain the solution for Eq. (12.33) for non-steady (i.e. transient) conditions, if we determine the contact forces and moments in the lubricated conjunctions between the piston skirt and the cylinder bore on the thrust and anti-thrust sides.

(b) *Contact forces and moments*

The contact reactions, f_{r1} and f_{r2}, and the moments, M_{fr1} and M_{fr2} are assumed to be generated by the lubricant film in the conjunction between the piston and the cylinder bore on its thrust and anti-thrust sides. The reactions are the instantaneous integrated lubricant pressure distributions. The moments are as a result of lubricant traction. Therefore, we need to obtain the pressure and shear stress distributions at any instant (see Chapter 10). The differencc here with the analysis in Chapter 10 is that there we imposed a known load, and obtained a solution for lubricant film thickness and pressure distribution to balance this within a specified limit (i.e. a quasi-static analysis, see the flowchart in Fig. 10.7). Here, the load (i.e. reaction) is generated by the lubricant film and no load convergence is carried out. Instead, the convergence criterion is based on the system dynamics (e.g. Eq. (12.34) and see the typical flow chart in Fig. 12.8). This linkage between the system dynamics and contact conditions is the essence of a transient analysis.

Neglecting side leakage, for a steady state case, the **Reynolds equation** is given by Eq. (10.10). For a transient study, we need to incorporate the time history of film thickness, which is given by $\frac{\partial \rho h}{\partial t}$, which we have already explained as the squeeze film term in Chapter 6 (also see Sec. 12.1), thus:

$$\frac{\partial}{\partial x}\left[\frac{\rho h^3}{\eta}\frac{\partial p}{\partial x}\right] + \frac{\partial}{\partial y}\left[\frac{\rho h^3}{\eta}\frac{\partial p}{\partial y}\right] = 12\left\{U\frac{\partial}{\partial x}(\rho h) + \frac{\partial}{\partial t}(\rho h)\right\}. \tag{12.35}$$

The **elastic film** equation for this case is:

$$h = c + e_l + s\tan\beta + \delta + \Delta, \tag{12.36}$$

where the effects of secondary motions of the piston are given by e_L and β. They represent respectively misaligned profile of the piston skirt, while δ is the elastic deflection due to the generated pressure and Δ is the global thermo-elastic distortion of the piston. This global deformation is quite small if thermal effects are not taken into account, and can thus be ignored in an isothermal analysis. The local deflection is given by Eq. (3.14), where: $\delta = w(x, y)$.

Tilt of piston, usually of the order of 0.1°, misaligns the profile, and is accounted for in Eq. (12.36). The profile of the piston skirt usually incorporates **relief radii** at its top and bottom to create a wedge effect (described in Chapter 6), which avoids generation of high pressures there (similar to that experienced at the edges of rollers, see Chapter 10). Lubricant density and viscosity variations are also taken into account (see Chapters 5 and 10).

Therefore, the contact forces described above are given as (subscript 1 denotes the **thrust side**, whilst 2 refers to the **anti-thrust side**):

$$f_{r1} = \iint p_1 dxdy, f_{r2} = \iint p_2 dxdy. \tag{12.37}$$

The corresponding moments are obtained as:

$$\begin{aligned} M_{fr1} &= r_{p1} \iint \left(-\frac{h_1}{2}\frac{\partial p_1}{\partial x} + \frac{\eta_0 \Delta_1 u}{h_1} \right) dxdy, \\ M_{fr2} &= r_{p2} \iint \left(-\frac{h_2}{2}\frac{\partial p_2}{\partial x} + \frac{\eta_0 \Delta_2 u}{h_2} \right) dxdy, \end{aligned} \tag{12.38}$$

where, r_{p1} and r_{p2} are the moment arms from the centers of pressure distributions p_1 and p_2 to the gudgeon pin axis, and $\Delta_1 u = \Delta_2 u \approx \dot{x}$, piston speed.

An example

Like the previous example for the case of thin shell journal bearings, it is clear that an analytical solution is not possible as we have to obtain a solution for piston inertial dynamics in a time marching procedure. This would involve solution of equations of motion (12.33), which requires forces and moments for secondary motion of the piston from the lubricated contacts between the piston skirt and cylinder wall on the thrust and anti-thrust sides. It is also clear that when a thick cylinder liner of high elastic modulus or the cylinder bore itself is used as the load bearing surface in contact with the piston, the deformation required for Eq. (12.36) must be obtained in the same manner as described in Chapter 10 for *hard* elastohydrodynamic conjunctions. This is the case for most vehicles, particularly the older generations, and certainly for larger commercial ones.

In motor sport, there has been a trend in use of lighter cylinder liners, made of high grade aluminium, which are usually coated with hard wear resistant alloys. A detailed analysis is reported by Balakrishnan and Rahnejat.[10]

The technology of advanced cylinder liners is gradually finding its way into the commercial sector, and is currently used in what is termed as **niche high performance vehicles**. For our example here we use the investigation of the contact of a piston skirt to a cylinder liner used in a motor sport application, where a typical cylinder liner is made of high grade aluminium with an **electrolytic nickel coating** of approximately 60 μm thick (known commercially as **Nikasil**). The surface of the liner is honed and cross-hatched to a finish of typically 0.1–0.4 μm Ra. In motor sport applications, in particular high performance engines which are often used for Formula 1 racing, the bore diameter is typically 85–90 mm (the piston clearance is typically in the range 10–50 μm), with piston skirt of 25–45 mm length, with axial relief radii at the top and bottom of the skirt. These are machined in order to mitigate high pressures generated at the contact extremities (similar to those in the contact of a roller against a semi-infinite elastic half-space, see Chapter 10). The relief radii also create a wedge effect to encourage entraining motion of the lubricant into the

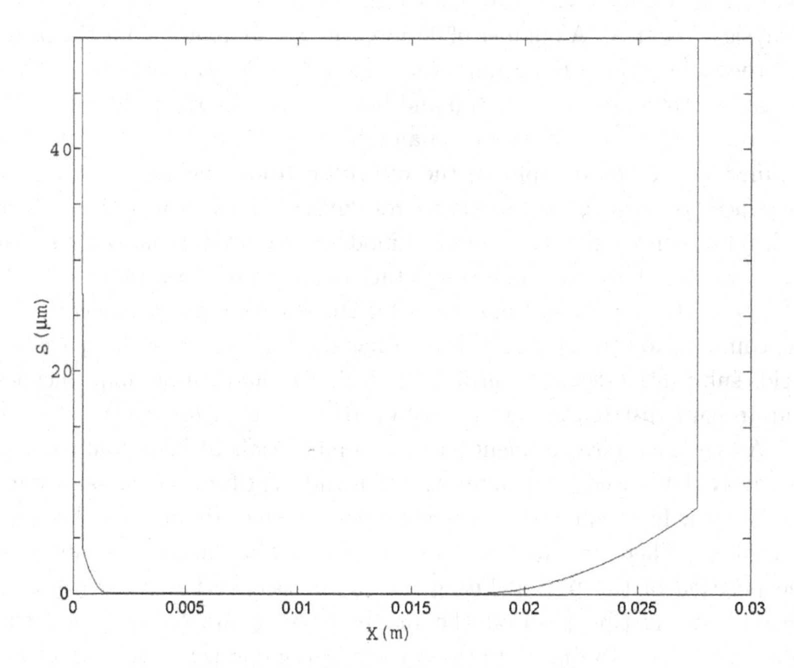

Fig. 12.13. Axial profile of a piston skirt.

contact (see Chapter 6). Figure 12.13 shows the axial profile of a typical piston skirt (S is the profile and X denotes the axial direction in the figure). The left hand side of the figure indicates the top of the piston skirt.

A typical high performance engine of this kind comprises 10 such cylinders in V-formation (i.e. a V10 engine), recently modified to a V8 formation. The engine is typically 3 liters and delivers 850–950 bhp, with a peak combustion pressure of 120 bars reached around 13° (in terms of the crank angle) past the position of TDC during its **power stroke** (downward sense of the piston). The above transient analysis shows that in this position the piston adheres to the thrust side by a tilt of 0.1°, the side force being approximately 2.5 kN.[11] The piston translational speed alters throughout its cycle from zero at TDC and BDC up to 42 m/s at mid-cycle. This means that with diminishing clearance due to piston tilt, cessation of entraining motion at the ends of the stroke and thermal distortion, causes the film thickness to be very small at **top dead center** (TDC) and **bottom dead center** (BDC). At mid-cycle, high speed of entraining motion, lower contact loads and larger contact areas due to aligned orientation of the piston, lead to a hydrodynamic regime of lubrication. At maximum combustion pressure, the side force reaches its maximum value. There is variation in the regime of lubrication throughout the power stroke. Thus, a transient analysis is required. A number of design features are required for the piston and the cylinder liner to optimise the varying tribological conditions. These include the relief radii at the top and bottom edges of the piston skirt, the material combination of the piston and liner, the choice of any coating and detailed surface topography of the **cylinder liner**, including the surface roughness or any modifications to its surface. This means that a very high mesh density for the numerical model is required so as to obtain the film contours. This very high computational mesh of a few million points, together with very small time steps for the dynamic analysis means that the computation times take at least a few days. However, such an analysis yields substantial detail, regarding the regime of lubrication, film thickness and pressure distribution at any instant during the piston cycle.

As we have already mentioned, a combination of high contact load, relatively slow speed of entraining motion and the tilt of the piston towards the TDC at high combustion pressure yield some of the worst tribological conditions. Therefore, we can look at the oil film thickness contours in the position of the piston during the power stroke. These are shown in Fig. 12.14. At this position the piston is doing its down-stroke, with lubricant being entrained into the contact in the upward sense, as shown in

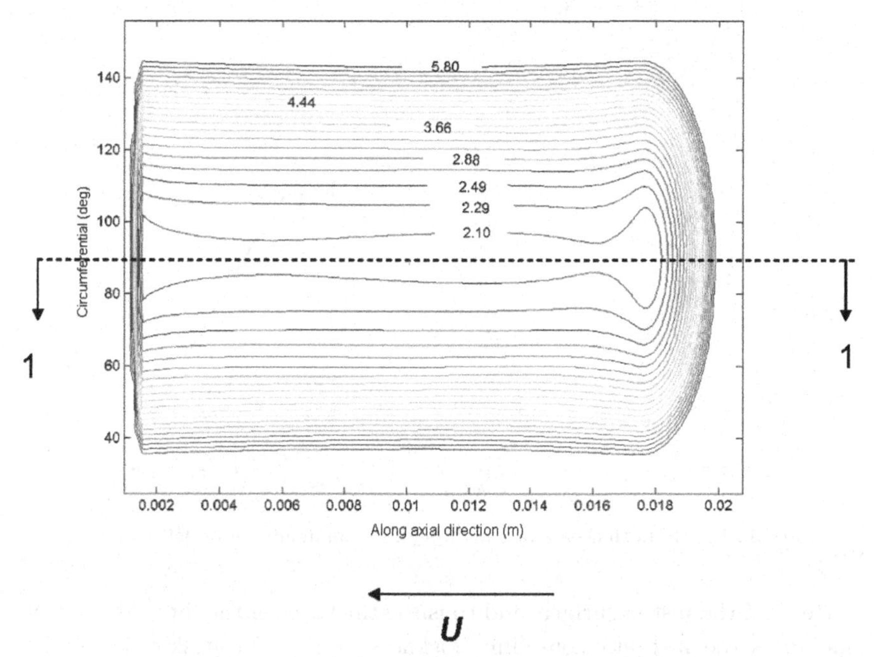

Fig. 12.14. Oil film contour of piston skirt-to-cylinder liner contact at maximum combustion pressure levels (Contour levels are in μm[11]).

the figure. The minimum film thickness occurs along the section 1–1, shown in the figure. A cut through this section of the oil film contour is shown in Fig. 12.15. We can see that the minimum film thickness is actually around 1.94 μm, and occurs at the inlet region of the contact, because of the sharper change in the relief radii of the piston at its lower edge (see Fig. 12.13), yielding the highest pressures there. The islands of high pressures at the leading and trailing edges of the contact can be seen in the isobar plot (a two dimensional presentation of the contact pressure distribution) of Fig. 12.16. Also note, from Fig. 12.15 that the film shape indicates deformation of the liner has taken place due to the high generated pressures in these locations (i.e. EHL).

It is important to ascertain the validity of this prediction. An engine test was carried out, where an ultrasonic sensor[12] was attached to the outside wall of the **cylinder liner** as shown in Fig. 12.17. The ultrasonic waves travel through the liner wall and the oil film and reflect from the piston surface. The wave speed alters through different media, from which we can ascertain the interfaces between the liner and the lubricant and the

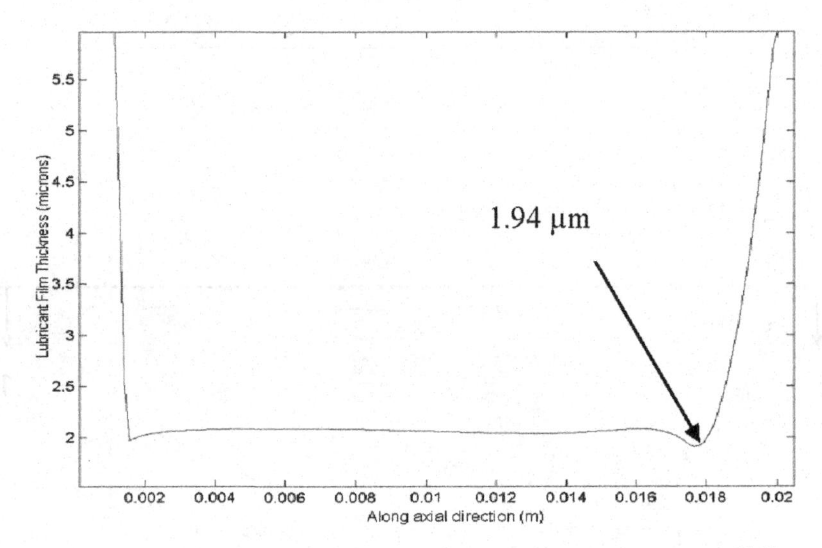

Fig. 12.15. Film thickness profile through the minimum film thickness.[11,12]

latter and the piston surface, and thus the thickness of the film. The sensor measures the instantaneous film thickness as the piston skirt passes its position on the liner wall.

The trace shown in Fig. 12.18 is taken under the same conditions as the predicted results in Fig. 12.16. Note that as the gap becomes large the measurements cannot be relied on. The accuracy in resolving the minimum film thickness depends on the sensor face-width. The value obtained for the minimum film thickness is 2 μm, which shows that the numerical predictions are very accurate indeed. If the surface roughness of both the cross-hatched and honed liner and the piston are 0.4 μm Ra, then the root mean square of the composite roughness (see Sec. 2.5.1) is: $\sigma_{rms} = \sqrt{2}(0.4) \approx 0.57\,\mu$m. Then, from Chapter 1: $\lambda = \frac{h_m}{\sigma_{rms}} \approx \frac{2}{0.57} = 3.51$, which according to **Stribeck analysis** points to an EHL condition, which is corroborated by the film shape in Fig. 12.15.

The conditions may even be worst at the TDC and BDC with no entraining motion and low load (thus no deformation). In practice, due to thermal distortion, shear thinning of the lubricant, degradation/ contamination of the lubricant and other such problems, the film thickness can reduce from those predicted. Therefore, some measures of surface modification have always been undertaken in an empirical manner to entrap a volume of lubricant. We do not intend to go into any great detail, and those interested should refer to Rahnejat *et al.*[13] and Etsion.[14] Here, we

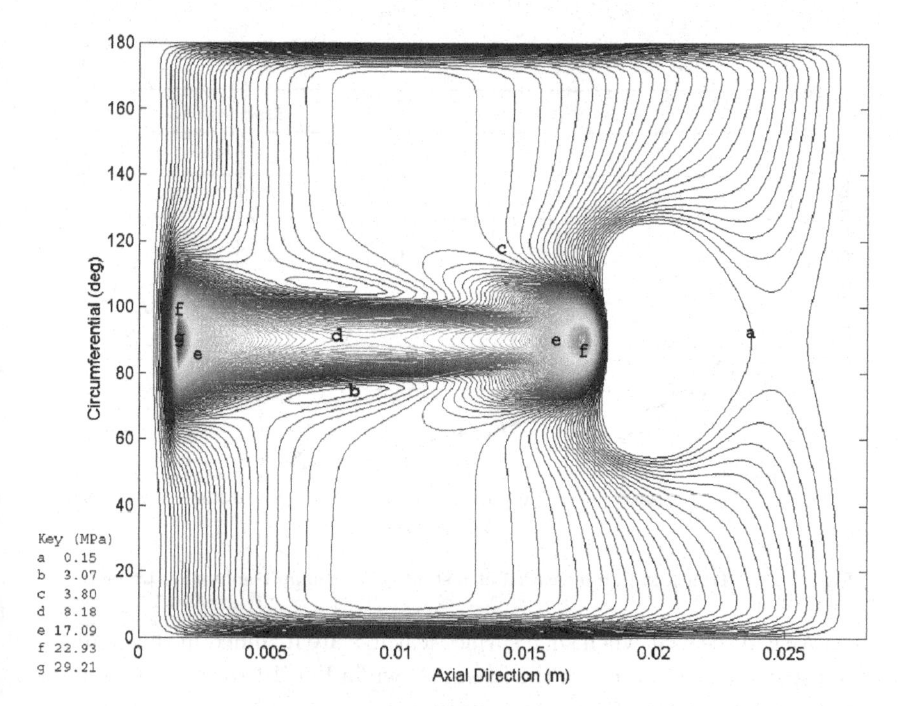

Fig. 12.16. Contact pressure isobar plot.[11]

Fig. 12.17. Measurement of film thickness by ultrasonic means.[12]

provide an example from Ref. 11, where grooves of depth of few micrometers are cut into the surface of the cylinder liner in specific areas to encourage retention of lubricant. In a micro-scale, such grooves actually create a wedge effect (see Chapter 6).

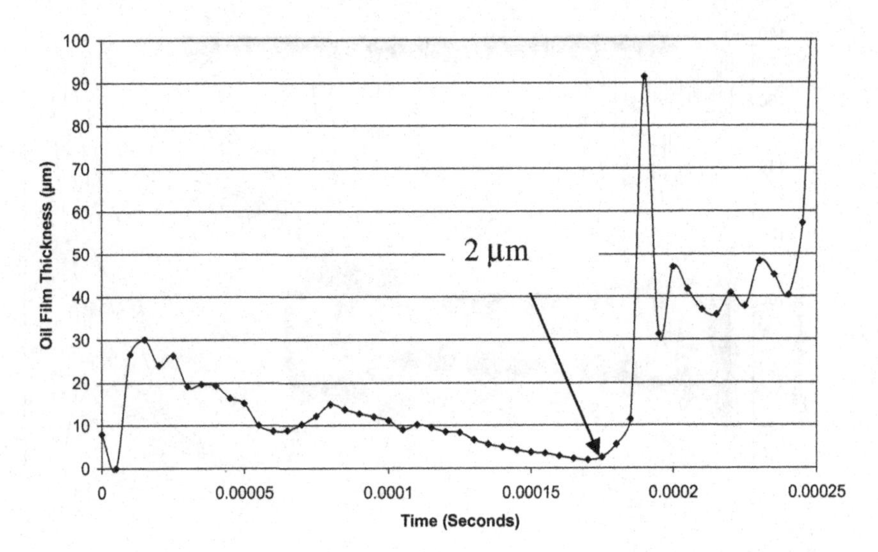

Fig. 12.18. Transient film measured at instant of maximum combustion pressure.[12]

We can see that when such surface features are included in the analysis, the predicted minimum film thickness, shown in Fig. 12.19 with the counter level **a**, is increased to 5.1 μm from its previous value of 1.94 μm (Fig. 12.14), thus improving the tribological conditions. The analysis carried out is, however, much more detailed and time consuming with such surface features included than for the case in Fig. 12.14, because one has to include the surface feature heights in the elastic film shape (Eq. (12.36)). This means that a **micro-EHL** analysis is carried out with even a greater number of mesh points than the previous case.

In practice, the surface features are introduced by laser etching techniques and their profiles (and pattern) may vary according to detailed analysis based on the flow around these features and local contact conditions. Analysis should take into account the adherence of lubricant to such micro-sized features, which would be governed by surface energy effects, topography and molecular behavior of the lubricant and its additives. We have provided some introduction to these issues in Chapter 13. Figure 12.20 shows the laser etched pattern of some typical features at the top dead center on a cylinder liner which is coated with **DLC** (Diamond Like Coating) which is a wear resistant very smooth material.

If such features act as small reservoirs of lubricant, they would promote the formation of a lubricant film between the cylinder liner and the piston skirt or ring-pack and reduce the chance of direct surface interactions.

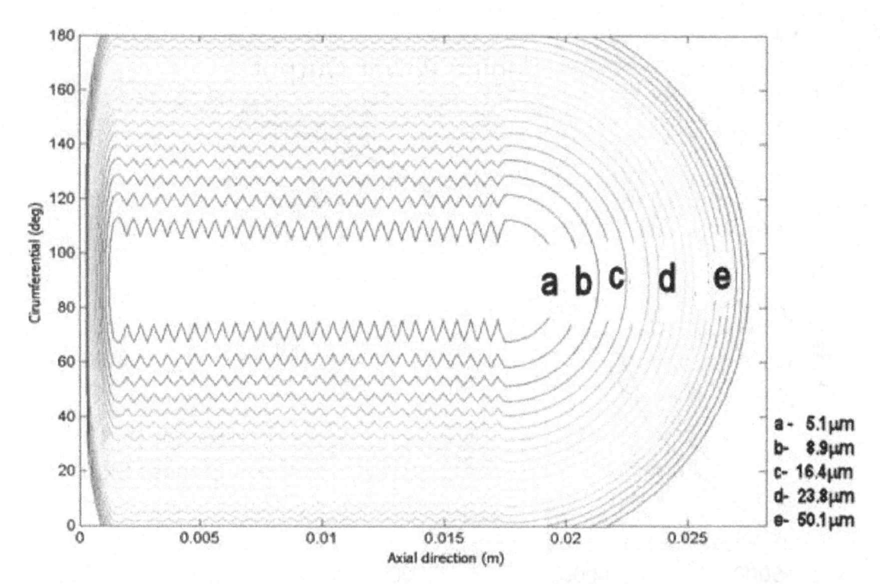

Fig. 12.19. Oil film contour of piston skirt-to-cylinder liner contact at maximum combustion pressure levels.[11]

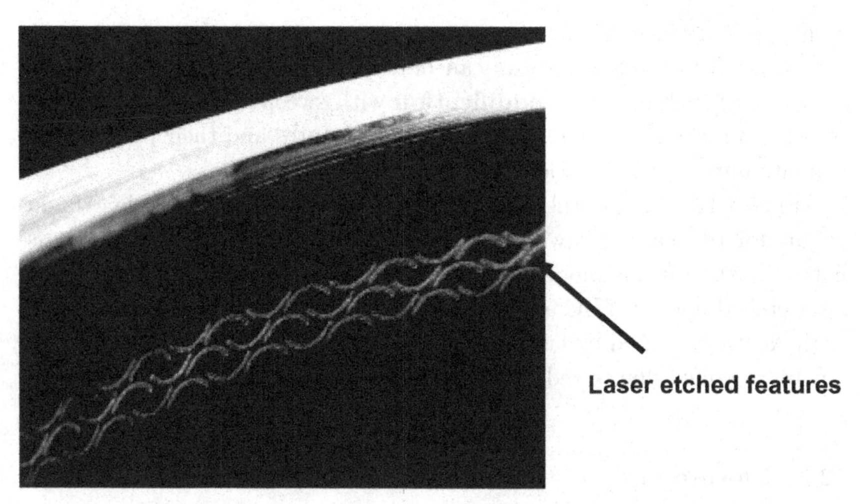

Fig. 12.20. Magnified pattern of laser-etched features at the TDC on a cylinder liner, coated with DLC.

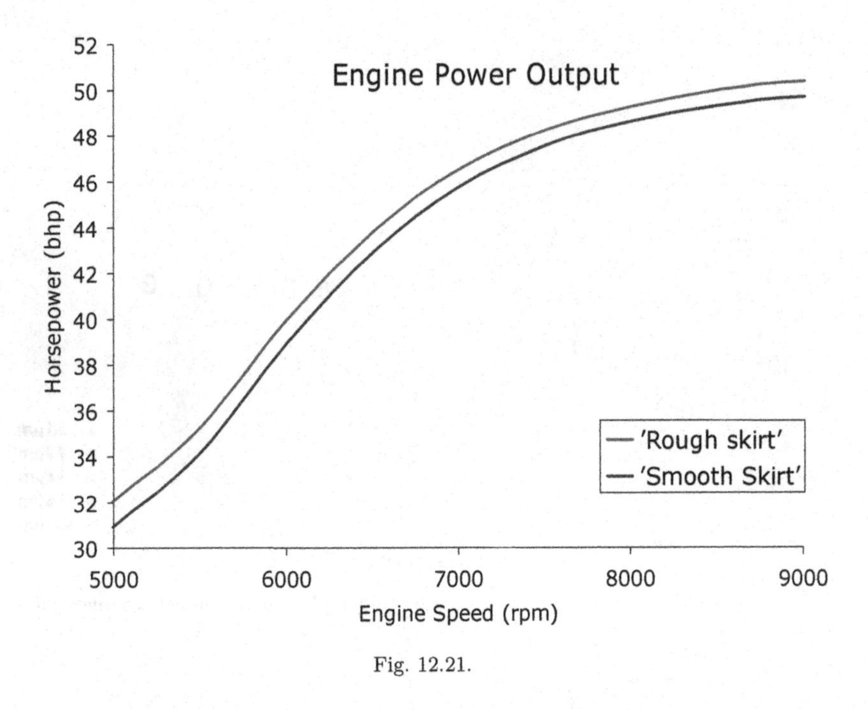

Fig. 12.21.

This would reduce friction by asperity interactions and improve engine output power. A series of tests can be carried out with and without the inclusion of such **surface modification** with sweeping engine speed tests to ascertain the effectiveness of introduced features and their pattern, and validate numerical predictions.

Figure 12.21 shows the results of tests carried out on a single cylinder engine for two cases: (i) with a standard cylinder liner (honed and cross-hatched, without any modifications, referred to as smooth), and (ii) with a laser etched liner at TDC and BDC, then honed to remove the sharp edges of the features etched by laser (referred to as rough). It is clear that 2–3 Hp has been gained due to reduced frictional losses.

12.5. Closure

In this chapter we have clarified the distinction between steady state and unsteady (transient) conditions in lubricated conjunctions. It is shown that the EHL conjunctions are sometimes subjected to either cyclic or other changing circumstances, which result in unsteady conditions. The cases highlighted are typical of many load bearing and transmitting conjunctions, which are often subject to accelerating or decelerating motions or vibration. Therefore, steady state or quasi-static solutions only provide an initial prediction, which in some circumstances may be significantly different from the actual conditions. We have also shown that transient solutions require the combined solution of tribology and system dynamics, which can lead to a computationally intensive effort. For solving most industrial problems, related to transient EHL, a sufficiently accurate solution, using extrapolated oil film thickness formulae, may be adapted as an alternative.

References

1. Mostofi, A. and Gohar, R., Oil film thickness and pressure distribution in elastohydrodynamic point contacts, *J. Mech. Eng. Sci.* **24** (1982) 173–182.
2. Rahnejat, H., Computational modelling of problems in contact dynamics, *Engineering Analysis*, **2** 4 (1985).
3. Rahnejat, H. and Gohar, R., The vibrations of radial ball bearings, *Proc. Inst. Mech. Engs. Part C, J. Mech. Eng.* **199(C)** (1985) 181–193.
4. Reeve, J., *Cams for Industry*, Mechanical Engineering Publications (1997).
5. Koster, M. P., Effect of flexibility of driving shaft on the dynamic behavior of a cam mechanism. *Trans. ASME, J. Engng for Industry* (May 1975) 595–602.
6. Kushwaha, M. and Rahnejat, H., Transient elastohydrodynamic lubrication of finite line conjunction of cam to follower concentrated contact, *J. Phys. D: Appl. Phys.* **35** (2002) 2872–2890.
7. Rahnejat, H. *Multi-Body Dynamics: Vehicles, Machines and Mechanisms* (UK: Professional Engineering Publishing), (USA: Society of Automotive Engineers), (1998).
8. Newmark, N. M., A method of computation for structural dynamics, *Trans. ASCE, J. Eng. Mech.* **85(EM3)**, 2094 (1959) 67–94.
9. Timoshenko, S., Young, D. H. and Weaver, Jr. W., *Vibration Problems in Engineering*, 4th ed, John Wailey and Sons, New York (1974).
10. Balakrishnan, S. and Rahnejat, H., Isothermal transient analysis of piston skirt-to-cylinder wall contacts under combined axial, lateral and tilting motion, *J. Phys. D: Appl. Phys.* **38** (2005) 787–799.
11. Balakrishnan, S., Howell-Smith, S. and Rahnejat, H., Investigation of reciprocating conformal contact of piston skirt-to-surface modified cylinder liner in high performance engines, *Proc. Instn. Mech. Engrs. Part C: J. Mech. Eng. Sci.* **219** (2005) 1235–1247.
12. Dwyer-Joyce, R. S. *et al.*, The measurement of liner-piston skirt oil film thickness by an ultrasonic means, *Trans. J. Engines, Pap.*, No. 2006-01-0648, 2006, 348–353.
13. Rahnejat, R., Balakrishnan, S., King, P. D. and Howell-Smith, S., In-cylinder friction reduction using surface finish optimisation Technique, *IMechE Part D* **220** (2006) 1309–1318 DOI: 10.1243/09544070JAUTO282.
14. Etsion, I., State of the art in laser surface, *J. Trib.* 1, 1 (2005) 248–253.

CHAPTER 13

NANO-TRIBOLOGY

13.1. Introduction

When Osborne Reynolds introduced his theory of fluid film lubrication in 1886, the minimum film thickness in many bearings was probably of the order of 10^{-5} m or even 10^{-4} m. When for certain applied loads the predicted film thickness due to hydrodynamic action was deemed to be below such values, surface interactions between contiguous bodies in contact were expected (see Chapters 3 and 4). However, in many cases the absence of wear was puzzling. The understanding of the mechanism of elastohydrodynamic lubrication (Chapter 10) provided the necessary explanation and paved the way for a steadily reducing minimum film thickness to provide greater load carrying capacity per unit area, improved stability and enhanced efficiency. In the latter half of the twentieth century many lubricated contacts were designed with film thicknesses, that are only a fraction of a micrometer (10^{-6} m) thick, such as in highly stressed components like gears, rolling element bearings and cam-follower pairs (Chapter 12). Indeed, as the understanding has advanced, the talk of sub-micrometre films in a number of situations; typically 10^{-7} m has become commonplace, as pointed out by Dowson.[1]

There is an increasing recognition that asperity or micro-elasto-hydrodynamic lubrication plays a major role in the effective lubrication of a number of components, involving both high (Chapter 10) and low elastic modulus materials (Chapter 12, thin shell bearings). Indeed, in very severe circumstances, perhaps involving plastic, as well as elastic deformation, the effective fluid films might well have thickness recorded in terms of nanometers (10^{-9} m) rather than micrometers (10^{-6} m). Nanometer thick films are promoted mostly in lightly loaded conjunctions, such as those in **magnetic storage media** (see Fig. 13.1).

A **hard disk drive system** of a computer, in which the flying height of a head-slider over a disk surface is approaching a few or several nanometers, gives rise to the formation of very thin lubricating films, currently of air. In

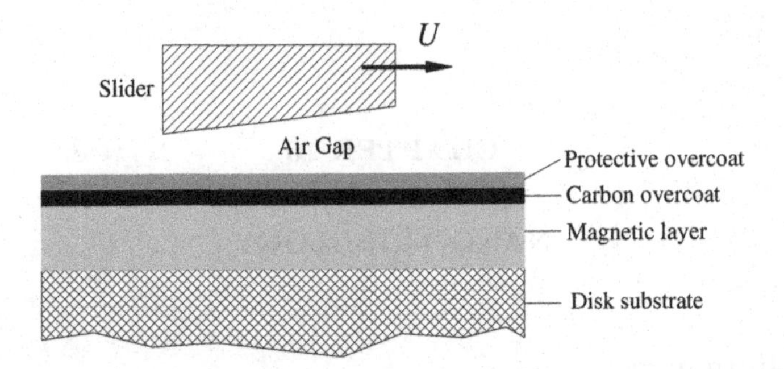

Fig. 13.1. Ultra-thin conjunction in magnetic storage media.

this system, the high storage density required in modern rigid disk drives can be achieved through the use of very smooth surfaces of thin-film rigid disks that allow ultra-low flying of read/write head sliders (see Fig. 13.1). A protective overcoat with a lubricant overlay is used over the soft metallic magnetic film in the construction of disks to maintain low friction, low wear and corrosion resistance in case of surface-to-surface contact, which can take place when the read/write head claps onto the disk platter with application of sudden shock loads.

Another emerging technology that requires fundamental understanding of film formation is **micro-electro-mechanical systems** (MEMS), where complete mechanisms are devised by etching/cutting system elements out of deposited layers of silicon.[2] The typical systems comprise many load bearing and transmission components such as gearing systems, some with diameters of a few micrometers (see Fig. 13.2). Thus, the contact between a pair of meshing teeth is of the order of a few to thousandths of μm^2, with separations of a few nanometers. The advantage of such devices is the thermal stability of materials such as silica. One problem, however, is that excessive wear is noted in load bearing dry contact conjunctions. Ingression of humidity into the contacts is guarded against (for example by inclusion of thin hydrophobic surface layers), but invariably takes place, which can lead to a form of ultra-thin film lubrication, with complex interactions, one of which is the meniscus effect which can lead to stiction and damage to the surfaces.

It is, thus clear that the thickness of effective lubricating films, as indicated by Dowson,[1] has been falling spectacularly throughout the 20th century. The working film has thinned by several orders of magnitude throughout this period as indicated above in the order of nanometers (i.e.

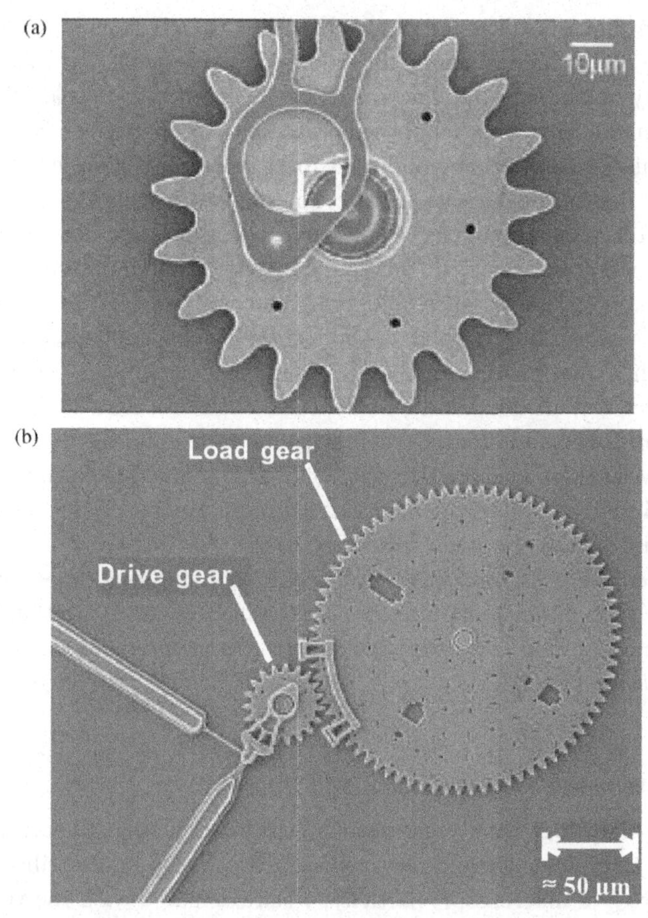

Fig. 13.2. SEM micrographs of a micro-engine operated in vacuum over 2,000,000 cycles (Tanner D. M. *et al.*[2]). (a) The micro-engine. (b) The engine drive gear shown with the load gear.

often of the order of the molecular diameter of the intervening fluid). The mechanism of lubrication in such narrow conjunctions is not well understood. In particular, the forces present in the contact are no longer dominated by the viscous shearing action, which is described according to the bulk rheological properties of the fluid film (Chapters 5, 6 and 10). Therefore, knowledge concerning the characteristics of very thin lubricant films will be indispensable as the basis of key technologies in the future. With such thin films of molecular lengths, the question arises, as to whether the conventional lubrication theory is still applicable to predict the minimum film thickness. The answer in short is no, since in such narrow gaps the continuum nature of fluid action is lost. We recall that this assumed continuity in laminar fluid flow was used in Chapter 6 to develop Reynolds' hydrodynamic flow equation, based on the bulk properties of the lubricant and the use of equilibrium conditions for a representative element of fluid.

Note that in the derivation of Reynolds equation from the more general Navier-Stokes equations (see Cameron[3]), one would neglect the effect of other forces, compared with the dominant viscous action. When the solid surfaces approach a distance of several nanometers we can no longer make such an assumption. In fact, we will observe in this chapter that some forces other than viscous force play a more significant role. Forces such as the attractive **Van der Waals force**, the oscillatory (attraction–repulsion) **solvation force** and **electrostatic repulsion** between the fluid molecules and between these and the adjacent solids also contribute to the formation of a fluid film, as well as the meniscus action in wetting of the two surfaces by building *bridges* between their asperities.

Indeed there is a multiplicity of interacting, and sometimes competing forces in such diminishing gaps, much of which is not as well understood as the more established viscous action. In this chapter, we briefly look at some of these forces.

13.2. Important Forces in Nano-Tribology

In **nano-tribology** we are interested in the forces acting at short distances (e.g. the intermolecular interactions), which can also be described as long range forces of macroscopic particles, when compared with the very short range forces in electrodynamics (between electrons and atomic nuclei) or at the nuclei of atoms or indeed the bonding forces such as those in covalent bonds. In particular, we are interested in the forces that act between fluid molecules and between these and the constraining surfaces (surfaces of contacting bodies), as well as surface forces between the adjacent contacting bodies themselves. These forces include Van der Waals forces that always exist, irrespective of the nature of molecules, be they polar or non-polar (as we shall describe in due course) and **electrostatic forces**. We will observe that the former are attractive in nature at long range (note that long range refers to a few nanometers) and can become repulsive when molecules get very close to each other. The latter (the electrostatic forces) are always repulsive. These two forces dominate in gaps of a few to several nanometer, but their effects are superseded by surface forces (also referred to as structural forces or **solvation force**) at a closer range, when the gap between the adjacent solids is of the order of several molecular diameters of the intervening fluid (in our case lubricants). The solvation force is one of the least fundamentally understood forces in very narrow conjunctions, and results from the constraining action of closely adhering solids, causing layering and re-ordering of the lubricant molecular layers.

There is also, of course, the **meniscus force**, which is responsible for the wetting action of the solid surfaces. This is an attractive force, which tends to bring the surfaces together by forming fluidic bridges between the surface asperities and causes a negative pressure (suction). Now we can look closer at some of these forces.

13.3. Van der Waals Forces

When two charges q^+ and q^- are separated by a small distance, r, they constitute an **electric dipole**, which is usually represented by a vector as shown in Fig. 13.3.

The magnitude of this vector, referred to as the dipole moment is qr, and the usual symbol used for this in physical chemistry is μ, which is rather unfortunate when studying tribology, where the symbol μ always denotes coefficient of friction. Thus, in this book we represent the **dipole moment** by the symbol: ξ. Its direction is from the negative charge towards the positive one. The unit of ξ is Cm (Coulomb meter), but the value of ξ is usually very small; of the order $10^{-30} - 10^{-29}$ cm, thus they are more conveniently expressed in the unit: Debeye (D), after Peter Debeye who was the pioneering scientist in the study of polarizability of molecules. Note that $1D = 3.336 \times 10^{-30}$ cm.

The electric properties of charges in the form of the dipole moment extends to ions and molecules. When a pair of such molecules, referred to as **polar molecules** are close to each other their dipole moments interact. The nature of interactions of their dipole moments depends on their relative orientation. Since dipole moments are represented by vectors, it is possible to resolve their contributions into various components. Those contributions

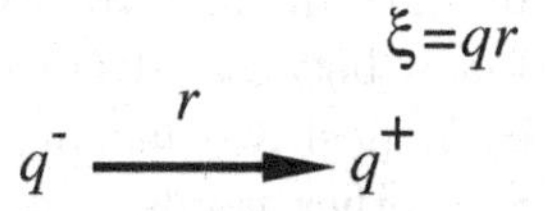

$$\xi = qr$$

$$q^- \xrightarrow{\quad r \quad} q^+$$

Fig. 13.3. An electric dipole.

in a head-to-head direction are repulsive, whilst those in a head-to-tail orientation are attractive. The latter correspond to a lower energy level, and thus on average are more prevalent in a fluidic medium, and give rise to the Van der Waals forces.

Now in a given medium, with a certain molecular structure and disposition there will be an assortment of molecular interactions due to the polarization effects. The Van der Waals forces arising from these can be obtained using the Lifshitz theory[4] for the interaction energy between two surfaces separated by a very thin fluid film. Taking into account the interactions between the atoms of the two surfaces and these with the molecules of the intervening fluid film the contact pressure distribution arising from this interaction energy can be obtained, depending on the geometry of the contact. Israelachvili[5] provides analysis for Van der Waals interaction energy per unit area for various contact geometries. Since the contact area is usually very small in nano-tribological contacts such as those examples cited in Sec. 13.1, we can assume the interaction energy to act between two flat parallel planes. For this case, the **Van der Waals pressure** between the contacting flat planes is obtained as:

$$p_{vdw} = -\frac{A_h}{6\pi h^3}, \tag{13.1}$$

where A_h is the **Hamaker constant**, usually in the range: $A_h = 10^{-20} - 10^{-19}$ J. Note that the Van der Waal's pressure is negative (i.e. attractive as in a suction effect).

13.3.1. *Concluding remarks on Van der Waals forces*

The importance of Van der Waals forces cannot be overstated as they act between atoms, particles and molecules and are particularly important in the interaction range 0.2–10 nm. In other words, unless dominated by other forces, such as solvation or meniscus action, as discussed later, they play an important role in the rapidly emerging new challenges in tribology of very narrow conjunctions, such as in the MEMS devices or data storage media (see Sec. 13.1). We have provided a very brief introduction to this important subject, and can make the following concluding remarks:

The Van der Waals forces between identical solids in a medium are always attractive (i.e. A_h is positive, making the generated pressures in Eq. (13.1) negative or in other words a suction effect). With water molecules being highly polar and MEMS devices mainly manufactured in all contacting parts from silica, presence of moisture can lead to stiction of parts, which also gives rise to adhesion through meniscus action.

The Van der Waals forces between different bodies in a medium can be attractive or repulsive. This means that care should be taken in design of contacts through use of appropriate combination of materials.

The Van der Waals forces between any pair of condensed bodies in vacuum or air are always attractive. This means that for applications such as read/write head slider contacts with a disk platter in advanced magnetic storage devices, highlighted in Sec. 13.1, the attractive action of Van der Waal's pressure in the air gap limits the gap size. This has the repercussion of lower storage density than otherwise would be expected if the system was to be immersed in a suitable fluid, in which case the effect of other forces such as the meniscus force and solvation should be carefully considered.

Finally, from Eq. (13.1) it is clear that as the film thickness decreases the Van der Waals pressure increases dramatically. In most cases this will lead to the stiction of surfaces and other forces should be encouraged in order to provide the balancing repulsive action. One such force is electrostatic repulsion and the other is the solvation force due to rapid oscillatory variation in fluid density near the contacting surfaces due to their constraining effects as barriers.

13.4. Electrostatic Forces

There are only a few situations in which Van der Waals forces alone would account for the intermolecular interactions. These are confined to interactions in vacuum or for non-polar films formed on inert surfaces. In almost all other systems the interplay between Van der Waals and electrostatic forces take place. This allows them to function properly. If only Van der Waals forces were operating, then because of their attractive nature in a medium, all dissolved particles and bodies would coagulate and form a solid mass. However, liquid media of high dielectric constant like water are polar and particles in them are prevented from adhering to each other and thus suspend. This is caused by the action of repulsive forces such as electrostatics and at shorter range by solvation. The acting short range repulsive potential energy dominates the interplay with the Van der Waals interactions, and is a function of a high power of distance between molecules: $w = \frac{C}{r^{12}}$, where C is a constant.

The interplay between electrostatic repulsion and attractive Van der Waals contribution is long recognised, such that the potential was originally

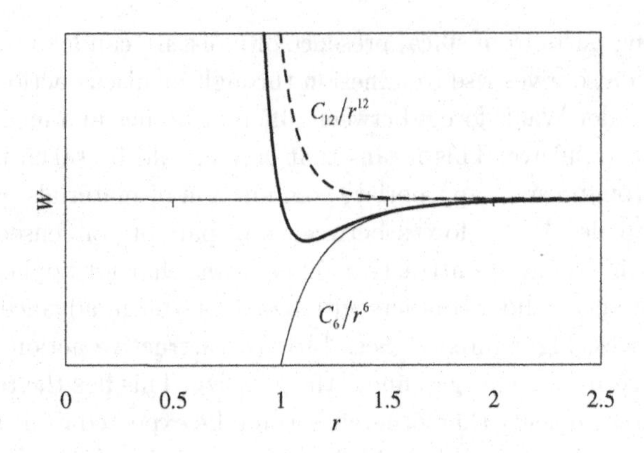

Fig. 13.4. Leonard-Jones potential.

given as an addition to these competing forces in the form called the two-parameter **Leonard-Jones potential**: $w = 4\varepsilon[(\frac{a}{r})^{12} - (\frac{a}{r})^6]$ (see Fig. 13.4), in which the overall potential is dominated by electrostatic potential at short range, below the separation $r = 2^{1/6}a$, where: $w = 0$. There is, however, plenty of evidence that the repulsive contribution in the form $1/r^{12}$ is rather poor, and that a gradual form $e^{-r/a}$ is more appropriate.

We are, therefore, suggesting that the electrostatic potential should have a negative exponential form. It is important to understand why, as we will see later that this forms the basis for repulsive action between two charged contacting surfaces in the presence of a polar liquid (as a lubricant).

Let us commence with the electrostatic force. The work done by this force is obtained as: $\int_{\infty}^{r} F dr = -\left(\frac{q_1 q_2}{4\pi\varepsilon_r\varepsilon_0}\right)\int_{\infty}^{r}\frac{1}{r^2}dr = \left(\frac{q_1 q_2}{4\pi\varepsilon_r\varepsilon_0}\right)\frac{1}{r}$. This can be represented as an increase in potential energy of the system: $q_1 w$, due to the potential arising from the charge q_2 as: $w = \frac{q_2}{4\pi\varepsilon_r\varepsilon_0}\frac{1}{r}$. Note that for solvents the value of permittivity ε_r is usually quite high (i.e. the interactions are much weaker than in vacuum, e.g. $\varepsilon_r = 78.5$ for water and $\varepsilon_r = 24.3$ for Ethanol). This means that ions interact with each other in a weak manner. Nevertheless, these weak interactions are crucial, otherwise the attractive Van der Waals forces would stick all bodies and particles together when in a solvent, such as water, which would have serious implications for example for human anatomy, made of at least 70% water. Therefore, in a solvent (such as water) the ionisation or dissociation of surface groups occur (this amounts to dissociation of protons from the surfaces, for details of the process see specialist texts on physical chemistry such as Atkins[6] or Israelachvili[5]). This process leaves the surfaces negatively charged. The solvent now embodies diffused positively charged counterions, which balance the attraction of charged surfaces as they approach each other. In nano-tribology electrostatic repulsion plays an important role between closely adhering charged surfaces, when separated by a polar liquid, which has a reasonably high dielectric constant. The charging of a surface in a liquid medium can take place by either ionisation, which results in a negatively charged surface, or by adsorption (i.e. binding) of ions from the solution to result in a positively charged surface.

The form of potential described so far does not include the effect of this ionic atmosphere. First, we can amend this potential to take into account the ionic atmosphere, and second, extend it for a distribution of charged particles/ions. If we consider ions as point-like entities, their potential decays more rapidly than that we have suggested thus far. The ions are then regarded as shielded, and the appropriate potential to be considered is termed: **shielded Coulomb potential**, where the term $\frac{1}{r}$ is replaced by $\frac{1}{r}e^{-r/\kappa}$, where κ is the **shielding length**, indicating the decaying (the damping effect) of the actual potential from its Coulombic value. Therefore, the shielded potential for an ion, i in a medium of **permittivity** or **dielectric constant** ε may be written as:

$$w = \left(\frac{z_i e}{4\pi\varepsilon_r\varepsilon_0}\frac{1}{r}\right)e^{-rk}, \tag{13.2}$$

where $q_i = z_i e$ is the charge of ion i, e is the electronic charge and z_i is valency. κ^{-1} is the shielding length (i.e. the **Debeÿe length**) and has the unit of nm (nanometers).

As already indicated in a medium, there would be, for example, a distribution of counterions, thus a charge density distribution must be assumed. If under equilibrium conditions we assume a uniform potential. This leads to the **Boltzmann distribution** of counterions at any given point within the solvent, then:

$$\rho = \rho_\infty e^{-z_i ew/KT}. \tag{13.3}$$

ρ_∞ is the concentration of electrolyte in the bulk lubricant/liquid, and can be stated as:

$$\rho_\infty = [X_i]N_A \times 10^3,$$

where $[X_i]$ is the concentration in mols per dm^3 of the ion X_i, $N_A = 6.02205 \times 10^{23}$ per mol is the Avogadro's number. Thus, ρ_∞ has the units of mol/m^3.

We clearly need to obtain w to quantify the charge distribution from (13.2). The **electrical potential** w arising from a charge distribution ρ (rather than a single ion, given by Eq. (13.2)) must be obtained by solution of **Poisson's equation**:

$$\nabla^2 w = -\frac{z_i e \rho}{\varepsilon_r \varepsilon_0}, \tag{13.4}$$

where,

$$\nabla^2 \equiv \left(\frac{\partial^2}{\partial x^2} + \frac{\partial^2}{\partial y^2} + \frac{\partial^2}{\partial z^2}\right) \text{ is the } \textit{Laplacian.}$$

Thus, solutions for w and repulsive **electrostatic pressure** can be obtained, using (13.4) for any charged distribution such as that given in (13.3). Israelachvili[5] derived the repulsive electrostatic pressure for a pair of flat contacting surfaces with a film thickness, h as:

$$p_{electro} = 64KT\rho_\infty \chi^2 e^{-\kappa h}, \tag{13.5}$$

where,

$$\chi = \tanh\left(\frac{zew}{4KT}\right).$$

13.4.1. *An example*

A simple example is to consider the contact of two smooth charged flat surfaces in electrolytic solutions, such as a salt (e.g. NaCl or $CaCl_2$) in water. The generated pressure in the conjunction is dominated by the competing actions of Van der Waals and electrostatic interactions between the two surfaces with this intervening fluid. More complex interactions of this kind can occur with base oils with electrolytic additives. The electrolyte ions (i.e. the dissociated inorganic salts) give rise to an electrostatic potential between the surfaces and generate a repulsive pressure, which competes with Van der Waal's force tending to close the gap. This is a simple example of nano-tribological conjunctions, which would normally be much more complicated, such as drug mixtures in inhalers, ingressing into the contact between the actuating plunger and the rubber o-ring seal, which is used to inhibit leakage of the mixture from the canister. Such drugs often contain various salts in water or ethanol.

Now to obtain the electrostatic pressure in our simple example, we need to use Eq. (13.5). For a simple electrolytic solution of NaCl in water, we need to determine the physical parameters required for use in Eq. (13.5), such as the valency, **Debeye length** and density. These depend on the concentration of the salt ions in the solution. The following procedure is used to obtain these physical data: Note that from the above relationships we can state that:

$$\kappa = \sqrt{\sum_i \frac{\rho_{\infty i} e^2 z_i^2}{\varepsilon_r \varepsilon_0 KT}} = \sqrt{\sum_i \frac{e^2 z_i^2 [X_i] N_A \times 10^3}{\varepsilon_r \varepsilon_0 KT}} = \sqrt{\frac{10^3 e^2 N_A}{\varepsilon_r \varepsilon_0 KT} \sum_i z_i^2 [X_i]}$$

$$= \sqrt{\frac{\varepsilon_r \varepsilon_0 KT}{10^3 e^2 N_A 10^3}} \frac{1}{\sqrt{\sum_i z_i^2 [X_i]}},$$

where **Boltzmann constant** is $K = 1.38066 \times 10^{-23}\,\text{J}$, **vacuum permittivity** is: $\varepsilon_0 = 8.85419 \times 10^{-12}\,\text{C}^2\,\text{J}^{-1}\,\text{m}^{-1}$, relative permittivity of water is: $\varepsilon_r = 78.5$, and electronic charge is: $e = 1.60219 \times 10^{-19}\,\text{C}$.

For NaCl, we have the ions: $[\text{Na}^+]_\infty + [\text{Cl}^-]_\infty$, both with a valency: $z_i = 1$.

For our example let us assume a negative surface charge density of $\phi = -0.2$ Coulombs and temperature of 25° C, $T = 298°$ K. Now we can determine the Debeye length for NaCl as follows:

$$\kappa^{-1} = \sqrt{\frac{\varepsilon_r \varepsilon_0 KT}{10^3 e^2 N_A}} \frac{1}{\sqrt{z^2 [\text{Na}^+]_\infty + z^2 [\text{Cl}^-]_\infty}} = \frac{1}{\sqrt{2}} \sqrt{\frac{\varepsilon_r \varepsilon_0 KT}{10^3 z^2 e^2 N_A}} \frac{1}{\sqrt{[\text{NaCl}]_\infty}}.$$

The term $[\text{NaCl}]_\infty$ represents the concentration of the salt, which is an input in mols per dm^3 as previously described for the generic case: $[X_i]$. Let the concentration be: $10^{-3}\,\text{mols/dm}^3$. This makes the Debeye length: $\kappa^{-1} = 9.6\,\text{nm}$.

Now we need to obtain the potential w, thus: χ and hence the electrostatic pressure from Eq. (13.5). You will note that we need to solve the Poisson's Eq. (13.4) to obtain a value for w. An approximate solution can be found, using a simplified form of Grahame's equation for small potentials (typically less than 25 mv (milli-volts)). We are not going to describe the Grahame's equation here (those interested should see Israelachvili[5] or Atkins).[6] This simplified form of **Grahame's equation** is:

$$\phi = 0.117 \sinh(w/51.4)\{[\text{NaCl}]\}^{1/2}.$$

Clearly, we can find w from this equation. For the data given thus far for this example: $w = -240$ Coulomb. Now if we assume a gap (i.e. a film thickness)

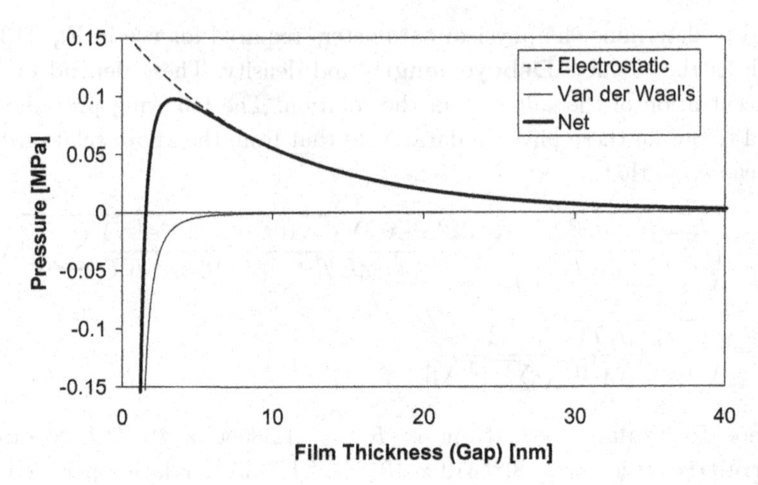

Fig. 13.5. Variation of electrostatic and Van der Waals pressure in the gap, filled with salty water.

we can obtain the corresponding pressure from Eq. (13.5). Figure 13.5 shows the variation of pressure with various assumed gaps. On the same figure the Van der Waals pressure is also plotted, with a **Hamaker constant** of 10^{-20} J. We can observe that as the gap is reduced the electrostatic pressure is increased, keeping the surfaces apart by repulsion. The Van der Waals pressure is increased at the same time, promoting convergence of surfaces. The net pressure is shown in the figure.

Note that the variation of pressure components (electrostatic and Van der Waals) in the contact conform to the behavior described by the Leonard-Jones potential. The effect of both phenomena become significant as the gap is reduced, and would account for the main pressure generating mechanisms as the gap becomes of the order of few to several nanometers. An equilibrium condition is reached around a couple of nanometers, being nearly equivalent to two molecular diameters of the solution. Thereafter, the Van der Waals pressure exceeds the electrostatic repulsion and contacting surfaces would tend to stick together. If the contiguous solid surfaces are considered to be perfectly smooth (molecularly smooth, of the order of tenths of a nanometer), then no menisci are likely to form as explained later, and the Van der Waals pressure would account for contact adhesion. For rough hydrophilic surfaces, the meniscus force would account for stiction (see later).

An important point to note is that the electrostatic pressure provides the load carrying capacity. If we were to assume a line contact, the hydrodynamic load carrying capacity is given as[7]:

$$W = \underbrace{\frac{2bU\eta_0 R}{h}}_{\text{pure rolling}} - \underbrace{\frac{3\pi b\eta_0 R^{3/2}}{\sqrt{2}h^{3/2}}\frac{\partial h}{\partial t}}_{\text{pure squeeze}}. \tag{13.6}$$

With no relative velocity of surfaces, $U = 0$, the only contribution will be due to squeeze, when $\frac{\partial h}{\partial t} < 0$ (mutually approaching of the surfaces). Large electrostatic pressures would usually inhibit this. Thus, for such nano-conjunctions there is no significant contribution from hydrodynamics. With such smooth surfaces as described in our simple example above, the density of the mixture near the surfaces undergoes oscillatory variations due to the constraining effect of the solid boundaries (surfaces in very close separation). This leads to another pressure generating effect, known as solvation, which is ignored in our example above (to be described in Sec. 13.6).

13.5. Meniscus/Surface Tension Force

(a) *Physics of surface tension*

The molecules of a liquid have intermolecular bonds or **cohesive forces** that hold them together. When the surface of the liquid comes into contact with another medium such as air or other material an interface is formed. We can easily observe such an interface, which is different in appearance to the bulk of the liquid. Often it looks like a *skin*, such as the top layer of water in a glass. Some energy is required to form this **interfacial layer**. This energy is called the **surface energy**, and in the case of liquids is referred to as surface tension. This is defined as the energy required to form a unit area of the *new surface* (i.e. the interface). When the liquid meets another substance, there is attraction between its molecules and that of this substance. For example, there is attraction between the molecules at the free surface of water and those of air. In general these are referred to as **adhesive forces**. These adhesive forces compete with the cohesive forces between the molecules of water on its free surface.

Liquids with weak intermolecular bonds or cohesive forces and strong attraction to another material tend to spread over the surface of that material. Let us consider water as an example. We can see that water spreads over certain surfaces. Such materials or substances are termed **hydrophilic**, literally meaning they love water. On the other hand, the molecules or atoms of some others do not have strong adhesive forces with water molecules. As a result water does not adhere easily to them. These substances are known as **hydrophobic** (i.e. this literally means that they have a phobia or fear of water). Nature has used hydrophilic or hydrophobic substances to its advantage. Duck feathers are oily and thus hydrophobic. This helps the ducks, not only to float easily on water (low adhesive force), but also to fly off the surface seemingly effortlessly. Mosquitoes with large feet (relative to the size of their bodies) generate large adhesive force with surface of ponds, which more than equates their weight, enabling them to walk on water!

Everyday experiences can remind us of the role of surface tension. When a wet glass is inverted and positioned on a rough draining board and left for a while, a film is formed between the glass lip and the board, and we sense that a force is required to separate the two. This is the **surface tension force**. If a surfactant (a substance reducing the surface tension of a liquid) is dissolved in water, our effort to separate the two surfaces is much reduced. **Teflon** is used as a non-stick substance in frying pans, because of its surface morphology. We can see that olive oil floats easily in the pan. Certain surfaces are even more hydrophobic for given liquids, to the extent that the liquid film breaks up to drop shapes, which is the natural shape with least surface area, indicating occasions that surface tension becomes the dominant force. We can see this when water meets an oily surface, such as the duck feathers.

Therefore, the properties: *hydrophobia* and *hydrophilia* extend to liquids other than water. This fact is not only exploited in Nature, but also in many man-made processes. For example, in ink-jet printing the droplets of paint are expected to cling to the surface of the paper after impact. Droplet dynamics is discussed later (see next section). It is, therefore, clear that surface material and topography as well as liquid composition play crucial roles in surface tension. These are critical factors in oil-surface combination in tribology.

(b) *Determining surface tension*

Lubricant is supplied to load bearing surfaces, for example in vehicle engines, through supply orifices by jet flows or sprays with the dual purpose of cooling and lubricating them. When a spherical droplet of oil falls onto a surface the inertial force tends to deform it rather like a pancake to flow along the surface. The inertial force required to deform the spherical drop is opposed by the surface tension force, which strives to maintain the drop's spherical shape. Upon impact, there is the additional viscous force, which resists the flow of the lubricant along the surface. Therefore, the inertial force must overcome surface tension to deform the droplet, and furthermore the viscous force of the fluid to spread it on the target surface.

The inertial force is a function of droplet mass and its acceleration. The viscous force is a function of lubricant rheology, kinematic conditions of surfaces and geometry of the conjunction (hydrodynamics, see Chapters 5 and 6).

To understand surface tension, we consider a simple experiment, shown in Fig. 13.6, in which a capillary tube of diameter D_c is filled with a fluid of density ρ. We will observe that the fluid droplet of diameter D_d forming at the end of the tube grows in size, before finally falling off it. The inertial force here is simply the weight of the droplet: $\rho g V$, V being the volume of the droplet. It is clear that a force Facts between the droplet and the rim of the capillary tube, which resists its fall. This force (surface tension force) acts along the contact line between the droplet tail and the tube rim as shown in the figure. This line of action makes an angle, θ with the tube wall referred to as the **contact angle**. We will describe this later. Note that we have ignored the effect of viscous force in this simple static balance analysis because conditions are considered at the equilibrium position. For now note that prior to the droplet's fall the following equilibrium of forces exist:

$$F\cos\theta = \rho g V. \tag{13.7}$$

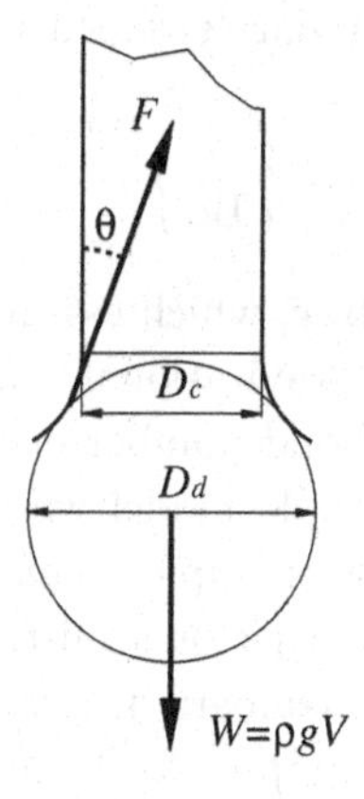

Fig. 13.6. Forces acting on a droplet at the end of a capillary.

The surface tension force, F is given as:

$$F = \pi\gamma D_c, \tag{13.8}$$

where γ is the surface tension in N/m.

Clearly the droplet continually grows in size before the inertial force finally exceeds the surface tension force, and the droplet falls off the tube. This is very similar to the physics of fall of a ripening fruit whose weight finally exceeds the tension holding it to the tree branch. Observation of this phenomenon in part resulted in the discovery of gravitational action by Newton in the 17th century. We can surmise, as in the case of the falling fruit, that the surface tension force tends to a vertical orientation (i.e. $\theta \to 0$, analogus to the tree branch being pulled vertically downwards by the increasing weight of the fruit) for the droplet to assume its maximum diameter. Now from (13.7) and (13.8), using this speculation, it follows that the maximum droplet size can be:

$$V_{\max} = \frac{\pi\gamma D_c}{\rho g}. \tag{13.9}$$

Referring to Fig. 13.6, the droplet may be considered to be nearly spherical in shape with diameter D_d, then its volume is: $V = \frac{\pi D_d^3}{6}$, and replacing for this in Eqs. (13.7) and (13.8), we obtain:

$$D_d = \left\{\frac{6\gamma D_c \cos\theta}{\rho g}\right\}^{1/3}. \tag{13.10}$$

A useful model, described by Middleman[8] is to non-dimensionalize the droplet diameter with the capillary size, thus (for $\theta \to 0, \cos\theta \to 1$):

$$\frac{D_d}{D_c} = \left\{\frac{6\gamma}{D_c^2 \rho g}\right\}^{1/3} = \left\{\frac{6}{\mathrm{Bo}}\right\}^{1/3} = 1.82Bo^{-1/3}, \tag{13.11}$$

where Bo is the **Bond number**, which essentially provides an insight into the interactions between the gravitational action and surface energy effect in fluid flow. Note that the Bond number decreases as the surface tension increases. Essentially, this simple model would enable us to measure the surface tension of a fluid, for example, engine oil. Clearly, it is desirable for the engine oil, sprayed as a jet or a mist, to adhere to the surfaces to form menisci for contacting members in relative motion, and also to flow and aid convection cooling (Chapter 6). If we measure the diameter of the droplet in a slow dripping experiment, from Eq. (13.11) we can obtain an estimate of the surface tension of the fluid. In such a simple experiment an estimate of the diameter of the spherical droplets are made by an average volume of so many falling into a container. Middleman[8] reports such an experiment, and it turns out that the simple model described here predicts drop volumes somewhat smaller than that observed in practice. This should be anticipated as careful observations show that some small proportion of each droplet continues to cling to the capillary post separation. In fact, the relationship is found through careful experimentation to be[8]:

$$\frac{D_d}{D_c} = 1.6\mathrm{Bo}^{-1/3}, \tag{13.12}$$

which gives

$$\gamma = 0.244\frac{\rho g}{D_c}D_d^3. \tag{13.13}$$

The difference between our simple model and the experimental measurements is because we have not considered the flow rate of the fluid through the tube, using the Bond number: $\mathrm{Bo} = \frac{D_c^2 \rho g}{\gamma}$. This is mainly because our simple model is static in nature, as already described. Dripping through a capillary, however, is a dynamic process, where the droplet falls from the tube when in contact with its outer diameter, whilst the flow takes place within the inner diameter of the capillary. This means that we should consider a more sophisticated model for droplet dynamics.

Fortunately, a substantial volume of work exists for droplet dynamics, mainly due to the efforts of Lord Rayleigh in the 19th century and more recently by Subrahmanyan Chandrasekhar. To fully appreciate the problem, consider a droplet impacting a surface. There are three forces involved. The inertial force is determined by the kinetic energy of the droplet. The surface tension force is governed by the energy required to deform the spherical droplet to adhere to the surface. The viscous force is due to the resistance of the droplet to flow (determined by the fluid viscosity, which yields the dissipation energy). The physics of the problem can be described adequately by two important non-dimensional numbers: the **Reynolds number** and the **Weber number**.

The Reynolds number is effectively the ratio of the inertial force to the viscous force, whilst the Weber number is that of the inertial force to the surface tension force. These are:

$$\mathrm{Re} = \frac{\rho U L}{\eta}, \tag{13.14}$$

$$\mathrm{We} = \frac{\rho U^2 L}{\gamma}, \tag{13.15}$$

where U and L are characteristic velocity and length of the system under consideration. For example, for the capillary tube problem, described above, U is the flow velocity through the inside diameter of the tube (say D_{ci}), and L is the outside diameter of the tube (D_{co}).

In lubrication problems the Reynolds number is usually quite low, indicating the dominance of viscous force. You should recall that one of Reynolds' assumptions (Chapter 6) was to ignore the effect of body and inertial forces. Thus, in lubrication typically is $\mathrm{Re} < 1$. Also remember that the lubricant is entrained into the contact by relative motion of contacting surfaces. However, lubricant is often supplied to the contacting surfaces by spray jets. This is often achieved by jet flow, in which droplets of lubricant must possess sufficient inertial force, not only to deform the *spherical* shape of the droplets, but also to overcome the viscous resistance to flow on the surface. This implies that: $\mathrm{Re} > 1$ and $\mathrm{We} > 1$.

It is clear that low Reynolds numbers represent highly viscous fluids, droplets of which dissipate all their kinetic energy upon impact and inhibit rebound. However, high viscosity fluids are difficult to atomize and pump, as well as cause excessive viscous dissipation, viscous heating and churning at higher temperatures. Most viscous lubricants with low Reynolds number also have low values of Weber number. This means that for the capillary problem described above the inertial force is too low to separate the drop from the capillary. The drop size becomes almost independent of the designed drip rate. Therefore, an additional problem with viscous fluids is detachment from the spray nozzles as well as larger droplets, which can rebound from the target surface. This is not ideal for lubrication. To keep a low Reynolds number, but control the drop size, we need to reduce the surface tension of the lubricant. For most spray jets a Weber number: $\mathrm{We} \leq 3$ translates to small droplet sizes which can become a part of an almost continuous jet, when: $\mathrm{We} > 3$.

This means that the ratio of the Weber number to the Reynolds number is quite critical in droplet dynamics. This ratio is called the **Capillary number**:

$$\mathrm{Ca} = \frac{\mathrm{We}}{\mathrm{Re}}. \tag{13.16}$$

By controlling this number we should be able to ideally design a spray system that possesses sufficient droplet inertia to maintain a steady dripping rate of lubricant/coolant onto a target surface, whilst reducing recoil and rebound of droplets from it. If the coolant used is water, the addition of surfactants to it reduces surface tension, as already mentioned above. One part of a typical surfactant molecule is attracted to water, whilst another part is repelled from it and forms an adhesive bond with air. Therefore, surfactants position themselves at the water-air interface and reduce water's surface tension (sometimes halving it) with the sprayed target surfaces.

For a lubricant spray, a practical solution would be for a combination of Reynolds and Weber numbers to enable ease of atomization, pumping and maintain a steady jet stream. Furthermore, we would wish to achieve small enough droplet sizes to have negligible gravitational effect and less chance of rebound from the target surface. This combination of requirements unfortunately calls for We ≥ 3 and $Re \gg 1$, which means that viscous shearing effect upon impact of the droplet with the target surface would be insufficient to inhibit its rebound for most base lubricants. This is one reason why additives, particularly polymeric molecules and esters are added to the base lubricant. These additives form droplets that dissipate some of the kinetic energy through **elongation viscous action** prior to and after impact.[9] This is a nonlinear viscous behavior of long-chain molecules, other than the usual shear viscous action we have described so far.

(c) *Formation of menisci*

Lubricant is supplied to almost all tribological conjunctions either through supply orifices (journal bearings, pad bearings) or by spray jets (piston ring-pack) or when the contacting members are either fully or partially immersed in a bath of lubricant (transmission systems, viscous dampers). For the lubricant to be entrained into the contact a film of it must be present on the contiguous free surfaces ahead of the contact (Chapter 6, boundary conditions). This free surface-adhered film acts as a reservoir that is drawn on by the surfaces in relative motion to form a contact film, separating them. Thus far, we have described the mechanisms of lubricant film formation, assuming that a plentiful supply of lubricant is available ahead of the contact (at the inlet) as the bodies undergo relative motion (Chapters 6, 7, 8 and 10). We have referred to this condition as a **fully flooded inlet**. Theoretically, the fully flooded condition means that a film of lubricant of sufficient thickness exists, even infinitely far ahead of the contact. In practice, flooded conditions arise when there is a sufficient thickness of film at a finite distance ahead of the contact. Otherwise, the supply of the lubricant falls short of the requirements governed by the speed of entraining motion. These circumstances lead to **starvation**, which has been extensively studied. For example, Birkhoff and Hayes[10] set the limiting case for fully flooded condition for point contact condition (Chapter 10), as:

$$\frac{h_i}{h_0} = \frac{x_i}{a} \approx 11.298, \tag{13.17}$$

where h_i is the inlet film thickness ahead of the Hertzian contact region, h_0 is the flat (parallel) film thickness in the Hertzian contact region, a is the Hertzian contact radius and x_i is the position of the inlet meniscus from the center of the Hertzian circle. According to Birkhoff and Hayes, film thickness ratios below that stated in Eq. (13.17) lead to partially flooded conditions. Clearly, the corresponding inlet pressure at the meniscus is atmospheric. Therefore, instead of Sommerfeld's inlet boundary condition: $p = 0$ (atmospheric) at $x = -\infty$, we should apply a condition such as: $p = 0$ at $x = -x_i$, as in practice fully flooded inlet is rarely achieved.

Note that unlike all the lubrication cases discussed so far we cannot use Reynolds equation to determine the position of the meniscus, the pressure inside it and consequently the height of the film there. Use of Reynolds equation is justified where the lubricant film is constrained by a solid boundary (a solid roof, see Chapter 6). However, a meniscus is formed as the boundary between two immiscible fluids, for example, the lubricant and air. The meniscus effectively forms a bridge between the contacting solids as shown in Fig. 13.7, and curves inwards as the pressure inside it (liquid side) is lower than that on the outside (air at atmospheric pressure).

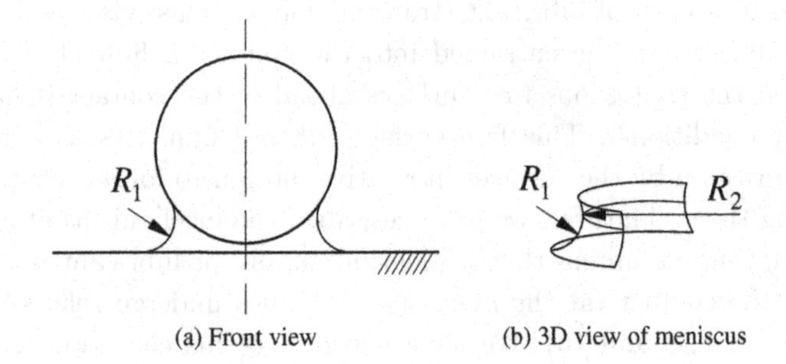

Fig. 13.7. Curved meniscus as a reservoir of lubricant.

Therefore, the negative pressure inside the meniscus creates a suction effect, which tends to close the gap between the contacting solids. The flow prior to the meniscus recirculates. Therefore, the meniscus can be regarded as the zero-reverse flow boundary (i.e. where there is no backward flow anymore). Boundary conditions that take into account the zero-reverse flow have been established by Prandtl and Hopkins. We do not propose to highlight these here. Interested readers should refer to Tipei.[11]

Normally, the hydrodynamic pressure generated by the relative motion of the contacting surfaces (Chapter 6) overcomes the aforementioned suction effect and maintains a gap of lubricant film. However, when the speed of entraining motion is considerably reduced or ceases altogether and the gap is of the order of few nanometers (of the order of molecular diameter of the lubricant), then this suction effect can lead to adhesion. This explains the reason for our earlier example of a force required to remove an inverted glass from a moist rough draining board. This **meniscus force** plays an important role in nano-tribology. To see its importance, we need to derive the equation, which governs the formation of fluid interfaces (menisci). This equation is known as the **Young-Laplace equation**, as it was first derived by them independent of each other.

Recall our example of the top layer of water in a glass, having an appearance different from the bulk of the water volume (rather similar to a formed skin). If we make the diameter of the glass much smaller (e.g. a capillary) and observe very carefully the top layer of water, we note a slightly concave shape, similar to Fig. 13.8(a). Figure 13.8(b) shows a small curvilinear differential element $dq_1 dq_2$ of the surface of the meniscus. The small dimensions dq_1 and dq_2 are considered to be a part of a surface with radii R_1 and R_2 respectively. The co-ordinates q_1 and q_2 are considered to be orthogonal. Small angles $d\theta_1$ and $d\theta_2$ subtend the extents of this small element. Forces act at the edges of this small element to keep its shape and are caused by surface tension, γ as shown in Fig. 13.8(c). These are distributed uniform forces on each edge.

We seek to determine the force balance on this differential element in the normal direction to its surface, indicated by the unit vector $\mathbf{n}$ in Fig. 13.8(c). The figure shows the surface of the element in the q_2 direction. The component of each force γdq_2 in the direction $\mathbf{n}$ is: $\gamma dq_2 \sin\left\{\frac{1}{2}d\theta_1\right\}$. Also note that for small dimensions: $\sin\left\{\frac{1}{2}d\theta_1\right\} = \frac{1}{2}d\theta_1$ and $d\theta_1 = \frac{1}{R_1}dq_1$. Therefore, replacing these relationships in the force equation and sum up

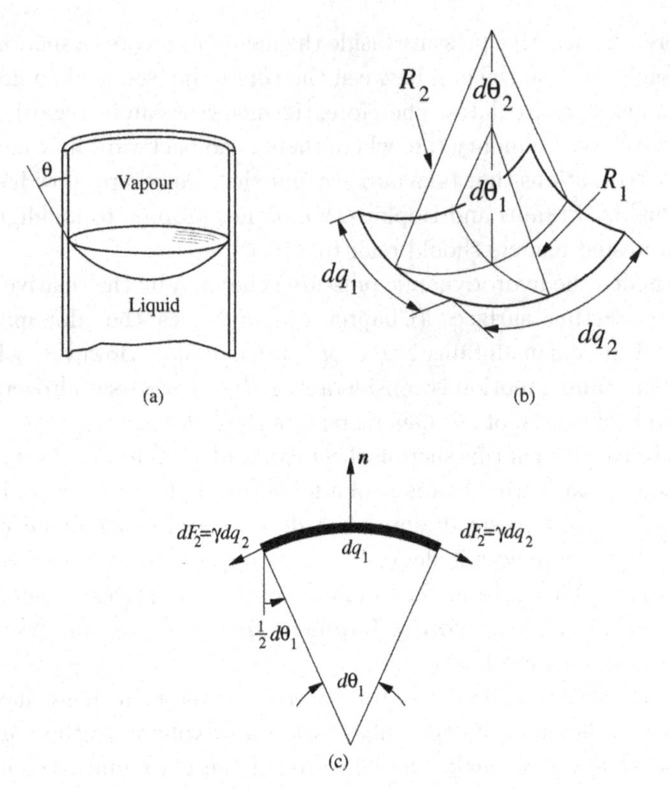

Fig. 13.8. Formation of a meniscus. (a) Concave meniscus in a capillary. (b) A small differential element of the interface. (c) Tensile forces on the small element.

for both forces, thus:

$$dF_2 = \frac{\gamma}{R_1} dq_1 dq_2. \tag{13.18}$$

Similarly, the forces in the q_1 direction also contribute in the normal direction to the surface of the element, **n**. Thus: $dF_1 = \frac{\gamma}{R_2} dq_1 dq_2$.

The net force due to interfacial tension is the addition of these components as: $dF = \gamma\left(\frac{1}{R_1} + \frac{1}{R_2}\right) dq_1 dq_2$.

Assuming static equilibrium condition, the interface should be upheld by the net pressure: $\Delta p = p_o - p_i$, where the p_o and p_i are the pressures on the outside and inside surfaces of the meniscus respectively. Thus, for equilibrium:

$$\gamma\left(\frac{1}{R_1} + \frac{1}{R_2}\right) dq_1 dq_2 + \Delta P dq_1 dq_2 = 0,$$

or:

$$\left\{p_o - p_i + \gamma\left(\frac{1}{R_1} + \frac{1}{R_2}\right)\right\} dq_1 dq_2 = 0. \tag{13.19}$$

This result is clearly true for any typical differential area, dq_1dq_2 on the surface of the meniscus. Since: $dq_1dq_2 \neq 0$, then the term in the bracket should vanish for the interface to be static, thus:

$$p_o - p_i + \gamma\left(\frac{1}{R_1} + \frac{1}{R_2}\right) = 0. \tag{13.20}$$

This is a fundamental equation, describing the behavior of all liquid interfaces, and is known as the **Young-Laplace equation**. Since, the meniscus (such as our example in Fig. 13.8(a)) is concave, the pressure outside (on the air side) is greater than that on the inside (the liquid side). When the pressure on the outside is considered to be atmospheric, the pressure inside the meniscus is negative, causing a suction effect, as we have already described above.

(d) *Attractive meniscus force*

Now consider the meniscus geometry formed by a liquid droplet underneath a sphere as shown in Fig. 3.9. Let us assume that the interfacial radius, $R_2 \gg R_1$, and γ_{lv} be the surface tension at the liquid-vapor interface (note that there are three such interfaces as we will describe later), then using the Young-Laplace equation, for $p_o = 0$ (atmospheric on the vapor (air)

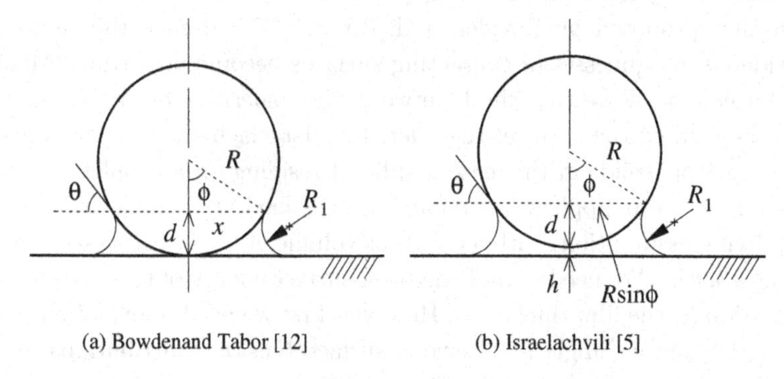

Fig. 13.9. Meniscus geometry of a drop trapped between a sphere and a flat plane.

side):

$$p_i = -\frac{\gamma_{lv}}{R_1}. \tag{13.21}$$

Using the geometry shown in Fig. 13.9(a), it is clear that the Laplace pressure acts on the surface area of the sphere immersed into the droplet, or: $\pi x^2 \approx 2\pi R d$. This pressure gives rise to the **meniscus force**:

$$F_m = -2\pi R d \frac{\gamma_{lv}}{R_1}. \tag{13.22}$$

For a small angle φ (see Fig. 13.9): $d = 2R_1 \cos\theta$, where θ is the contact angle. Thus,

$$F_m = -4\pi R \gamma_{lv} \cos\theta. \tag{13.23}$$

We observe the following:

The meniscus force is attractive (note the negative sign) and tends to bring the contacting surfaces together.

As the contact angle reduces, the cosine of the angle tends to unity. This also corresponds to very thin films, where contact angle is usually a few degrees only. Then, the meniscus force becomes large for very thin films. In fact, Bhushan[13] has shown that when the film thickness is of the order of two molecular diameters of the intervening fluid adhesion takes place.

The derivation given above follows that of Bowden and Tabor,[12] and as can be seen from Fig. 13.9(a) proposes a meniscus geometry that does not include a film of lubricant existing in the contact. It also does not take into account the surface tension effect at other interfaces; between the liquid and solid, and solid and vapor.

To include the effect of any separation (usually an ultra-thin film of molecular dimension) we need to consider a more thorough analysis than that proposed by Bowden and Tabor.[12] When such thin films are considered, roughnesses of contacting surfaces become important. Menisci are formed by a wetting fluid between the asperities of the contacting surfaces, pulling the surfaces together. Like Israelachvili,[5] we can consider a macroscopic solid (in this case a sphere) residing on a droplet of liquid, which is in turn applied to a flat surface, having a single asperity (a simplifying assumption) with a constant volume of fluid. This case is shown in Fig. 13.9(b). We need to find the total surface energy of this system with separation h (the film thickness). However, first we need to explain in more detail the contact angle and various surface tension contributions at the solid-liquid, solid-vapor and liquid-vapor interfaces.

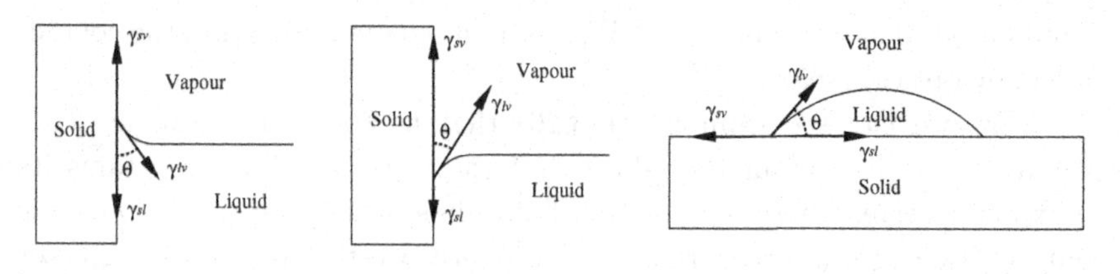

(a) Positive meniscus (b) Negative meniscus (c) Interfacial forces on a drop on a surface

Fig. 13.10. Contact angle in solid-liquid interfaces.

The **contact angle** is defined as the angle formed by the solid-liquid or liquid-vapor interface, measured through the liquid, as shown in Fig. 13.10. Note that we have shown the surface tensions at these interfaces to be different. These are energies per unit area (force per unit length of the contact line), denoted as γ_{sl} for solid-liquid interface, γ_{lv} for the liquid-vapor interface, and γ_{sv} for the solid-vapor interface. These surface tensions act tangentially to their respective contact lines. Now refer to Fig. 13.10(c) to find a static force balance (per unit length) in the horizontal direction as:

$$\gamma_{sl} + \gamma_{lv} \cos\theta = \gamma_{sv}. \tag{13.24}$$

This equation is often referred to as the **Young's law**. Since the vapor phase is regarded as not having any significant influence, the surface tension γ_{lv} may be considered to be a property of the liquid itself.

For solid-liquid-vapor interfaces (as in Fig. 13.10), the total free energy change is:

$$\gamma_{sl} = \gamma_{sv} + \gamma_{lv} - W_{svl}, \tag{13.25}$$

where W_{svl} is the work of adhesion required to separate unit areas of the two media of solid and liquid in a vapor medium from their contact to infinity (remote from the contact). In fact, this accounts for our effort to separate the inverted glass from the wet draining board per unit length of its perimeter. Therefore, it can be carefully measured.

Equation (13.25) is known as the **Dupree equation**. Combining Eqs. (13.24) and (13.25) after minor re-arrangement yields:

$$W_{svl} = \gamma_{lv}(1 + \cos\theta). \tag{13.26}$$

This equation is referred to as the **Young-Dupree equation**, and is normally used to calculate the value of surface tension, γ_{lv}. To do so, we would need to measure our effort W_{svl} as pointed out above, and the contact angle by optical means.

Now we can see from Eq. (13.26) that as the contact angle, $\theta \rightarrow 0, \cos\theta \rightarrow 1$ (its maximum value), and the work of adhesion attains its maximum value. Therefore, for thin fluid films, where the contact angle is only a few degrees, the surfaces are properly wetted and have a greater tendency to stick together. This can be a major problem in ultra-thin contacts such as **magnetic storage devices** (Fig. 13.1) and MEMS (Fig. 13.2). Therefore, good surface wetting requires the contact angle to be much less than 90°. For $\theta < 90°$, γ_{sl} and γ_{lv} are in the same direction (see Fig. 13.10(a)). The meniscus geometry in this case is defined as positive and the liquid wets the surface. From Eq. (13.26) we can observe that as the contact angle increases the degree of hydrophilicity decreases (i.e. the wetting action becomes less effective). Since the work of adhesion is also a function of γ_{lv}, it is clear that it also depends on the type of liquid (in our case the lubricant composition) and the solid surface topography (in fact roughness amplitude and geometry).

When γ_{sl} and γ_{lv} are in the opposite direction (Fig. 13.10(b)), then the contact angle, $\theta > 90°$ and the meniscus geometry is defined as negative. Note that in this case, using Eq. (13.26) indicates a decreasing value for W_{svl} as the contact angle increases. The liquid does not properly wet the solid surface (**hydrophobicity**). When: $\theta \rightarrow 180°$, $W_{svl} \rightarrow 0$, the liquid fails completely to adhere to the solid surface, forming droplets that fall off it. There are very few such completely hydrophobic cases in Nature. One example is lotus leaves. Close examination of the surface of a lotus leaf through a microscope shows a topography strewn by sharp *hills* with breadths of around 10 μm, each of which appears to be covered by inserted tubes of 1 μm diameter. It is clear that droplets of water from a spray cannot adhere easily to such a surface and can be easily dislodged. This physical morphology, to a lesser extent, is also apparent on many other leaf types, and is due to the surface waxes, containing large hydrocarbon molecules, which are quite hydrophobic. Silicon and Teflon surfaces also have similar characteristics due to applied chemical treatments.

Now we can return to the development of our more detailed meniscus force model, based on the approach of Israelachvili.[5] In this approach we need to determine the variation of the total free surface energy of the system (solid-liquid-vapor), W_{svl} (i.e. **work of adhesion**) with the film thickness, h, using the Young's law that we have already described above. For the system shown in Fig. 13.10(b), according to Israelachvili[5]:

$$W_{svl} = 2\pi R^2 \sin^2 \varphi(\gamma_{sl} - \gamma_{sv}). \tag{13.27}$$

Now, use the Young's law (Eq. (13.24)) and for small $\varphi : \sin\varphi \approx \varphi$, then:

$$W_{svl} = -2\pi R^2 \varphi^2 \gamma_{lv} \cos\theta. \tag{13.28}$$

Therefore, the meniscus force due to a single asperity is obtained as:

$$F_m = -\frac{dW_{svl}}{dh} = 4\pi R^2 \gamma_{lv} \varphi \left(\frac{d\varphi}{dh}\right) \cos\theta. \tag{13.29}$$

To find the quantity $\frac{d\varphi}{dh}$, we can assume a constant volume for the meniscus, or: $\frac{dV}{dh} = 0$. Israelachvili[5] approximated the assumed constant volume of fluid to be (see Fig. 13.10(b)):

$$V = \pi R^2 (h+d) \sin^2 \varphi - \frac{\pi R^3}{3}(1-\cos\varphi)^2(2+\cos\varphi). \tag{13.30}$$

For small value of φ, $\sin\varphi \approx \varphi$, $1-\cos\varphi = 2\sin^2 \frac{\varphi}{2} \approx \varphi^2$ and $2+\cos\varphi = 3$. Also, $h \gg d$, thus: $h + d \approx h$. Therefore:

$$V = \pi R^2 \varphi^2 \left(h + \frac{R\varphi^2}{4}\right). \tag{13.31}$$

If one assumes a constant volume for the formed meniscus, then: $\frac{dV}{dh} = 0$. Using Eq. (13.31), and noting that $\varphi = f(h)$, then:

$$\frac{dV}{dh} = \pi R^2 \left(\varphi^2 + 2\varphi \frac{d\varphi}{dh} h\right) + \frac{\pi R^3}{4}\left(4\varphi^3 \frac{d\varphi}{dh}\right) = 0,$$

which simplifies to:

$$\frac{d\varphi}{dh} = -\frac{1}{\left(\frac{2h}{\varphi} + R\varphi\right)}. \tag{13.32}$$

Now substituting into Eq. (13.27), and after some manipulation:

$$F_m = -\frac{4\pi R \gamma_{lv} \cos\theta}{1 + \frac{2h}{R\varphi^2}}.$$

Since:

$$d = R(1-\cos\varphi) = 2R\sin^2\frac{\varphi}{2} \approx 2R\frac{\varphi^2}{4} = \frac{1}{2}R\varphi^2,$$

then:

$$F_m = -\frac{4\pi R\gamma_{lv}\cos\theta}{1+\frac{h}{d}}. \tag{13.33}$$

Note that for films of molecular dimension (in nano-tribology), $\frac{h}{d} \ll 1$, thus Eq. (13.31) in fact simplifies to that given by Bowden and Tabor.[12]

13.5.1. *An example*

We have shown that the meniscus force is strongly dependent on the topography of surfaces, contact angle that a fluid makes with the surface and the surface tension γ_{lv} (see Eqs. (13.33) and (13.23)). For water: $\gamma_{lv} = 72.75\,\text{mN/m}$ (milli-Newton per meter). These equations assume a single asperity case, thus for real problems they may be regarded as simplified cases for practical surfaces with many asperities of different geometry. For our example, we use Eq. (13.23) and water as the fluid, but with surfaces of different materials. Figure 13.11 shows the variation of the meniscus force for the geometry of Fig. 13.9(a), where $R = 1\,\mu\text{m}$ is considered to be the radius of an asperity tip. Four specific cases are highlighted in the figure, being Silica, Diamond Like Coating (DLC), which is a hard wear resistant

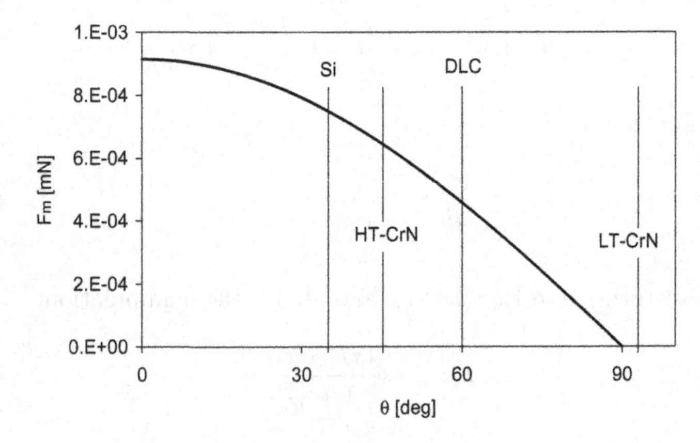

Fig. 13.11. Meniscus force variation with contact angle for various surface materials.

coating, and two types of Chromium Nitride coatings, also wear resistant. The contact angle that water makes with surfaces often depends on the way the coatings are deposited or grown on the substrate material (such as at high temperature, HT or low temperature, LT) or the resulting crystalline structure is etched or cut, such as in the case of Silica (Si).

The menicus force is clearly very small for a single asperity contact. Its total value in a contact depends on the number of asperity contact pairs, and the average asperity contact tip radius, discussed later in Sec. 13.7. Note that as the contact angle increases the meniscus action is diminished, as we have already discussed (hydrophobic surfaces). Also, if the contact angle becomes negative, water droplets have a tendency to fall off the surface. This effect is exploited in cases, where it is intended to inhibit water condensation on contacting surfaces, such as in MEMS devices, as pointed out in Sec. 13.1. In such cases, a mono-layer (single molecular layer) of a hydrophobic material, such as Octadecyltrichlorosilane (OTS) is assembled to/grown on the surface. These are known as **self-assembled mono-layers** (SAMS).

13.6. Solvation Effect

(a) *Introduction*

We have discussed interaction pair potential between a pair of particles or molecules in free space, given, for example, by w in Secs. 13.3 and 13.4. This is, in fact, often referred to as the free energy. However, when the interaction between, for example, two solute molecules in a solution is taken into account, then the effect is not just described by the pair of molecules themselves, but also many of the solvent molecules nearby. For example, ionic crystals often dissolve in solvents, which consist of molecules that form an electrostatic association with them. This effect is called **solvation**. When the solvent is water, then the effect is referred to as **hydration**. Therefore, two molecules in free space would not behave in the same manner as they would in a solution, irrespective of their polar or non-polar nature, because they are also affected by the presence of the solvent molecules. In fact, the nature of their interactions may completely alter.

The continuum of the solvent itself is also affected by the presence of solute or guest molecules, which are accommodated by the host solvent molecules. The structure of the continuum is, therefore, disturbed and an additional force comes into play between the solute molecules, referred to as the solvation force. The structure of the solvent medium alters by molecular re-ordering of its molecules around a solute molecule. We can regard this as the manner in which molecules re-organize their packing to adhere around an introduced guest molecule. It is now clear that re-ordering also takes place as a medium of molecules meets a solid boundary (i.e. a wall). The ordering of molecules near a solid surface is dependent upon the molecular geometry, their packing formation and the local geometry of the surface (i.e. its topography). This means that the solvation force between the fluid molecules and the solid boundary dictates different ordering near its surface than elsewhere within the bulk of the fluid. This indicates that the density of fluid molecules near a solid surface is quite different to that in the bulk. The attractive interactions between the wall and the fluid molecules, as well as the constraining effect of the wall, forces the fluid molecules to re-order themselves into quasi-discrete layers, several molecular layers into the depth of the bulk of the fluid. The same effect has not been observed at interfaces between liquids, and liquids and vapors (see Abraham[14]). It has also become clear that the solvation effect occurs at very smooth surfaces (generally referred to as molecularly smooth, of the order of tens of Angstrom — tenths of nanometer, such as some mica or silica surfaces).

Before we proceed further, it is important to state the growing importance of solvation effect in tribology. As already stated, there are emerging advanced technologies that employ contacts of vanishing separation of very smooth surfaces, with intervening fluids, such as Octamethylcyclotetrasiloxane (OMCTS). The fact that solvation effect can occur merely due to the constraining effect of a pair of smooth solid walls means that at small enough separations this can play a very significant role. A study by Chan and Horn[15] has shown that as the separation between two such surfaces is reduced, fluid drainage from the contact occurs in discrete steps (ejection of rows of molecules). This is evidence of fluid film layering due to solvation. It also shows that the drainage does not follow the usual viscous effect that we have come to expect in tribological contacts (see Chapter 6 on hydrodynamics: principle of continuity of flow).

(b) *The oscillatory solvation force*

The layering effect, briefly described above, is reflected in an oscillatory density profile of molecules, extending several molecular diameters deep into the bulk of the liquid. This is as the result of attractive interactions between the liquid molecules and the constraining effect of solid walls. The effect of the latter is found to be more dramatic. Constraint geometry

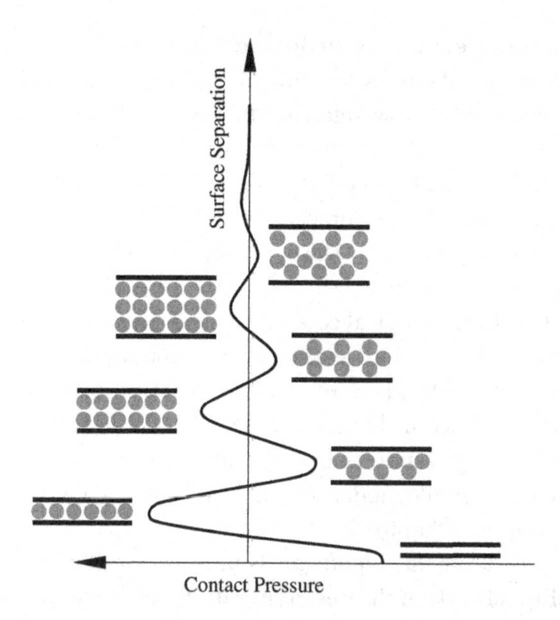

Fig. 13.12. Molecular re-ordering due to solvation effect in very small separations.

of the contact dictates the re-ordering of liquid molecules so as to be accommodated within a separation h. The variation in this ordering with changes in h gives rise to the solvation force (see Fig. 13.12). Because the solid structures play such an important role, the resulting solvation force is also referred to as the **structural force**.

To understand solvation, it is necessary to understand the density profile of the liquid molecules near solid surfaces. Observations and simulation (usually Monte Carlo) studies have shown that a higher density profile of molecules takes place near the adjacent solid surfaces than into the bulk of the liquid film. Henderson *et al.*[16] and Evans and Parry[17] proposed a contact density model for the constraining effect of **solvation pressure** as:

$$p_s = KT[\rho_s(h) - \rho_s(\infty)], \tag{13.34}$$

where $K = 1.381 \times 10^{-23}$ J/°K is the Boltzmann constant, T is temperature in °K, and P_s is the solvation pressure with separation h. $\rho_s(h)$ is the contact density of liquid molecules at each contact surface for finite surface separations and $\rho_s(\infty)$ is the corresponding value for a single surface (i.e. when $h \to \infty$). Clearly, Eq. (13.34) shows that solvation pressure arises when there is a change in the liquid density at the surfaces as they approach each other (i.e. **molecular re-ordering** takes place). For an inert hard wall, this is brought about as a result of changes in molecular packing as the separation h varies. Now referring to Fig. 13.12, we can observe that $\rho_s(h)$ will have a high value at values of h, which are integer multiples of the molecular diameter a, and falls off at intermediate separations. Both density and solvation pressure (thus force) are oscillatory functions of distance, of periodicity roughly equal to the molecular diameter of the intervening liquid. At large values of separation, h: $\rho_s(h) \to \rho_s(\infty)$, and the solvation effect diminishes. In tribological contacts, this means that the density of the liquid tends to its bulk value, leading to the dominance of hydrodynamic effect. Note that film thickness of order of tenths to several micrometers, which are commonplace in hydrodynamics and EHL (see Chapters 6–12) are considered as large separations, with no significant constraining effect. Thus, the density profile under those conditions tends to $\rho_s(\infty)$ (bulk density of the liquid, Chapter 5).

As the last layer of liquid molecules is eventually squeezed out of contact, Eq. (13.34) indicates a negative solvation pressure (i.e. an attractive force): $p_s(h \to 0) = -KT\rho_s(\infty)$. This accounts for the limit of adhesion due to the solvation effect alone. However, in practice there are other forces that should be taken into account at such limiting separations, such as the meniscus effect, described above. Al-Samieh and Rahnejat[18] have shown that at the separation limit of two molecular diameters of the intervening fluid, meniscus pressure inhibits the oscillatory behavior due to solvation. Bhushan[13] has also shown that stiction takes place for such thin films due to the meniscus force. In fact, to observe the effect of oscillatory solvation force, Chan and Horn[15] had to apply a force to break the meniscus bridge between a pair of mica surfaces to allow the solvation effect to be observed, causing molecular re-ordering, layering of film and step-wise drainage from the contact. These observations show that in lightly loaded contacts of vanishing gap size there are many intermolecular and surface interaction effects that cannot be taken into consideration in an isolated manner, and the contact model described here is rather idealistic, particularly at limiting conditions.

Israelachvili[5] has suggested that, as a first approximation, the solvation pressure between any two surfaces separated by a very thin film or in any highly restricted space may be described by an exponential decaying cosine function of the form:

$$p_s = -KT\rho_s(\infty)e^{-h/a}\cos\left(\frac{2\pi h}{a}\right) = -Ce^{-h/a}\cos\left(\frac{2\pi h}{a}\right), \qquad (13.35)$$

where a is the molecular diameter of the liquid. Note that both the oscillatory period and decay length are close to a, as also shown in Fig. 13.12. Also, note that this exponential form is similar to electrostatic repulsion form, discussed in Sec. 13.4. An important point to bear in mind is the multiplicity of forces in very narrow conjunctions and the similarity of their physical mechanisms. We must, therefore, be prudent that many of these effects may be different manifestation of the same phenomena, which are not fully understood in a deterministic manner.

13.6.1. *An example*

A simple example is the contact of a rigid roller of diameter 0.02 m and length 10 μm with a semi-infinite elastic half-space made of mica at very low load (i.e. no deformation takes place: hydrodynamic condition). The roller is rotating, entraining the lubricant into the contact at $U = 0.2\,\text{mm/s}$. The lubricant is OMCTS, with its rheological data provided as: $C = 172\,\text{MPa}$, $\eta_0 = 2.35\,\text{mPas}$, $a = 1\,\text{nm}$.

In very small gaps of the order of a few to several nanometers, and with molecularly smooth surface of mica (roughness in the order of few tenths of nanometer), the pressure between the solids is assumed to be generated by a combination of hydrodynamic viscous action and the solvation effect. The effect of electrostatic repulsion is ignored due to the non-polar nature of the OMCTS, and the contribution of the Van der Waals force is considered to be negligible.

The total contact load, W, is assumed to be due to combined effect of hydrodynamics and solvation, as:

$$W_{total} = W + \int p_s dx \quad \text{for } p_s > 0, \tag{13.36}$$

where p_s is given by Eq. (13.35) and W is the hydrodynamic reaction, given here by Eq. (13.6). However, it is more appropriate to obtain W as integrated pressure distribution through numerical solution of Reynolds equation at any instant of time. This approach is described by Teodorescu and Rahnejat.[19]

Figure 13.13 shows the film thickness variation with an increasing total contact load. We can see that the contribution to film thickness from hydrodynamics decreases sharply (shown by the thin line) as the contact

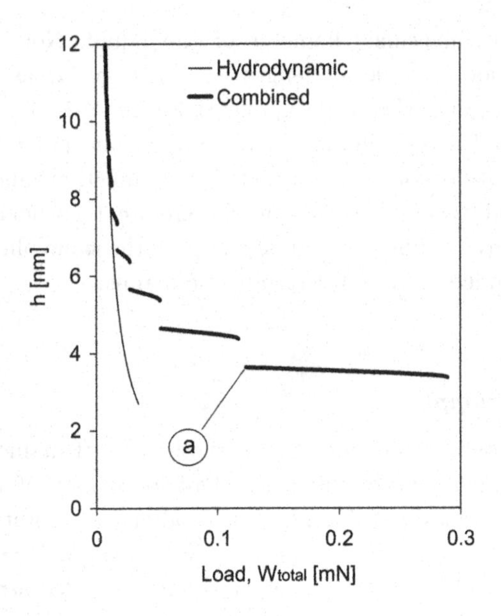

Fig. 13.13. Layering of film due to solvation with an increasing contact load.

load increases. Also lubricant film layering effect is observed in discrete steps with increasing load due to the solvation pressure.

We should also note no layering effect is observed above a film thickness of 8 nm, in other words a film of several molecular diameters of OMCTS ($a = 1$ nm). Hydrodynamic effect dominates in this region. The important points to remember are: use of extrapolated or analytically obtained film thickness formulae for hydrodynamics is inappropriate for gaps of several molecular diameters of the intervening fluid. Also, that solvation effect operates for molecularly smooth surfaces only. Additionally, we have assumed that the contacting surfaces are not charged, as well as the fact that OMCTS is non-polar, thus no electrostatic repulsion occurs.

Solvation pressure is oscillatory (attractive-repulsive) in nature. The negative pressures (attractive) tend to bring the surfaces together. This reduces the load carrying capacity of the contact. These are ignored in Eq. (13.36).

Figure 13.14 shows the pressure distributions due to hydrodynamics and that combined with solvation (i.e. the total pressure distribution). These are the contributions for the separation level, marked by letter (a) in Fig. 13.13. The dominance of solvation pressure is evident in Fig. 3.14.

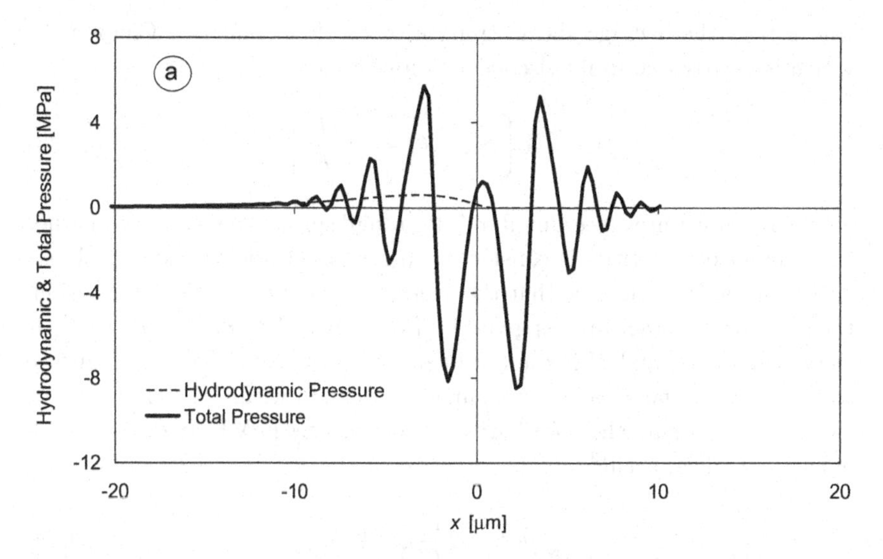

Fig. 13.14. Pressure distribution at a separation of $h \approx 4\,\text{nm}$.

13.7. Mechanisms of Friction in Nano-Conjunctions

The fundamental aspects of friction are already covered in Chapter 4. For dry contacts, the mechanisms giving rise to contact friction were shown to be due to adhesion of surface asperities and their deformation. Thus, for dry contact of rough surfaces the total friction force is given as addition of these two contributions, as indicated by Eq. (4.2) (i.e. $F = F_a + F_d$).

The adhesive contribution is given by Eq. (4.3) as: $F_a = A\tau_s$. Note that A is the real area of contact (made by all the asperity tips, rather than the apparent area of contact). In most contacts a film of fluid is intended to separate the surfaces, and in fact in most cases such a film is often interrupted by asperity interaction (we have already described this resulting regime of lubrication as **mixed lubrication**, see Chapter 1). Thus, we must account for the existence of a fluid film for some of the contact area, which is not taken into account by Eq. (4.2). Bowden and Tabor[12] used a simple method by using a fraction coefficient for dry contact, $0 \leq \psi \leq 1$. Thus, Eq. (4.3) can be written as:

$$F_a = A\{\psi\tau_s + (1 - \psi)\tau_\ell\}, \tag{13.37}$$

where τ_ℓ is the average shear strength of the fluid film (see Chapter 4), which is clearly due to its viscous action (see Chapter 6), thus:

$$F_a = A\left\{\psi\tau_s + (1-\psi)\frac{\eta U}{h}\right\}. \tag{13.38}$$

The real contact area (see Chapter 4) for rough surfaces depends on whether the contact of asperities is considered to remain elastic or undergo plastic deformation. If we assume that the asperities undergo plastic deformation, then the real contact area is given by Eq. (4.10): $A \approx F_n/H$, where F_n is the normal load, and H is the hardness. If, on the other hand, the surface asperities with the area A are assumed to remain within the elastic limit, as we wish to assume here for hard ceramic surfaces in MEMS devices, such as mica or silica, then[13]:

$$A \approx 3.2\frac{F_n}{E^*}\sqrt{\frac{r_p}{\sigma_p}}, \tag{13.39}$$

where r_p is the average value of asperity tip radius in the contact, σ_p is the standard deviation from this average value and E^* is the reduced elastic modulus of the two contacting solids, given by Eq. (3.7).

Before replacing for A into (13.38), it is necessary to determine the normal (i.e. the net contact) force. Referring to Fig. 13.15, it can be observed that the net contact force would be usually as a result of a number of forces. These can include any applied force W, a meniscus force, F_m in the presence of an intervening fluid between the rough surfaces (with bridges forming between the asperity tips as shown in the figure), and depending on the nature of surfaces and the fluid: electrostatic and Van der Waals forces. For simplicity, let us assume that the only significant force is due to

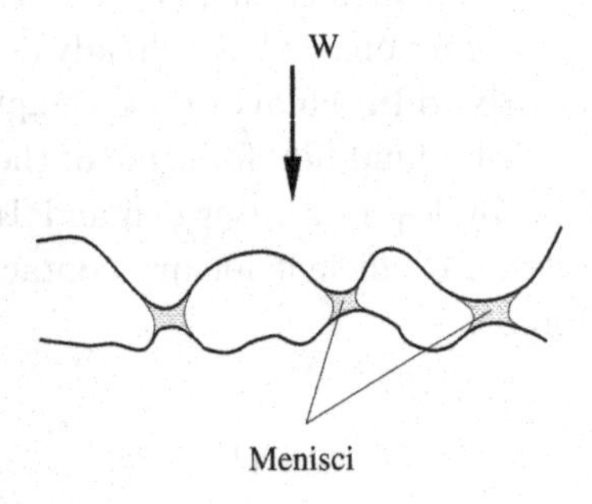

Fig. 13.15. Wet asperity tip contacts with formed menisci.

the meniscus effect, thus:

$$F_n = W + F_m, \tag{13.40}$$

where F_m can be given by either Eqs. (13.23) or (13.33). Using the simpler Eq. (13.23), and replacing for F_n in (13.40), and subsequently into (13.38), it yields the adhesion force as:

$$F_a = 3.2\frac{1}{E^*}\sqrt{\frac{r}{\sigma_p}}(W + 4n\pi r\gamma_{lv}\cos\theta)\left\{\psi\tau_s + (1-\psi)\frac{\eta U}{h}\right\}. \tag{13.41}$$

n is the number of asperity tips in contact (in Fig. 13.15, 3 such pair of contacting tips are shown). We have assumed that the fluid makes the same contact angle with both the rough surfacc asperities. Note also that the meniscus force is negative (i.e. tends to bring the surfaces together).

The other contribution to dry friction is ploughing friction (sometimes referred to as deformation friction), F_d, which we have already described in Chapter 4. This is simply the force required to deform the asperities in contact. The asperities may be considered to deform either elastically or plastically. We have already described the simple case of a hemispherical asperity ploughing a groove of a certain depth in the counterface in contact of rough surfaces in Sec. 4.2.3.1.

Clearly in a nominal contact area of asperity contacts in relative motion some asperities undergo elastic deformation, whilst others are plastically deformed. Therefore, one would need to assume a distribution for asperity peak heights to begin with, and then consider elastic, plastic or a proportion of either of the mechanisms in the ploughing action. A negative exponential distribution is often assumed for asperity peak distribution, such as e^{-z}, where z is the peak height. We have to also take into account a number of peaks, n, and their geometry such as average radius of such peaks, r and its standard deviation σ_p.

The deformation takes place as the result of conversion of some of the kinetic energy due to relative sliding motion of rough surfaces expended to deform the asperities, elastically or plastically. If the equivalent mass is considered as m, and the relative sliding velocity of the two rough surfaces as V and the proportion of this kinetic energy used to deform the asperity as β, then we can state that:

$$\frac{1}{2}\beta m V^2 = (\pi d^2)Hd, \tag{13.42}$$

where d is the extent of asperity in the ploughing action as shown in Fig. 4.7 (Chapter 4) and H is the hardness of surface. This relationship is for a plastic contact. Similarly, for an elastic contact, we obtain:

$$\frac{1}{2}\beta mV^2 = E^* d\left(\frac{\sigma_p}{r}\right)^{1/2}\pi d^2 \tag{13.43}$$

The right hand side of Eqs. (13.42) and (13.43) is the work done to deform an asperity through the distance d. We can denote this by w_{def}. The force to deform such an asperity is, therefore, given as: $f_d = w_{def}/d$. Thus:

$$f_d = \pi d^2 H \qquad \text{(for \textbf{plastic ploughing} of an asperity)}, \tag{13.44}$$

$$f_d = \pi d^2 E^* \left(\frac{\sigma_p}{r}\right)^{1/2} \qquad \text{(for \textbf{elastic ploughing} of an asperity)}. \tag{13.45}$$

If we now assume n such asperities with a negative exponential distribution as we have explained above (also see Chapter 4), then the contribution due to deformation friction becomes:

$$F_d = \int_0^{z_0} f_d e^{-z} dz, \tag{13.46}$$

where z_0 is the maximum wear depth of asperities. Note that we can use either Eqs. (13.44) or (13.45) to assume fully plastic or fully elastic deformation of these asperities. Alternatively we can assume that n asperities are plastically deformed and m others undergo elastic deformation, and thus have two contributions to the overall value of F_d. You will have noted by now that we have not yet related the deformation of the asperities to the normal load F_n (see Eq. (13.40)) and the sliding velocity V. We need to do this, by assuming a wear model. You will recall that in Chapter 4 we described simple models for both adhesive and abrasive wear, and found that very similar relations result in both cases for the wear rate (see Eqs. (4.27) and (4.30)). These relationships were given for unit sliding distance. Here we use a very similar relationship, but take into account the effect of sliding velocity, thus:

$$Q = kF_n X. \tag{13.47}$$

This is known as the **Lancaster's equation** for sliding wear, in which Q is the wear volume, usually given in mm^3, F_n is the normal contact load in N, X is the sliding distance, given in m, and k is the wear factor with the resulting units: mm^3/Nm. Of course this happens to be the traditional convention for these units, one that is not particularly suited to nano-tribology. Therefore, we can measure the quantities in any suitable unit, as long as we make sure that we are aware of the resulting unit for the wear factor.*

We can re-present the equation in terms of sliding velocity as:

$$\frac{dQ}{dt} = \dot{Q} = k' F_n \frac{dx}{dt} = k' F_n V, \tag{13.48}$$

where X the total sliding distance replaced here by its changing value, the rate of which is the sliding velocity. Note that: $U = V/2$, if a film of lubricant forms in the contact.

k' can be considered as a non-dimensional wear factor, if $\dot{Q}$, defined as wear rate, is given in units Nm/s (i.e, in Watts), which is the required power to plough through the asperities.

If the rate of wear depth for a typical asperity is denoted by: $\dot{z}$, and the force to deform it (as we have already stated above) as: f_d, then we can also state: $\dot{Q} = f_d \dot{z}$, and substituting for this in Eq. (13.48), and re-arranging:

$$f_d = \frac{k' F_n V}{\dot{z}}. \tag{13.49}$$

Now substituting in Eq. (13.46):

$$F_d = nk' F_n V \int_o^{z_0} \frac{1}{\dot{z}} e^{-z} dz = \frac{nk' F_n V}{\dot{z}} \int_o^{z_0} e^{-z} dz = \frac{\pi nk' F_n V}{\dot{z}} (1 - e^{-z_0}). \tag{13.50}$$

when assuming that the rate of wear depth, $\dot{z}$ is independent of the actual wear depth, z. If the asperities are considered to be hemispherical of radius R as in Fig. 4.27 (Chapter 4), and note that z_0 here is the same as h in that figure, then: $z_0 = d^2/8R$. We can now replace this for z_0 in Eq. (13.48), thus:

$$F_d = \frac{\pi nk' F_n V}{\dot{z}} (1 - e^{-\Phi V^{4/3}}), \tag{13.51}$$

where $\dot{z}$ can be replaced in terms of f_d (using Eq. (13.49)), which is simpler to specify in terms of geometrical and mechanical properties of the rough

*If we now consider a unit sliding distance and take into account the assumptions made in chapter 4 for wear of n asperities we end up with $k = k_{ad}$, a non-dimensional value used in Eq. (4.27).

surface, as shown in Eqs. (13.44) and (13.45). Thus:

$$F_d = \pi n f_d (1 - e^{-\Phi V^{4/3}}), \tag{13.52}$$

where for plastic deformation:

$$\Phi = \frac{1}{8R}\left(\frac{\beta m}{2\pi H}\right)^{2/3} \quad \text{with } f_d \text{ given by Eq. (13.44),} \tag{13.53}$$

and for elastic deformation:

$$\Phi = \frac{1}{8R}\left\{\frac{\beta m}{2\pi E^*}\left(\frac{r}{\sigma_p}\right)^{1/2}\right\}^{2/3} \quad \text{with } f_d \text{ given by Eq. (13.45)} \tag{13.54}$$

Now the total friction force is obtained by the addition of F_a and F_d as indicated by Eq. (4.2). Thus:

$$F = F_a + F_d = 3.2\frac{1}{E^*}\sqrt{\frac{r}{\sigma_p}}\,(W + 4n\pi r\gamma_{lv}\cos\theta) \times \left\{\psi\tau_s + (1-\psi)\frac{\eta U}{h}\right\} + \pi n f_d (1 - e^{-\Phi V^{4/3}}). \tag{13.55}$$

We note that other contributions to friction such as micro-stick-slip have been ignored, as well as other normal forces which would contribute to friction under certain circumstances, such as electrostatics, when a thin film of a polar fluid adheres to a charged surface. Nevertheless, the derived friction model can be considered as a good basic model of friction for nano-tribological contacts.

13.7.1. *An example*

Ceramic surfaces are fairly smooth compared with metal surfaces that we are mostly used to in tribology, the typical roughness of which vary according to the mechanical process (see Table 2.2, Chapter 2). The very smooth surface of ceramics means that asperity interactions are less likely to take place in presence of thin films than for rougher metal surfaces, an example being the use of ceramic femoral heads in hip joint replacements, as described in Chapter 14. However, with very thin films of nanometer thickness asperity interactions are more likely. Therefore, we look at a number of examples, where a rigid roller or hemispherical tip of radius $R = 0.2\,\mu$m and mass, $m = 20.41\,\mu$gm slides against (i) a surface coated with hard wear resistant **DLC** (diamond like coating) or (ii) a flat silicon substrate, in both cases at a sliding speed of 0.15 mm/s, whilst the applied contact load, $W = 2\,\text{mN}$. At such a low speed and in the presence of water condensation only, a coherent hydrodynamic lubricant film is not likely to form. For the same reason contribution due to electrostatic pressure can also be ignored, even though water molecules are polar.

Solvation pressure can act in molecularly smooth conjunctions (in the range of few tenths of nanometer roughness) at very close range. Therefore, for our example, we can choose average roughness values which inhibit the solvation effect: 100 nm is chosen for average asperity tip radius in both examples, with DLC roughness being 10 nm Ra and that of silicon 20 nm Ra. Menisci bridges, however, can form between these surface asperities and the ploughing hemispherical tip or roller, when such surfaces are exposed to the environment. Thus, meniscus forces would act wherever a meniscus bridge is formed. The equation we have used for meniscus force does not include the speed with which a meniscus bridge can be formed, which is a function of the sliding speed of the contact and condensation of water on the surfaces. Riedo *et al.*[20] give a more detailed analysis of this, but in general time taken for water molecules to form on any surface is around 25 μs. Let us assume that such condensation has taken place for all the surface asperities in this example. Therefore, we need to consider the overall meniscus action, which is already included in Eq. (13.55). Other required input data include:

For the case of DLC surface:

Let: $\tau_s \approx 100\,\text{MPa}$ and water makes a contact angle of $\theta = 85°$ with DLC.

For the case of silicon surface:

Let: $\tau_s \approx 100\,\text{MPa}$ and water makes a typical contact angle of $\theta = 45°$ with silicon.

As we have already indicated in the example in Sec. 13.5: $\gamma_{lv} = 72.75\,\text{mN/m}$.

Now the following procedure can be used to evaluate the overall friction force for the examples chosen here:

- We ignore viscous friction as a coherent film of lubricant is unlikely, ignoring the term:$(1-\psi)\frac{\eta U}{h}$ in Eq. (13.55). This means that: $\psi = 1$.
- The number of asperities, n must now be determined. We need to determine the real contact area, A. Assuming elastic deformation of asperities only, we can use Eq. (13.39), where $E^* = \frac{E}{1-\upsilon^2}$, where the sliding tip/roller is considered to be rigid (for DLC, $E = 800\,\text{GPa}$,

$\upsilon = 0.1$ and for silica, $E = 161\,\text{GPa}$, $\upsilon = 0.23$). However, we would need to know the value of F_n in order to evaluate A from Eq. (13.39). This is given by Eq. (13.40), which includes the meniscus force, F_m, depending on the number of asperities n. An iterative procedure can be set with these two equations (both having F_n), such that values of A and n can be determined, satisfying the force balance in (13.40) for certain pre-specified accuracy. This procedure yields the results: (i) for DLC: $A = 1.8 \times 10^{-2} \mu\text{m}^2$, $n = 302$ and (ii) for silicon: $A = 8.8 \times 10^{-2} \mu\text{m}^2$, $n = 1502$.

- Now we can evaluate the meniscus force, and having the values of ψ, τ_s and W the adhesion contribution to the overall friction (i.e. the first term on the RHS of Eq. (13.55) can be obtained).
- To calculate the contribution due to ploughing (deformation) friction (the last term in Eq. (13.55)), we need to calculate f_d for a single asperity and the exponent Φ. Since we have assumed that the asperities undergo elastic deformation, then we will use Eqs. (13.45) and (13.54). We first need to determine values for d and β. Let us assume that the proportion of kinetic energy expended to elastically deform the asperities, $\beta = 0.5$. Now using Eq. (13.43), with all the data given: (i) for DLC: $d = 80\,\text{nm}$ and (ii) for silicon: $d = 57$ nm. Putting all the data in Eqs. (13.45) and (13.54) and subsequently into the last term in the RHS of Eq. (13.55), the contribution due to ploughing friction is obtained.

The total friction, F is thus obtained as; (i) for DLC: 1.79 μN and (ii) for silicon: 9.32 μN. DLC has a very smooth hard wearing surface, thus generates a lower friction force. Note that the meniscus force is quite small (i.e. 17.6 μN for DLC and 87.5 μN for silica) compared to the applied load $W = 2\,\text{mN} = 2000\,\mu\text{N}$. Under the conditions described, the coefficient of friction becomes: 0.0045 for silicon and 0.0009 for DLC. These are very small values, since the surfaces are really quite smooth and the contribution due to ploughing friction is negligible. If there is no applied load ($W = 0$), then the coefficient of friction becomes the ratio of adhesive friction to the meniscus force. This will be 0.1 in both examples, indicating the force required to break the menisci, and initiate motion (i.e. a form of static coefficient of friction, representing the worst case). The point to note is that for MEMS conjunctions, friction plays an important role if water molecules condensate on their "rough" surfaces, causing stiction. This is the reason for maintenance of an inert environment for such devices, as well as introduction of hydrophobic monolayers.

References

1. Dowson, D., Developments in lubrication-the thinning film, *J. Phys. D: Appl. Phys.* **25** (1992) 334–339.
2. Tanner, *et al.*, *MEMS Reliability: Infrastructure, Test Structures, Experiments, and Failure Modes*, SANDIA REPORT SAND 2000–0091.
3. Cameron, A., *Basic Lubrication Theory*, Ellis Horwood Limited (1976).
4. Lifshitz, E. M., The theory of molecular attractive forces between solids, *Sov. Phys. JETP* **2** (1956) 73–83.
5. Israelachvili, J. N., *Intermolecular and Surface Forces*, New York: Academic (1992).
6. Atkins, P. W., *Physical Chemistry*, Oxford University Press, New York, 3rd Edition (1986).
7. Rahnejat, H., Computational modelling of problems in contact dynamics, *Engineering Analysis* **2**(4), (1985).
8. Middleman, S., *An Introduction to Fluid Dynamics: Principles of Analysis and Design*, John Wiley and Sons, New York (1998).
9. Bergeron, V. *et. al.*, Controlling droplet deposition with polymer additives, *Nature* **405**, 772.
10. Birkhoff, J. and Hays, D. F., Free boundaries in partial lubrication, *J. Math. and Phys.* **32** (1963) 2.
11. Tipei, N., Boundary conditions of a viscous flow between surfaces with rolling and sliding motions, *Trans. ASME J. Lubn. Tech.* (January 1968).
12. Bowden, F. P. and Tabor, D., *The Friction and Lubrication of Solids*, Clarenden Press, Oxford (1950).
13. Bhushan, B., *Handbook of Micro/Nanotechnology*, 2nd Ed., CRC Press, Boca Raton, Florida (1999).
14. Abraham, F. F., The interfacial density profile of a Leonard — Johns Fluid in contact with a (100) Leonard–Johns wall and its relationship to idealised fluid/wall systems: A Monte-Carlo simulation, *J. Chem. Phys.* **68**(8), (1978) 3713–3716.
15. Chan, D. Y. C. and Horn, R. G., The drainage of thin liquid films between solid surfaces, *J. Chem. Phys.* **83** (1984) 5311–5324.
16. Henderson, D., Compressibility route to Solvation structure, *Mol. Phys.* **59**(1), (1986) 89–96.
17. Evans, R. and Parry, A. O., Lubricant at interfaces: What can a theorist contribute, *J. Phys. Condens. Matter* **2** (1990) SA15–32.
18. Al-Samieh, M. F. and Rahnejat, H., Ultra-thin lubricating films under transient conditions, *J. Phys. D: Appl. Phys.* **34** (2001) 2610–2621.
19. Teodorescu, M., Balakrishnan, S. and Rahnejat, H., Physics of ultra-thin surface films on molecularly smooth surfaces, *Proc. Inst. Mech. Engrs., J. Nano-Technology* **220**(1), (2006) 7–19.
20. Riedo, E., Lévy, F. and Brune, H., Kinetics of capillary condensation in nanoscopic sliding friction, *Phys. Rev. Lett.* **88** (2002) 185505.
21. Johnson, K. L., Kendall, K. and Roberts, A. D., Surface energy and the contact of elastic solids, *Proc. Roy. Soc., Series A* **324**, 1558 (1971) 301–313.
22. Fuller, K. N. G. and Tabor, D., The effect of surface roughness on the adhesion of elastic solids, *Proc. Roy. Soc., Series A* **345**, 1642 (1975) 327–342.

9. [illegible]
10. [illegible]
15. [illegible]
16. [illegible]
17. [illegible]
22. [illegible]

CHAPTER 14

BIO-TRIBOLOGY

14.1. Introduction

The term "*bio-tribology*" was first used in 1973[1] and defined as ...

> "*those aspects of tribology concerned with biological systems.*"

Tribology itself had been defined only seven years earlier[2] and **bio-tribology** is now one of the most rapidly developing aspects of the larger subject. In 1973 it was possible to cite only a few areas in which tribological studies of biological systems were active, but the subject now embraces topics such as; Animal Joints/Bearings.

Each of these topics brings important sub-sections of tribological interest. For example, it is impossible to consider joint replacements without developing an appreciation of the associated bio-mechanics, joint geometry, articulation, bio-materials, manufacturing, rheology, lubrication, friction, wear, implant fixation and surgical requirements. The list of topics in biotribology shown in Table 14.1 is not exhaustive, but a few examples from this expanding field will be presented in more detail to illustrate the developments.

Much attention has understandably been devoted to human bearings, but some remarkable examples of tribological solutions designed to provide attachment and yet permit locomotion of small animals over vertical surfaces have emerged in recent times.

The universal and troublesome problem of arthritis has ensured that studies of the lubrication, friction and wear of natural synovial joints and the development of total joint replacements have continued to dominate activity in biotribology. Natural synovial joints are remarkable bearings. Joint replacement is a major aspect of orthopaedic surgery today, with about one million hip prostheses being implanted throughout the world each year. Furthermore, satisfactory knee joint replacements have been developed, such that the numbers of knee replacement procedures have now caught up those of hips in several countries.

Table 14.1. Topics in biotribology.

Tribology in the animal world	Natural synovial joints	Joint replacements
Skin	Hair/Shaving	Textiles Clothing and Shoes
Ocular tribology	Teeth/Oral tribology/	Heart valves
Cardiovascular flow	Biological attachment	Swimming

Skin provides an interface between the body and the external world. It is a flexible, tough layer of tissue which protects the body organs, provides temperature control and water retention. The tribological properties of the nerve rich skin determine our tactile response to the environment. Locomotion and our ability to grasp and hold objects is determined by friction, as is our ability to avoid accidents by slipping in baths and showers. Likewise, lesions of the skin in the form of friction blisters, which are almost exclusively confined to the human species, are troublesome and not uncommon reactions to trauma. Work in the cardiovascular field has focused upon the lubrication of corpuscles in narrow capillaries and the contribution of wear to replacement heart valve deterioration.

14.2. Tribology in the Animal World

An intriguing relationship between the loads applied to the joints of mammals and their mass emerged from a combination of **allometry** and biomechanics.

14.2.1. *Allometry*

The lengths and diameters of animal bones would be proportional to (body mass)$^{0.33}$ if they were geometrically similar for animals of different sizes. Measurements of bones from the limbs of a large range of mammals from Shrews ($2.9 - 3.1 \times 10^{-3}$ kg) to Elephants (2.5×10^{3} kg),[3] with body masses differing by six orders of magnitude indicted that:

$$\text{Length} \propto (\text{body mass})^{0.35} \quad \text{Diameter} \propto (\text{body mass})^{0.36}$$

The two powers on body mass are so impressively close to the predicted values of 0.33 that they are entirely consistent with the concept of geometrical similarity.

14.2.2. *Forces and stresses*

Determination of ground reaction forces during activity is normally necessary in the biomechanical analysis of forces on animal joints. In addition, estimates have to be made of the size and disposition of the bones in the skeleton, body segment mass distribution and the location of muscle and tendon insertion points.

For relatively small animals, with body masses similar to or less than those of man, it is often possible to record the ground reaction during activity by means of a Force Platform.[4,5] The ground reactions for small animals such as frogs and dogs, were determined directly from force plate measurements,[5,6] but cinematograph and video analysis alone was used for larger animals such as kangaroos (*Macropodidae*),[7] ostriches (*Struthio camelus*),[8] buffalo (*Syncerus caffer*) and elephants (*Loxodonta Africana*).[9] These studies enabled relationships between body mass and joint forces to be developed for an impressive range of animals.

The maximum joint forces for vertebrates were found to be proportional to $(\text{body mass})^{0.747}$. The nominal areas of load bearing regions of joints would be proportional to $(\text{bone diameter})^2$ or $(\text{body mass})^{0.72}$. The stresses in the nominal load bearing regions of joints in vertebrates are thus related to animal body mass as follows:

$$\begin{aligned} \text{Joint stress} &\propto (\text{maximum load})/(\text{bone diameter})^2 \\ &\propto (\text{body mass})^{0.75}/(\text{body mass})^{0.72} \\ &\propto (\text{body mass})^{0.03} \end{aligned}$$

The mean stresses in animal joints are thus reasonably independent of their overall body mass or size. It appears that the operating characteristics of load bearing synovial joints reflect the ability of the bearing material, articular cartilage, to sustain the same mean stress level in a wide range of animals. This result is of considerable significance in studies of the tribology of synovial joints in man.

14.3. Attachment and Locomotion of Small Animals on Smooth Surfaces

Numerous systems of attachment of plants and small creatures such as barnacles, fishes, slugs, insects and spiders to smooth surfaces have been recognized in recent times.[10] Studies of the diverse systems adopted by nature have revealed valuable information on nano-scale tribology. Tarsal hairs, soft pads, suction cups and mechanical systems such as barbs and hooks have been recognized, but the outstanding tribological feature is the intimate contact between very large numbers of minute, soft, flat tips on the extremities of **Gecko** toes.

The remarkable Gecko[11] can move at speed up smooth vertical walls or even upside down over ceilings. More dignified speeds are evident through a different system of locomotion with slugs. The lengths and masses of a Tokay Gecko are generally in the ranges (0.3–0.4) m and (0.25–0.3) kg. It has long been known that attachment of the gecko toes to almost any material is non-sticky. The detailed structure of the tiny hairs on the gecko toes and the associated attachment process have been studied for over a century, originally by light microscopes but more recently by electron microscopy. The findings have revealed quite a remarkable system of attachment of these highly mobile creatures.

Each Geckos toe pad is covered by ridges which in turn support fine hairs or **setae** of length (30–130) μm and diameter (4–10) μm. The impressive density of *setae* is about 14,000 per mm^2, but this is not the end of the division. Each *seta* divides further into (100–1000) *spatulae* about 200 nm long and wide as shown in Fig. 14.1. The ends of millions of these

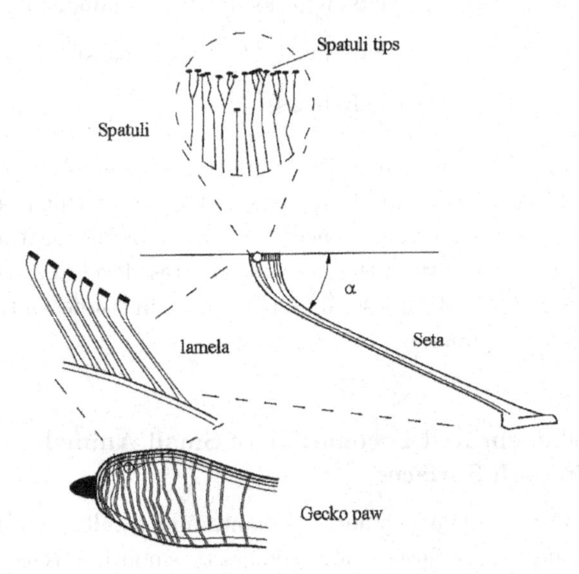

Fig. 14.1. Representation of a Gecko's foot.

exceptionally fine, soft spatulae contact an external surface with a total adhesive force several times greater than the weight of the Gecko.

It is now widely recognized that the *'dry'* adhesion mechanism adopted by geckos is attributable to **van der Waals forces**,[12] which arise from systems of electric charges associated with atoms and molecules. In the theory of elastic contact between two solids it is usual to assume that no tensile stresses can exist at the true interface when the applied load is removed. However, the intermolecular forces at very small separations exhibit both attractive and repulsive components,[13] such that the resultant potential energy w can be expressed approximately as (see also Chapter 13):

$$w = -\frac{A}{r^m} + \frac{B}{r^n}. \tag{14.1}$$

And since the interacting force (F)is given by $F = \frac{\partial w}{\partial r}$,

$$F = \frac{mA}{r^{m+1}} - \frac{nB}{r^{n+1}}. \tag{14.2}$$

The variation of potential energy w, the forces of both repulsion F_r and attraction F_a and the total force of interaction between the solids F, with distance between the solids r are depicted in Fig. 14.2 (see also Fig. 13.4).

As two solids approach each other the competing forces of attraction and repulsion balance each other at a small separation r_o. Perfectly smooth flat solids would, therefore, never touch, but in equilibrium would be separated by a gap r_o. For pure metals the equilibrium separation r_o is of the order of a few (Å). However, in the real world all solid surfaces are rough on the atomic scale and for hard materials, such as metals, contact is associated with asperity interactions. Only materials like mica approach the theoretical representation of flat surfaces. If the asperities are represented by hemispherical tips the contact mechanics analyses by Maugis-Dugdale (MD); **Derjaguin-Muller-Toporov** (DMT) or **Johnson-Kendall-Roberts** (JKR) models based upon the usual parabolic approximations to the unloaded surface profile can be applied. It has been pointed out,[12] however, that it is important to model the surface separations carefully if such studies are to succeed in predicting the limiting van der Waals interaction at a separation r_o. The force of adhesion F_a (tension) for a single hemispherical attachment is given by[14]:

$$F_a = -\frac{3}{2}\pi\gamma r_p, \tag{14.3}$$

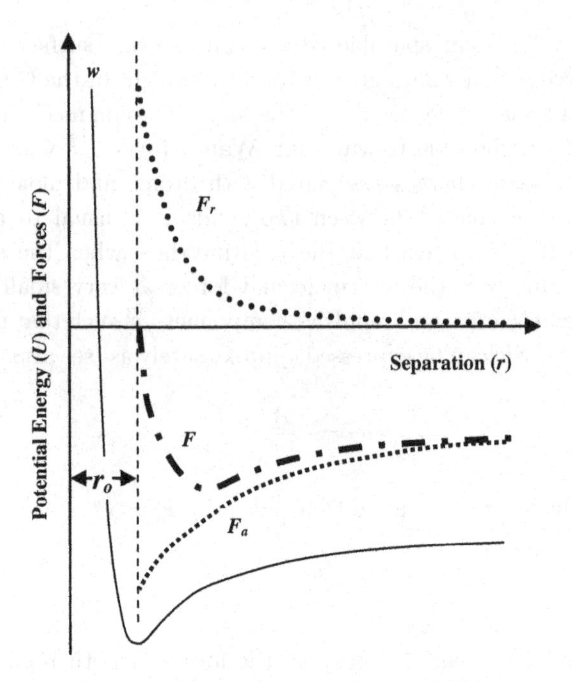

Fig. 14.2. Interacting forces between solids.

where γ is the surface energy (mJ/m^2) and r_p the radius of the hemisphere (m).

There are 100–1000 spatula on each seta and since the number of spatula on each seta n is related to the radius of the spatula tip (r_p) and the total radius of each seta by ($n = (R/r_p)^2$), **adhesion** with the counterface, of each seta (F_a) is:

$$F_a = -\frac{3}{2}\pi\gamma\frac{R}{\sqrt{n}}. \tag{14.4}$$

It has been estimated theoretically[15] that in general the intermolecular van der Waals force of adhesion range from (10–100) mJ/m^2. Measurements of the normal pull-off force for individual **gecko spatulae**[16,17] of nanometer proportions have resulted in values of about 10 nN. The value of the surface energy γ for a spatula of radius 100 nm can thus be found from Eq. (14.3) to be about 21 mJ/m^2. This is well within the theoretical range (10–100) mJ/m^2 reported in.[15]

With very low elastic modulus materials the compliant biological surfaces readily wrap themselves around asperities to achieve good proximity to the counterface. With the Gecko the very soft tips of the spatulae achieve this, such that the ratio of real to apparent contact area is much closer to unity than in the traditional models of interacting hard rough surfaces. Estimates of the effective elastic modulus of spatulae tips have ranged from about 0.1 MPa to 2 GPa.[18,19]

There are 10^2–10^3 spatulae on each seta and about 10^9 spatulae on each Gecko foot. If the measured force of adhesion for each spatula is accepted to be about 10 nN, the maximum force of adhesion for each Gecko foot must be about 10 N. Not all seta on every foot will be in contact and not all feet will necessarily be in contact with the counterface at any one time in the walking process. Since the weights of Geckos are typically (50–100) gf (0.49–0.98) N, there is nevertheless a substantial safety margin in the impressive climbing activity of Geckos.

The above assessment of theoretical and experimental information clearly endorses the view that van der Waals forces play a major role in the remarkable adhesion process attributable to Geckos. Other possible mechanisms, such as capillary action, have been proposed, but for the Gecko intermolecular forces appear to play the dominant role, at least under dry contact conditions.

Once the attachment mechanism of the Gecko had been understood, it was necessary to explain how the attachment forces could be overcome to enable swift, smooth climbing action. It appears that the contacts are detached by a peeling process, which requires considerably less force than pulling normal to the counterface. To achieve this action the gecko has developed special muscles and a joint design which permits digital hyperextension.[12,19]

Other creatures such as flies, grasshoppers and beetles have developed alternative forms of attachment surfaces. These include very low elastic modulus spherical tips, flat-ended or torroidal attachment pads and suction pads.

14.4. Natural Human Synovial Joints

Human joints normally function satisfactorily as self-contained plain bearings with quite low friction and wear for three score years and ten. However, some fail early due to accident or disease and if, in due course, clinical procedures and medication fail to relieve the arthritic symptoms, surgical intervention may be considered.

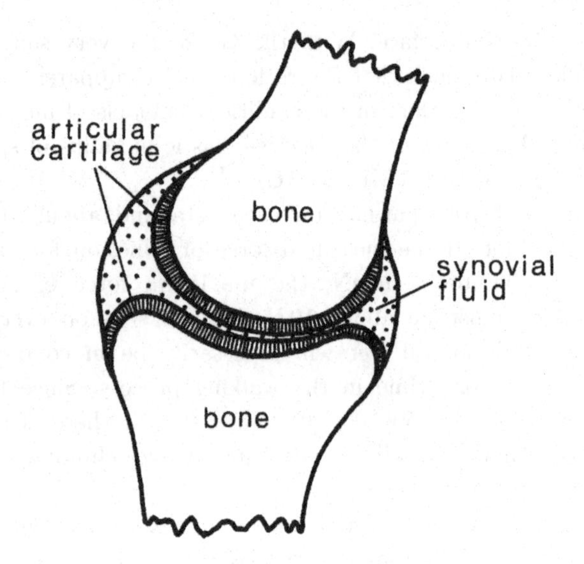

Fig. 14.3. Simple representation of natural human synovial joint.

A simple engineering representation of a natural synovial joint is shown in Fig. 14.3.

All the essential features of a plain bearing are evident in Fig. 14.3. The bearing material (**articular cartilage**) is mounted on a relatively hard backing material (bone) and the fluid lubricating the cartilage surfaces (**synovial fluid**) is contained within a synovial membrane.

14.4.1. *Measurement of friction*

Many attempts have been made to measure the coefficient of friction of finger, hip and knee joints. The experiments are not easy to perform and the results often difficult to interpret. One of the major problems is to isolate the frictional torque of the joint being tested, exhibiting coefficients of friction of about 0.002 to 0.02, from the support bearing friction. Pendulum machines have been popular for this purpose and a valuable understanding of the frictional characteristics of hip joints emerged in the 1970s.[20,21] A new form of pendulum machine was introduced in which the mounting cradle was supported on hydrostatic bearings. In the initial experiments[20] carried out under constant load, the powerful influence of squeeze-film action was clearly evident, but fluid-film lubrication characteristics arose only when high viscosity lubricants were used. When dynamic loads were applied to reflect walking cycle conditions,[21] clear evidence emerged of fluid film lubrication. The measured coefficients of friction ranged from 0.005 to 0.025, well within the range associated with hydrodynamic bearings. Equally important was the finding that the coefficient of friction increased as the ratio of (speed/load) increased. This observation was entirely consistent with the characteristics of fluid-film lubricated plain bearings, as depicted on a **Stribeck diagram** (see Chapter 1).

14.4.2. *Lubrication of synovial joints*

While the general features of the **synovial joint** depicted in Fig. 14.3 reflect those of a plain bearing, suggesting that fluid film lubrication must prevail, a complete understanding of human joint lubrication is still emerging. A common approach to the determination of the mode of lubrication in bearings is to calculate the film thickness h for perfectly smooth bearing surfaces and to compare the result with a measure of the combined average surface roughnesses defined as $\{(R_{a1})^2 + (R_{a2})^2\}^{1/2}$ and known as the composite surface roughness. The ratio of these two quantities is known as the lambda ratio λ_s and is defined as (see also Chapter 1):

$$\lambda_s = (h)/\{(R_{a1})^2 + (R_{a2})^2\}^{1/2}. \tag{14.5}$$

The lambda ratio was found to be very useful in elastohydrodynamic lubrication analysis (see Chapter 10) of gears. If $\lambda_s \geq 3$, fluid film lubrication offered good protection against surface damage, including fatigue, but if $\lambda_s \leq 1$ boundary lubrication prevailed and excessive wear could occur.

14.4.3. *Analysis of fluid-film lubrication in joints*

Surface roughness of **articular cartilage** appears to be in the range (1–5) μm, depending upon the age and condition of the natural bearing material. The lubricating film thickness would have to exceed at least ($3 \times \sqrt{2} = 4.23\,\mu$m) to ensure that satisfactory fluid film lubrication prevailed with the smoothest cartilage considered, if the bearing material was rigid. However, the elastic modulus of cartilage is relatively low, typically about (10–50 MPa), while the viscosity of synovial fluid has been shown to be sensibly independent of joint pressure.[22] It is, therefore, appropriate to estimate the film thickness for iso-viscous-elastic elastohydrodynamic lubrication conditions (see Chapters 10 and 12).

The appropriate **elastohydrodynamic lubrication** film thickness equation has been shown to be[23]:

$$h_{\min} = 1.40 \left(\frac{D^2}{c_d}\right) \left(\frac{\eta \Omega c_d}{2E'D}\right)^{0.65} \left(\frac{4Wc_d^2}{E'D^4}\right)^{-0.21}, \tag{14.6}$$

where,

c_d = diametral clearance
D = diameter of femoral head
E' = effective elastic modulus $\frac{1}{E'} = \frac{1}{2}\left[(1-\nu_1^2)/E_1 + (1-\nu_2^2)/E_2\right]$
W = load
η = viscosity of synovial fluid
Ω = angular velocity

Few attempts have been made to quantify diametral clearance in **natural hip joints**, but for the present purposes values in the range (0.1–2) mm will be adopted for a 50 mm diameter femoral head. The effective elastic modulus for cartilage is typically 25 MPa, while the average load on the hip during the walking cycle is about 1750 N. A representative angular velocity is 1.5 rad/s and the viscosity is about 0.002 Pas at high shear rates. Calculated values for the film thickness are shown in Table 14.2.

It is evident that none of the calculated film thicknesses achieve the value of 4.23 μm required to ensure satisfactory fluid film lubrication in natural hip joints. It is also clear that neither the equivalent elastic modulus nor the load materially affect the film thickness. The most dominant effect for a femoral head of fixed diameter is the clearance between the unloaded **femoral head** and the **acetabular cup.** It is concluded that the film

Table 14.2. Calculated film thicknesses in natural hip joints.

Diameter (mm)	50	50	50	50	50	50	50
Diametral clearance (mm)	2	1	1	1	0.5	0.25	0.1
Viscosity (Ns/m^2)	0.002	0.002	0.002	0.002	0.002	0.002	0.002
Angular velocity (rad./s)	1.5	1.5	1.5	1.5	1.5	1.5	1.5
Equivalent elastic modulus (MN/m^2)	25	50	50	25	25	25	25
Load (N)	1750	3000	1750	1750	1750	1750	1750
Film thickness-μm	0.30	0.34	0.38	0.51	0.87	1.49	3.01

thickness in natural joints is unlikely to exceed the composite roughness of the unloaded articular cartilage surfaces.

Experimental and functional experience nevertheless suggests that an excellent lubrication mechanism provides good protection to the cartilage surfaces over many years in most load bearing human joints. Studies of micro-elasto-hydrodynamic lubrication[24] brought new evidence of a significant mechanism of fluid film lubrication in human joints. The asperities on the elastic articular cartilage surfaces are capable of perturbing the pressure distributions predicted for smooth surfaces. The locally generated pressure peaks partially flatten the asperities and effectively produce a much smoother surface in the load bearing region than that suggested by the unloaded topography. The mechanism is shown diagrammatically in Fig. 14.4.

The result of micro-elasto-hydrodynamic action is that the surfaces become much smoother as they pass through the load carrying conjunction than they are in the initial unloaded state. The significant role of pressure perturbations in squashing the asperities on rough, low elastic modulus layers of cartilage can be illustrated by considering a simple column model for elastic deformation. If it is assumed that the local deformation δ is proportional to the local pressure p and the layer thickness d, $\delta \propto pd$ (see also Chapter 12).

The local deformation for flattened asperities thus attains a magnitude $\pm a$ and this requires a pressure perturbation of only $(a/d)(p)$. Since the

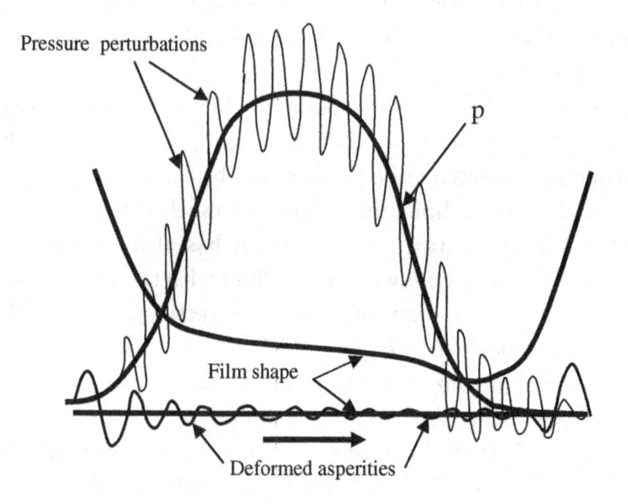

Fig. 14.4. Micro-elastohydrodynamic lubrication.

ratio a/d is typically of order 10^{-3} a modest perturbation to pressure of similar magnitude readily smoothes the initially rough cartilage surfaces.

For the example considered[24] the amplitude of the partially flattened asperities was less than 10% of the assumed amplitude of the unloaded cartilage roughnesses of $1\,\mu$m. The potential for developing adequate fluid film lubrication should thus be assessed upon a modified composite roughness of $0.14\,\mu$m rather than the originally accepted value of $1.4\,\mu$m. Four of the seven computed film thicknesses shown in Table 14.2 now satisfy this condition and the remainder should also be able to provide quite a significant reduction in friction and wear from partial fluid film lubrication.

The conclusion to be drawn from this analysis of fluid film lubrication in natural hip joints is that there is a good possibility of effective hydrodynamic action throughout much of the walking cycle. This does not mean that it is the sole mechanism of lubrication, particularly under high loads and low speeds. Under these and similarly tasking conditions, protection of the articulating surfaces must rely upon mixed or boundary lubrication.

14.4.4. *Evidence of boundary lubrication in synovial joints*

The nature of boundary lubrication in human joints is still the subject of active consideration. Two of the major constituents of **synovial fluid** thought to be associated with articular joint lubrication are **hyaluronic acid** and **glycoproteins**. The former is now known to be associated with the fluid viscosity and hence with fluid film lubrication rather than boundary lubrication.

Early studies[25–27] of the boundary lubricating ability of various constituents of synovial fluid indicated that a particular glycoprotein named **lubricin** exhibited good boundary lubricating characteristics. More recently[28,29] the role of lipids has attracted much attention, with lubricin and hyaluronic acid identified as carriers. It has also been postulated that hyaluronic acid not only endows synovial fluid with the viscosity required for effective fluid film lubrication, but also as a wetting agent for the strongly hydrophobic biological surfaces.

A thin oligolamellar boundary lubricating layer of surface active phospholipid has been identified on both cartilage and pleural surfaces as shown in Fig. 14.5. It is also interesting to note that similar layers can be associated with the pericardium, synovium, ocular and gut surfaces; all of which require low friction and ease of movement.

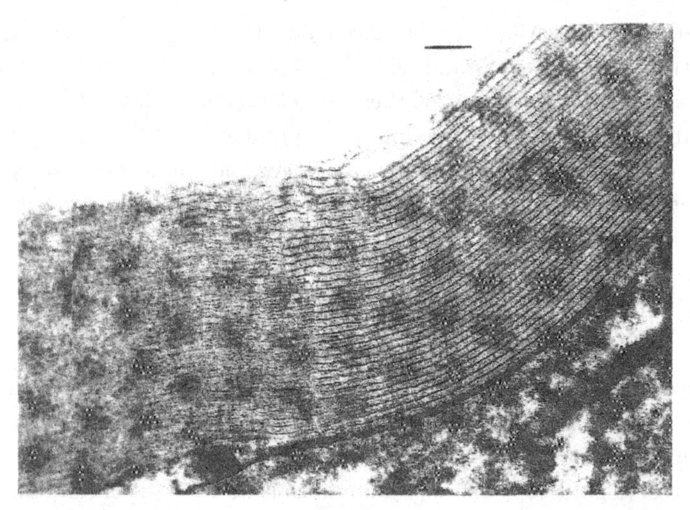

Fig. 14.5. Oligolamellar structure of surface active phospho lipid on pleural epithelium. (Hills, B. A., *J. Appl. Physiol* (1992), 73, 1034–1039).

14.4.5. *Overall view of human joint lubrication*

Disparate theoretical and experimental studies of **synovial joint** lubrication have provided clear indications of both fluid film and boundary lubrication, depending upon the test conditions. It seems clear that human joints, like many lubricated engineering systems, operate for much of the time in a mixed lubrication regime. Micro-elastohydrodynamic lubrication is capable of maintaining very low friction and good protection to the articular cartilage bearing surfaces over a wide range of operating conditions. For particularly severe operating conditions an effective boundary lubrication mechanism protects the surfaces.

Engineers would be hard pressed to design and make self-contained plain bearings to operate under tasking dynamic conditions in an environment as hostile as that of the body for three score years and ten. The human synovial joint is a truly remarkable illustration of effective biotribology.

14.5. Total Joint Replacements

Damaged or diseased synovial joints may have to be replaced by man-made bearings. **Arthroplasty** (total joint replacement) has developed impressively, particularly during the past half century. Joint replacement has often been described as the major advance in orthopaedic surgery in the 20th Century. Understanding of friction, wear, lubrication and bearing materials has guided many of these developments. Most of the synovial joints in the body now receive man-made replacements, including the hip, knee, ankle, toe, shoulder, elbow, wrist, finger and spine. The hip represents the best known arthroplasty and this is used as an example of replacement joint tribology in the next section.

14.5.1. *The hip*

Arthroplasty of the hip has been practiced for almost two hundred years.[30] Surgeons experimented by interposing between the articulating bone ends in the damaged or diseased joint various materials including wood, ivory, gold foil, glass, celluloid, Bakelite, cobalt-chromium alloys, gold, platinum, white oak, stainless steel, acrylic, polymers (nylon, polytetrafluoroethyene (ptfe or teflon), polyethylene) and ceramics, together with natural tissues such as fat and muscle. These early experiments led to the first metal-on-metal hip replacements in the late 1930s, when Philip Wiles[31] replaced both the femoral head and the acetabular cup with stainless steel components at the Middlesex hospital in London.

Since the middle of the 20th century total hip replacements have been manufactured almost exclusively from metals, polymers and ceramics. In the 1950s McKee[32] developed one of the best known metal-on-metal hip replacements of the 20th Century. Initially both components were made from stainless steel, but both friction and wear were excessive and so chrome-cobalt alloys were introduced. The **acetabular cup** was originally screwed into the bone, but later fixed by acrylic cement. A success rate of fifty percent was reported by the end of the 1950s. In due course it became clear that most of these early metal-on-metal implants survived for unsatisfactorily short periods, but equally significant was the observation that some survived for two decades or more. These pioneering empirical steps took place before there was an adequate understanding of the tribological principles upon which metal-on-metal hip replacements had to be designed.

When identical materials are used for both components in bearings the friction and wear can be excessive owing to the ease with which adhesion takes place at the interacting asperities on the loaded interface. High friction and subsequent loosening of the components were major issues in hip arthroplasty in the middle of the twentieth century. This prompted Charnley[33] to combine polymeric acetabular cups with relatively small diameter (7/8 inch or 22.225 mm) metallic femoral heads to minimize frictional torque. Polytetrafluoroethylene (ptfe or teflon) was used initially, but wear was excessive and after a mere three years the metallic heads had penetrated through many of fhe polymeric cups. The subsequent introduction of **ultra-high molecular weight polyethylene** (UHMWPE) cups and metallic femoral heads provided the basis for a form of hip replacement which became the '*gold standard*' for at least half a century.

The use of ceramic femoral heads reduces the volumetric wear of the UHMWPE by about fifty percent. This configuration, alongside ceramic-on-ceramic hip replacements, exhibiting very low wear rates, introduced a third important material into the story of total hip replacement in recent decades. Concerns for the brittleness of ceramic bearing materials have greatly diminished with improvements to bio-ceramic materials and their manufacture.

These three material combinations are depicted in Fig. 14.6.

In due course it was recognized that a biological response to the accumulation of wear debris from metal-on-polymer articulations promoted osteolysis and loosening of the joint components. This generated renewed interest in the use of ceramic femoral heads and also in hard-on-hard material combinations such as ceramic-on-ceramic and metal-on-metal

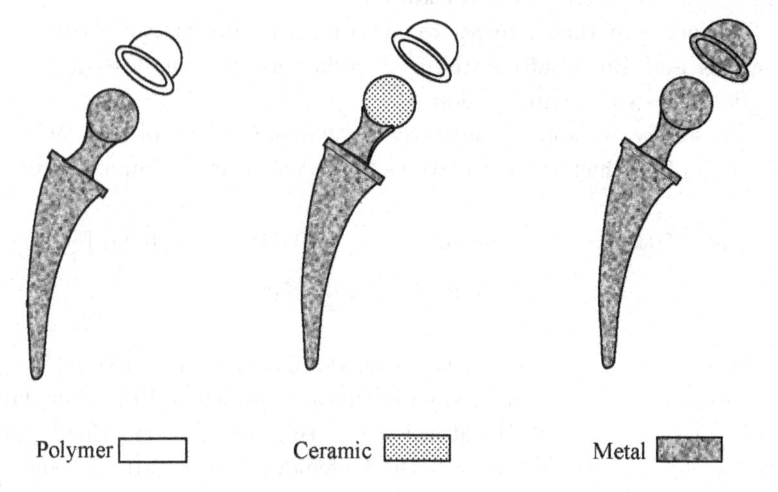

Fig. 14.6. Principal hip joint replacement materials.

total hip replacements. Reappraisal of the metal-on-metal combination of materials pioneered by McKee-Farrar in the 1950s has been a fascinating feature of recent studies.

14.5.2. *Metal-on-metal total hip replacements*

This combination of materials for the femoral head and acetabular cup offers a considerable reduction in volumetric wear compared with metal-on-polymer or ceramic-on-polymer implants. This is potentially attractive since the problem of **osteolysis**, afflicting metal-on-polymer joint replacements, appears to be related to the accumulated volume of polyethylene wear debris. Larger diameter femoral heads are also attractive since they offer greater stability and the opportunity to introduce surface replacement shells on the femoral head rather than solid, monolithic heads.

It has clearly emerged that large diameter metal-on-metal hip replacements enjoy improved lubrication and very low wear rates.[34] They can be analyzed by established elastohydrodynamic lubrication analysis outlined in Chapter 10. The shear rates in the thin fluid films which may be developed during at least part of the dynamic loading and sliding motion cycle associated with activities such as steady walking are typically of order 10^6 1/s. At these high shear rates the viscosity of **synovial fluid** is little affected by shear rate and is but a few times greater than the viscosity of water. The lubricant can thus be treated as being iso-viscous in initial estimates of the average film thickness.

The problem thus reduces to one of iso-viscous-elastic elastohydrodynamics and Eq. (14.6) introduced earlier for the analysis of natural synovial joints can again be adopted.

For steady walking, representative average values of the variables involved in film thickness calculations for metal-on-metal implants are;

$$D = 50\,\text{mm}, \quad c_d = 200\,\text{nm}, \quad E' = 250\,\text{GPa}, \quad \eta = 0.001\,\text{Pas},$$
$$\Omega = 1.5\,\text{rad/s}, \quad W = 2500\,\text{N}.$$

The average minimum film thickness for these conditions is 18 nm.

This is very small compared with the range of 300–3010 nm calculated in Sec. 14.4.3 for compliant natural joints. However, the remarkably good initial average surface finish on both the femoral heads (≈10 nm) and the acetabular cups (≈5 nm) of modern prostheses offers encouragement for the view that fluid film lubrication might provide some relief to the severity of friction and wear that would otherwise be encountered.

The composite surface roughness for this example is $\{R_{a1}^2 + R_{a2}^2\}^{\frac{1}{2}} = 11.2\,\text{nm}$, yielding a lambda ratio (see Eq. 14.5) of 1.6. This value suggests that fluid film lubrication will support at least some of the applied load, leaving a reduced proportion of the total load to be carried by highly stressed asperities on the opposing, sliding solids. Selection of the femoral head diameter and control of the diametral clearance will thus play a significant part in the tribological performance of the prosthesis. Values of the lambda ratio for implants with roughness average values of 10 nm and 5 nm, heads of diametral clearances of 200 μm and diameters ranging from 22 mm to 60 mm are shown in Table 14.3.

The results show that boundary lubrication action will govern tribological behavior of the smaller diameter prostheses, but that there is every possibility that a substantial proportion of the load applied to the larger heads will be supported by fluid film (**iso-viscous-elastic**) lubrication action. Experimental support for this conclusion was presented in Refs. 35 and 36. Results for the total running-in volumetric wear for metal-on-metal prostheses with broadly similar surface finish, but different diameters, were measured on joint simulators. The results are shown in Fig. 14.7.

Table 14.3. Influence of femoral head diameter upon lambda ratio. ($R_{a1} = 10\,\text{nm}$ $R_{a2} = 5\,\text{nm}$) ($c_d = 200\,\text{nm}$; $\eta = 0.001\,\text{Pas}$; $\Omega = 1.5\,\text{rad/s}$; $E' = 250\,\text{GPa}$; $W = 2500\,\text{N}$).

Diameter (mm)	22	28	36	45	50	55	60
Lambda (Λ)-ratio	0.27	0.46	0.79	1.29	1.63	2.01	2.43

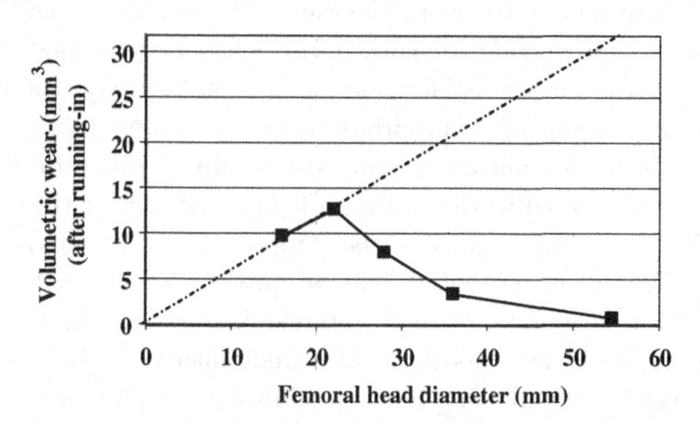

Fig. 14.7. Influence of femoral head diameter upon running-in wear (simulated walking cycle).

A broken line has been drawn from the origin through the running-in wear values for the two smallest diameter heads. The ratio of the wear volumes at 22.225 mm and 16 mm was almost the same as the ratio of the head diameters and the slope of this line, as would be expected as the volumetric wear increased under boundary lubrication conditions. The wear then decreased as the head diameter increased, relative to this linear relationship, such that at 54 mm the measured volumetric wear was only slightly more than 2% of the projected wear that would have occurred under boundary lubrication conditions. As the diameter increased beyond 22.225 mm, mixed lubrication was encountered owing to the generation of thicker effective lubricating films and reduced asperity contact. While full fluid film lubrication and zero wear throughout the dynamic loading and motion associated with simulated walking conditions was never achieved, the protection offered by the elastohydrodynamic films was nevertheless most impressive.

14.5.3. *Alternative material pairs*

In the previous section the role of tribology in total hip joint replacement was illustrated for metallic femoral heads and acetabular cups. This combination of materials has attracted much attention in recent years for the treatment of damaged or diseased joints in younger, active patients. The joints generally work well as bearings, although it will be some time before the biological response to the millions of metallic wear particles of nano-metre proportions and the substantial release of metal ions is known.

It is sometimes deemed to be desirable to use a thin hemispherical metallic shell to replace the worn femoral head cartilage, rather than using a traditional, solid metallic femoral head. The previous analysis applies equally well to the surface replacement and monolithic designs of prostheses. Likewise, by introducing appropriate values of material properties, the film thickness in ceramic-on-ceramic **total hip replacements** can be ascertained and related to the surface roughness offered by these materials which offer exceedingly good surface topography and attractive wear resistance. A further potential pair of prosthetic materials is offered by differential hardness metal-on-ceramic materials. This again takes advantage of the very smooth surface and inherent wear resistance of ceramics. While the higher elastic modulus of ceramic will lead to a slightly reduced theoretical fluid film thickness, the lambda ratio, and hence, effectiveness of lubrication may be enhanced owing to the improved composite roughness of pairs of materials.

None of these pairs of hard-on-hard materials mimics the properties of bearing materials used in natural synovial joints. The possibility of using man made, low modulus bearing layers capable of promoting effective lubrication continues to appeal. However, as yet no material structures with adequate long term integrity and suitable bearing properties have emerged.

14.6. Further Examples of Bio-Tribology

While natural and replacement synovial joints represent major topics in the field of biotribology, there are many other significant examples of tribology in biological systems.

14.6.1. *Skin*

This tissue represents about one eighth of the body mass. It exhibits a stratified structure, consisting of three distinct layers. The lower, fatty connective tissue (*hypodermis*) supports a relatively thick middle layer (*dermis*) while the latter supports a thinner outer layer (*epidermis*). The outermost layer of the epidermis, the stratum corneum, consists of flattened scales of **keratin** skin some 10 μm thick. Cells formed deep in the epidermis constantly divide and as the daughter cells move towards the surface they are keratinised before reaching the surface some two to four weeks later. These dead cells are constantly being rubbed off the surface, thus forming an unusual illustration of a biological wear process.

Skin is rough by engineering standards, the R_a being in the range 4–40 μm. When initially stretched it exhibits substantial strains with low values of elastic modulus in the range (0.04–1000) kPa. At higher strains the modulus can rise to about 3 GPa.

Hydration plays a dominant role in determining both the elasticity and the coefficient of friction of skin. The former can be reduced by factors of about 10^3 when the skin is wet. Pharmaceutical and cosmetic companies undertake considerable research into the friction of skin, with tribologists being increasingly involved.[37–39]

Friction is the tribological feature of greatest interest for skin, since the tissue forms the interface between our bodies and the environment. The coefficient of friction varies with site and particularly with hydration. The average value for untreated skin is just under 0.5, being highest on the fingers and soles of the feet ($\approx$1.2) and lowest on the back, ankle and upper arm ($\approx$0.2). Laboratory measurements of the friction of skin are important since touch, locomotion, the comfort of clothes and the grasping tools and safety rails all depend upon appropriate values of the coefficient being achieved. Accidents in the home, particularly on stepping into or out of a bath and on wet pavements are particular hazards for the elderly and infirm.

14.6.2. *Hair*

The **protein keratin** also forms the major constituent of hair. With diameters of about (50–70) μm, each strand of hair exhibits two or three distinct layers. A thin outer layer (*cuticle*), a much thicker inner region (*cortex*) and in thicker hairs a central (*medulla*). The modulus of elasticity is about 3 GPa.

Once again a biological wear process is evident in the continuous growth and shedding of hairs. Mechanical or chemical damage to the outer cuticle can sometimes be viewed as a wear process.

Friction is again the tribological characteristic of major interest with this keratinous tissue. A most distinctive difference between friction during motion from *root-to-tip* and *tip-to-root* is observed during combing. This *differential friction effect* is attributable to the mechanical structure of **cuticle scales**, which overlay the cortex like the tiles on a roof as shown in Fig. 14.8.

The recorded coefficients of friction for sliding towards the tip are similar to those of untreated skin ($\approx$0.5), but when combing towards the root the values almost double to about 0.9.

14.6.3. *Shaving*

Friction plays a part in the wet shaving process, with the tribological characteristics of both skin and hair contributing to the action. The

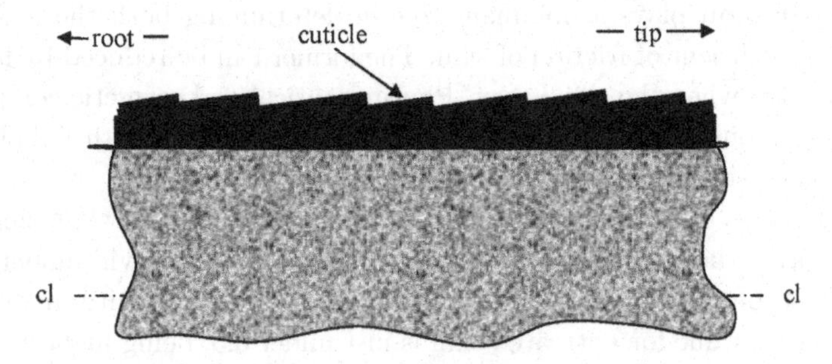

Fig. 14.8. Representation of the cross-section of a strand of hair.

principal objective during shaving is to hydrate the skin and hairs in order to soften the latter and hence reduce their elastic modulus and resistance to cutting. At the same time, specially formulated shaving soaps, creams or gels reduce friction between the razor and skin and razor blade and hair. Treatments of skin and hair and the formulation of preparations to assist shaving continue to attract the interest of pharmaceutical and cosmetic companies.

14.6.4. *Teeth*

Wear of both natural tissues and restorative materials are primary concerns in the tribology of teeth. In the former category the properties of enamel and dentine have been explored extensively, while in the latter area there is interest in existing and potential restorative materials. In both cases toothbrushes and dentifrice/toothpaste must be sufficiently abrasive to clean teeth effectively, but they must not cause excessive wear which could expose the underlying dentine.

Enamel is the hardest tissue in the body, with measurements of the modulus of elasticity ranging from (30–130) GPa. The underlying **dentine** is much softer with values of the elastic modulus ranging from (10–20) GPa. **Physiological wear** is linked to masticatory processes with food acting as the third body, while **Pathological wear** arises from conditions like bruxism. **Prophylactic wear** is associated with systems of cleaning teeth such as tooth sticks, brushes, toothpaste and flossing.

Physiological wear rates of (0.04–0.2) μm/day have been reported, while there is much more variation in pathological wear from one individual to another. These daily wear rates yield wear factors of order $10^{-3}\,mm^3/Nm$. Several tribology research groups have contributed to studies of the wear of teeth and the role of the toothbrush and toothpaste.[40–42] The totality of work in this field has contributed to the development of standards for both manual and electric toothbrushes and toothpaste abrasivity.

14.7. Closure

This brief outline of major topics in biotribology is by no means exhaustive. Other active areas include ocular tribology (lubrication of eyelids and contact lenses), cardiovascular tribology (blood flow and heart valves), and the lungs (membrane lubrication).

Biomimetics, in which engineers endeavour to understand well developed tribological systems in nature and possibly to apply the governing principles to the solution of engineering problems, has not yet had the same impact upon tribology as it has in other branches of physics and engineering. However, interest in this field is growing as solutions are sought to new classes of problems.

References

1. Dowson, D. and Wright, V., Bio-Tribology, *Proceedings of the Conference on The Rheology of Lubrication*, organised by The Institute of Petroleum, The Institution of Mechanical Engineers and the British Society of Rheology (1973) 81–88.
2. Department of Education and Science, *Lubrication (Tribology) Education and Research*, A Report on the Present Position and Industry's Needs, HMSO, London (1966) 1–80.
3. Alexander, R. McN, Jayes, A. S., Maloiy, G. M. O. and Wathuta, E. M., Allometry of the limb bones of mammals from shrews to elephants, *J. Zool., Lond.* **189** (1979) 305–314.
4. Alexander, R. McN, *Animal Mechanics*, Sidgwick & Jackson, London (1968).
5. Calow, L. J. and Alexander, R. McN., A mechanical analysis of a hind leg of a frog, *J. Zool., Lond.* **171** (1973) 293–321.
6. Alexander, R. McN., The mechanics of jumping by a dog (Canis familiaris), *J. Zool., Lond.* **173** (1974) 549–573.
7. Alexander, R. McN. and Vernon, A., The mechanics of hopping by kangaroos (Macropodidae), *J. Zool., Lond.* **177** (1975) 265–303.
8. Alexander, R. McN., Maloiy, G. M. O., Njau, R. and Jayes, A. S., Mechanics of running of the ostrich (Struthio camelus), *J. Zool., Lond.* **187** (1979) 169–178.
9. Alexander, R. McN., Maloiy, G. M. O., Hunter, B., Jayes, A. S. and Nturibi, J., Mechanical stresses in fast locomotion of buffalo (Syncerus caffer) and elephant Loxodonta africano), *J. Zool., Lond.* **189** (1979) 135–144.
10. Scherge, M. and Gorb, S. N., *Biological Micro-and Nano-tribology-Nature's Solutions*, Springer-Verlag, Berlin, Heidelberg, (2001) 1–304.
11. Arzt, E., Gorb, S. and Spolenak, R., From Micro to Nano Contacts in Biological Attachment Devices, *Proc. Natl. Acad. Sci. USA* **100**, 19 (2003) 10603–10606.
12. Gao, H., Wang, X, Yao, H., Gorb, S. and Arzt, E.,Mechanics of hierarchical adhesion structures of geckos, Mechanics of Materials **37** (2005) 275–285.
13. Akhmatov, A. S., Molecular physics of boundary lubrication, *Gosudarstvennoe Izdatel'stvo Fiziko-Matematicheskoi Literatury*, Moskva (1963) 1–480.
14. Johnson, K. L., Contact mechanics, Cambridge University Press, Cambridge, (1985) 1–452.

15. Israelachvili, J. N., *Intermolecular and Surface Forces*, Academic Press, London (1992) 179–204.
16. Autumn, K., Chang, W.-P., Fearing, R., Hsieh, T., Kenny, T., Liang, L., Zesch, W. and Full, R. J., Adhesive force of a single gecko foot-hair, *Nature* **405** (2000) 681–685.
17. Autumn, K., Sitti, M., Liang, Y. A., Peattie, A. M., Hansen, W. H., Sponberg, S., Kenny, T. W., Fearing, R., Israelachvili, J. N. and Full, R., Evidence for van der Waals adhesion in gecko setae, *Proc. Natl. Acad. Sci. USA*, **99**, 19 (2002) 12252–12256.
18. Spolenak, R., Gorb, S., Gao, H. and Artz, E., Effects of contact shape on the scaling of biological attachments, *Proc. R. Soc.* **461** (2005) 305–319.
19. Autumn, K. How Gecko toes stick, *American Scientist* **94** (2006) 124–132.
20. Unsworth, A., Dowson, D. and Wright, V., The frictional behavior of human synovial joints-Part I: Natural Joints, *Transactions of the ASME, F, Journal of Lubrication Technology* **97**, 3 (1975) 369–376.
21. O'Kelly, J. O., Unsworth, A., Dowson, D., Hall, D. A. and Wright, V., A study of the role of synovial fluid and its constituents in the friction and lubrication of human hip joints, IMechE, *Engineering in Medicine* **7**, 2 (1978) 73–83.
22. Cooke, A. F., Dowson, D. and Wright, V., The rheology of synovial fluid and some potential synthetic lubricants for degenerate synovial joints, IMechE, *Engineering in Medicine* **7**, 2 (1978) 66–72.
23. Hamrock, B. J. and Dowson, D., Elastohydrodynamic lubrication of elliptical contacts for materials of low elastic modulus, Part I-fully flooded conjunction, *J. Lubr. Technol.* **100**, 2 (1978) 236–245.
24. Dowson, D. and Jin, Z.-M., Micro-elastohydrodynamic lubrication of synovial joints, *Engineering in Medicine* **15**, 2 (1986) 63–65.
25. Swan, D. A., Hendren, R. B., Radin, E. L., Sotman, S. L. and Duda, E. A., The lubricating activity of synovial fluid glycoproteins, *Arthritis and Rheumatism* **24**, 1 (1984) 22–30.
26. Swan, D. A., Bloch, K. J., Swindell, D. and Shore, E., The lubricating activity of human synovial fluids, *Arthritis and Rheumatism* **27**, 5 (1984) 552–556.
27. Swan, D. A., Silver, F. H., Slayter, H. S., Stafford, W. and Shore, E., The molecular structure and lubricating activity of lubricin isolated from bovine and human synovial fluids, *Biochem. J.* **225** (1985) 195–201.
28. Hills, B. A. and Butler, B. D., Surfactants identified in synovial fluid and their ability to act as boundary lubricants, *Annals of the Rheumatic Diseases* **43** (1984) 641–648.
29. Hills, B. A., Boundary lubrication in vivo, Proc. I.Mech.E., Part 'H', *Engineering in Medicine* **214**, H1 (2000) 83–94.
30. Scales, J. T., Arthroplasty of the hip using foreign materials: a history, in 'Lubrication and wear in living and artificial human joints, *Proceedings of the Institution of Mechanical Engineers, 1966–1967* **181**, Part 3J, (1967) 63–84.
31. Wiles, P., The surgery of the osteoarthritic hip, *British Journal of Surgery 1957–1958* **45** (1958) 488–497.

32. McKee, G. K. and Watson-Farrar, J., Replacement of arthritic hips by the McKee-Farrar prosthesis, *Journal of Bone ad Joint Surgery* **48B** (1966) 245–259.
33. Carnley, J., *Arthroplasty of the Hip-a New Operation,* Lancet (1961) 1129.
34. Dowson, D. (Ed.), Special issue on the development of monolithic and surface replacement metal-on-metal hip replacements, *Proc. IMechE, part H, Journal of Engineering in Medicine* **220**, H2 (2006) 1–412.
35. Dowson, D., Hardaker, C., Flett, M., and Isaac, G. H., (2004), A hip joint simulator study of the performance of metal-on-metal joints, Part I: the role of materials, *J. Arthroplasty* **19**, (8), Suppl. 3 (2004) 118–123.
36. Dowson, D., Hardaker, C., Flett, M., and Isaac, G. H., A hip joint simulator study of the performance of metal-on-metal joints, Part II: design, *J. Arthroplasty* **19**, (8), Suppl. 3 (2004) 124–131.
37. Adams, M. J., Briscoe, B. J. and Wee, T. K., The differential friction effect of keratin fibres, *J. Phys. D.* **23** (1990) 406–414.
38. Bhushan, B., Wei, G. and Haddad, P., Friction and wear studies of human hair and skin, *Wear* **259** (2005) 1012–1021.
39. Gitis, N. and Sivamani, R., Tribometrology of skin, *STLE Tribology Transactions* **47**, 4 (2004) 461–469.
40. Wright, K. H. R., The abrasive wear resistance of human dental tissues, *WEAR* **14** (1969) 263–284.
41. De Gee, A. J., ten Harkel-Hagenaar, H. C. and Davidson, C. L., Structural and physical factors affecting the brush wear of dental composites, Journal of Dentistry **13**, 1 (1985) 60–70.
42. Powers, J. M. and Bayne, S. C., *Friction and wear of dental materials, ASM Handbook, Friction, Lubrication and Wear Technology,* (Blau, P. J.), (1992) 665–681.

QUESTIONS

We have set a number of questions here for you to try and consolidate what you have learnt by reading through the text. These questions supplant examples provided in the various chapters. There are few questions for some of the chapters. Solutions for some of these are given in the Book of Solutions at the end of the book. Some solutions make use of little programs written in Mathcad or similar popular mathematical solvers. In all cases you are encouraged to attempt to solve the problems first, before referring to the Book of Solutions.

Chapter 2 Questions

2.1. Find the analogue values of R_a and σ for the triangular roughness profiles (a) and (b) shown in Fig. 2.1. Compare them with the sinusoidal profile in the worked Example (2.1).

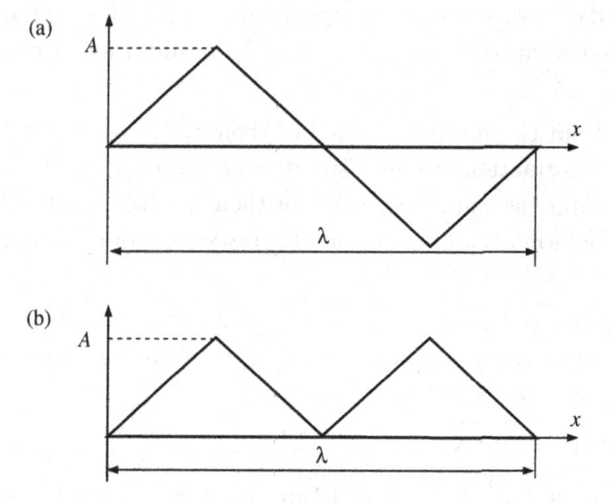

Try to minimize your effort by considering the definitions of R_a and σ, finding their values in question (a) by only integrating over $\lambda/4$. Justify that the same approach applies to case (b) as well.

Ans: (a) and (b): $R_a = A/2$, $\sigma = A/\sqrt{3}$

2.2. Find the Root Mean Square height of the parabolic surface roughness below if the amplitude $= a =$ 03 micron, and the mean radius of curvature $= R = 150$ micron.

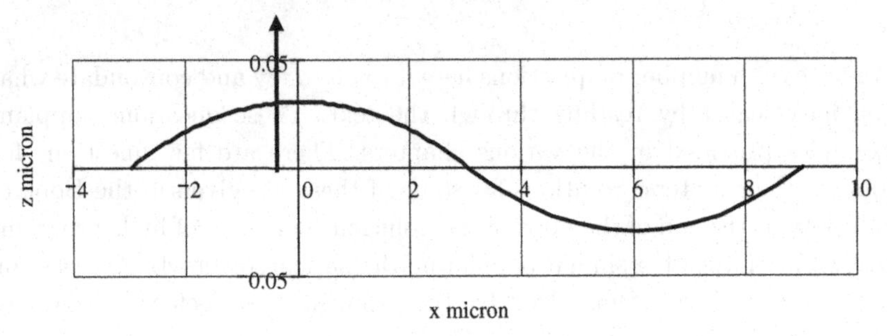

Parabolic profile

Ans: $\sigma = 0.024 \times 10^{-6}$ m

Chapter 3 Questions

3.1. Two flat steel surfaces are lapped to give each a standard deviation of their surfaces peak heights of $0.2\,\mu$m and a mean peak radius of $5\,\mu$m.

(a) Find the *minimum* value of their hardness that will make the deformation of their asperity peaks predominantly elastic

(b) Find the *maximum* value of their hardness that will make the deformation of their asperity peaks predominantly plastic
$E^* = 110\,$GPa

3.2. In the worked example (1) in Chapter 3 (Sec. 3.4.3), if the ball and the races have $E^* = 110\,$GPa and $H = 8\,$GPa. Find the load capacity before yielding commences.

Ans: $W = 2381\,$N

3.3. A ceramic ball of diameter 12 mm is pressed onto another steel ball of diameter 15 mm under a normal load of 400 N. Calculate the minimum hardness of the steel ball for the contact to (a) remain fully elastic, (b) just become fully plastic

Note: $E_{cer} = 460\,\text{GPa}, E_{st} = 200\,\text{GPa}, \sigma_{st} = \sigma_{cer} = 0.3, H_{cer} > H_{st}$.

3.4. In a contact mechanics experiment, a steel ball of 0.01 m radius is loaded against a cylinder of equal radius. If the normal force is 1 kN, find the dimensions of the contact footprint and the maximum pressure

Note: $E^* = 1.1 \times 10^{11}\,\text{Pa}$

Ans: $a = 4.57 \times 10^{-4}\,\text{m}$ and $b = 2.88 \times 10^{-4}\,\text{m}$

Chapter 4 Questions

4.1. The drawing by Leonardo, shown in Fig. 1.2, is a bobbin type separator for a thrust ball bearing. It was discovered in the Madrid National Library in 1967. Leonardo's original description is given below.

> *'I affirm that if a weight with a flat surface moves on a circular surface, the movement will be facilitated by interposing between them balls or rollers, save the fact that the balls have uniform motion whereas rollers can move only in one direction. But if the balls or rollers touch each other in their motion, then movement will be more difficult than if there was no contact between them because when they touch the friction causes a contrawise motion and, for this reason the movements will contradict each other. However, if the balls or rollers are pitched apart, they will touch only at one point between the load and its resistance and consequently it will be easy to produce the movement'.*

With the aid of sketches, describe more fully the friction principles he refers to. Elaborate on why he believed that it would work. Try inventing your own more comprehensive designs of rolling separators for radial and thrust ball bearings.

Assume that the balls themselves are guided by annular grooves cut in the top and bottom races.

Reference

Gohar, R. Low friction rolling element bearings *Wear* **104** (1985) 309–333.

4.2. We need to design a short inclined conveyor that allows a pallet, of convenient shape, to slide down it on two identical parallel cylindrical rods with minimum friction. The rods are supported by journal bearings attached to the conveyor and can be driven about their axes of rotation at speeds well in excess of the desired sliding speed of the pallet. There is no lubricant available for the pallet and rod surfaces.

Assuming that total friction force is independent of speed, sketch a possible arrangement and explain why the friction forces on the pallet *along* the conveyor may be much lower than if the rods did not rotate. What is a disadvantage of such an arrangement?

4.3. A single hard conical asperity of apex semi angle θ, ploughs a groove under a load W, of depth h and width $2r$ at the groove rim, in a softer surface. Show that the ploughing friction coefficient is

$$\mu_p = \frac{4h}{\pi d} = 2\cot\theta/\pi.$$

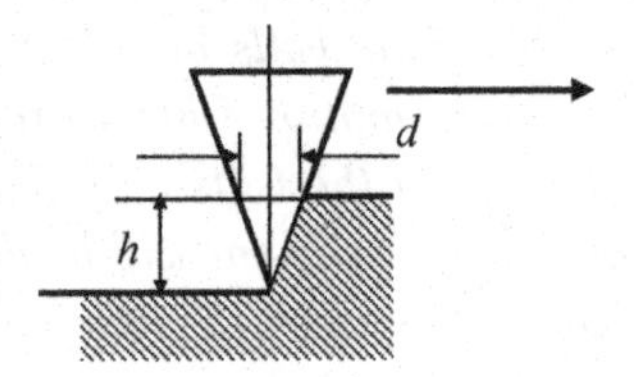

4.4. A wear test uses a face of a bronze annulus, of outside diameter 0.025 m and inner diameter 0.015 m, loaded against a flat steel plate and rotated about its spin axis at 5 rev/s for 20 hours under a force of 100 N. Upon completion of the test, the difference between the pre and after test masses of the specimens shows that the bronze one has lost 250 mg and the steel one 8 mg. Estimate the wear coefficients for the bronze and steel.

The hardnesses in *MPa* are 2400 (steel) and 800 (bronze) and the densities are 7800 $\mathrm{Kgm^{-3}}$ (steel) and 8400 $\mathrm{Kgm^{-3}}$ (bronze).

Ans: $K_s = 1.08 \times 10^{-6}$ and $K_b = 10.64 \times 10^{-6}$.

4.5. In friction tests, two hard conical sliders 1 and 2, of semi angles 70 and 80 degrees respectively, were slid along a metal surface under a constant load.

In the first test the surface was dry and it was found that the measured coefficients of friction for each slider were such that $\mu_{1d}/\mu_{2d} = 1.1$.

In the second test, the experiment was repeated but the surface was lubricated and the measured coefficients for each slider were such that $\mu_{1l}/\mu_{2l} = 1.3$. From these results, find the coefficient of adhesive friction in each case assuming the ploughing components are not affected by addition of the lubricant.

$$\therefore \mu = 2 \cot \theta / \pi.$$

Chapter 5 Questions

5.1. A mineral oil has the following viscosity-temperature characteristics

(a) Using Vogel's formula, find the constants a, b and c. Hence find the viscosities at 30°C and −5°C. Plot the result on ASTM paper or, with suitable scales, employ computer graphics

Temperature °C	Viscosity cSt
20	170
40	46
100	6.97

(b) Choose two of the temperatures from the data supplied and compare the results from (a) using Walther's formula.

5.2. A special lubricant is required with a viscosity of 115 cSt at 35°C. This is to be achieved by blending two standard lubricants, 1 and 2, with their properties shown in the table.

Standard lubricant	Kinematic viscosity (cSt) at 25°C	Kinematic viscosity (cSt) at 100°C
1	100	8.5
2	350	17

Using Walther's equation

(a) Find the values of its constants from the above data.
(b) Determine the viscosity of each standard fluid at 35°C.
(c) Using these results, on a blending chart you make, or ASTM chart, determine the weight percent of the two standard lubricants that must be blended to create the special lubricant.

The aim of the procedure, outlined below, is to find a solution to Question (2) assuming an ASTM viscosity-temperature chart is not available.

Procedure

(a) From the data, find the values of the constants, d and e, in Walther's formula for each of the standard lubricants
(b) Insert the values of these constants in the respective equations and find their viscosities at 35°C. This produces the vertical line through C in Fig. 5.4.
(c) The chart you should produce will have $\log(t + 273)$ as the abscissa and $\log(\log(\alpha))$ as the ordinate (remember that α is in cP). To do this, its a good idea to arrange Walther's equation with α only on the left hand side. The two lines you will get should be parallel. Excel or Math Cad are the softwares of choice
(d) Print the chart and draw a vertical line through log $(35 + 273)$ and a horizontal line through 115 cP.
(e) Measure CA/AB to find the solution. The required characteristic for the blended should pass through point C and will be parallel to those of the two standard oils.

Chapter 6 Questions

6.1. Calculate the peak pressure between 2 long rigid discs each 0.1 m diameter each rolling at 1000 rev/min and 2.54×10^{-4} m apart with a lubricant of viscosity 0.1 Pa s. Use half Sommerfeld boundary condition.

6.2. If the minimum pressure for the contacting disks in Example (1) cannot exceed $-68950\,\text{Nm}^{-2}$ gauge, and there is a flooded inlet, discuss the modified boundary conditions for Eq. (6.27) and sketch the pressure distribution. What is the maximum positive pressure in this case?

6.3. Figure 3.1 shows a *long* wedge of width B having a stationary top surface, and a geometry that produces an exponential lubricant film of the form $h = h_0 e^{\alpha x}$

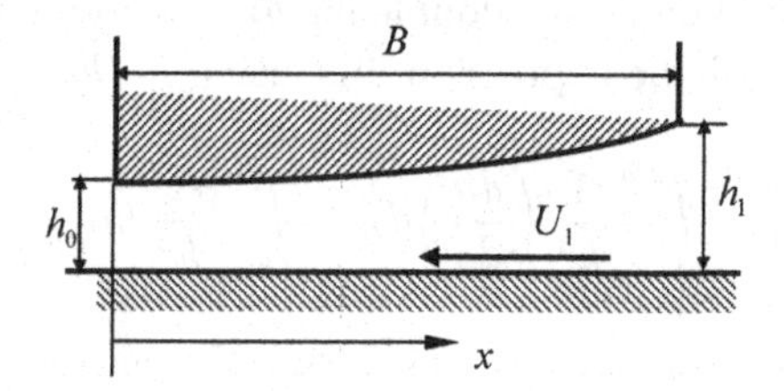

(a) If the bottom surface moves to the left at speed U_1, starting with Eq. (6.22), show that the film thickness where $dp/dx = 0$ is given by:

$$h_c = \frac{3h_0}{2}\left(\frac{1 - e^{-2\alpha B}}{1 - e^{-3\alpha B}}\right).$$

(b) Derive an expression for the pressure distribution at a coordinate x in the film if the pressure is zero at either end of the wedge.

(These are the 2 boundary conditions necessary for the solution of Reynolds equation (6.22) for long bearings.)

6.4. Figure 4 shows a *narrow* wedge of length L, where the lubricant film thickness is given by

$$h = h_m(1 + m\cos\pi x/B) \quad \text{and} \quad dh/dy = 0.$$

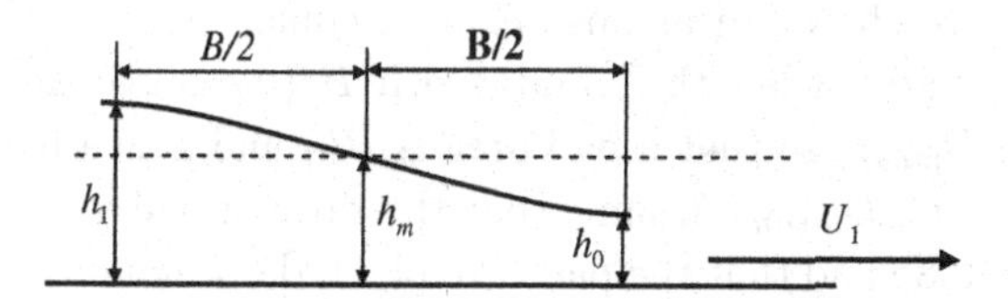

If $h_1/h_0 = 2$, show that the load capacity of the arrangement can be written as

$$\frac{Wh_0^2}{\pi\eta U_1 LB^2} = \frac{3}{16}\left(\frac{L^2}{B^2}\right).$$

Hint: use Eq. (6.35) and do the double integral to obtain the load, *ignoring* the x direction film shape expression, but using $h_1/h_0 = 2$. Note that:

$$\int_0^{B_0} \frac{1}{h^3}\left(\frac{dh}{dx}\right)dx = \int_{h_1}^{h_0} \frac{1}{h^3}dh.$$

As long as there is zero slope at either end of a narrow bearing, does the film shape matter, if h_1/h_0 only is given?

Chapter 7 Questions

7.1. A high speed plane pivoted thrust bearing pair is required to support a load of 0.4 mN, with a minimum film thickness of 50 microns and a runner speed of 50 m/s. The effective temperature of the test oil is 50.5°C operating at its effective viscosity.

The viscosity properties are: viscosity 19.5 cS at 100°C, viscosity index = 95, density = 870 kg/m^3.

Find:

(a) The effective viscosity of the test oil
(b) The minimum pad dimensions needed to fulfil the above specification, assuming that the bearing pair will operate at its optimum convergence ratio (K).
(c) The approximate power consumed by the bearing under these conditions

The ASTM data table gives: for $\nu = 19.5$ cSt at 100°C, $L = 471.3$ cS at 40°C and $H = 221.1$ cSt at 100°C

Ans: (a) $\eta_e = 0.113$ Pas, (b) $B = L = 0.14$ m and (c) 81 kW.

7.2. The figure shows an arrangement of equally spaced pivoted thrust bearing pads of length L and width B (measured along the mean radius, D_{mean}, so that a pad area is BL) and L is a fixed dimension equal to 0.3 D_{mean}. Assume that the ratio of pad area to swept area is to be 0.85 and that the pads are pivoted optimally so that the load on each pad is given by

$$W = \frac{6U_1\eta B^2 L}{h_0^2} k_p S_l,$$

where k_p is a wedge factor equalling 0.85 for the optimum position used here and S_l is a side leakage factor defined by curve fitting as $S_l = 0.370L/B$.

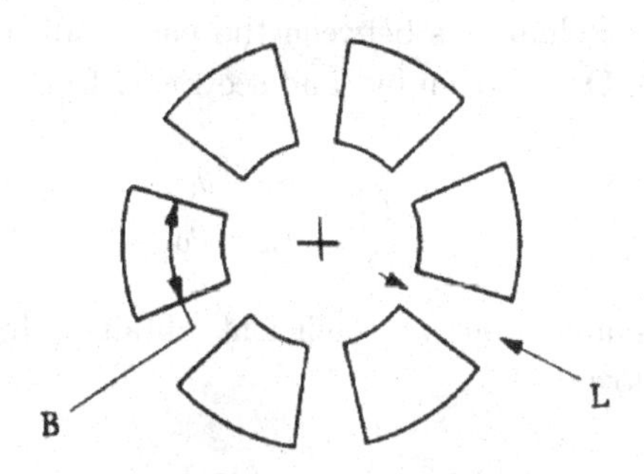

Show that the load carried by the whole assembly, W_t, is independent of the number of pads, N, and find its magnitude for the following specification

$$U_1 = 50\,\text{m/s}, \ D_{\text{mean}} = 0.4\,\text{m}, \ \eta_e = 30\,\text{cP}, \ h_0 = 51\mu\,\text{m}$$

Ans: $W_t = 0.0141^{U_1\eta D^2_{\text{mean}}}/h_0^2 = 0.52\,\text{MN}$.

7.3. A pivoted plane thrust bearing has a square plan form with $B = L = 50\,\text{mm}$ and has its pivot close to its optimum position at $K = 1$. The bearing is required to support a maximum load of 7 kN at a runner speed of 10 m/s. The same mineral oil as in worked Example 7.1 is to be used at a supply temperature of 30°C. Assuming that 60% of the heat flow in the film is carried away by convection, find the minimum film thickness, effective temperature and effective viscosity under these conditions. Comment on these results when comparing them to those of worked Example 7.1.

Ans: $\theta_{se} = \theta_o + 45.1 = 75.1°\text{C}, \ \therefore \ \eta_e = 19.8\,\text{cP}$ and $h_0 = 1.56 \times 10^{-5}\text{m}$.

7.4. A fixed inclination plane thrust bearing is to have a square plan form with $B = L = 50\,\text{mm}$, and a minimum film thickness of $2.2 \times 10^{-5}\,\text{m}$

0.03 mm. The bearing is required to support a maximum load of 7 kN at a runner speed of 10 m/s. The same mineral oil as in example 7.1 is to be used at a supply temperature of 30°C. Assuming that 60% of the heat flow in the film is carried away by convection, find the taper, t, the effective temperature and the effective viscosity under these conditions. Assume $\frac{k_1}{\rho\sigma} = 0.376 \times 10^{-6}$.

7.5. Using data from Question 4, find the temperatures at the leading edge (θ_1) and trailing edge (θ_2) of a thrust bearing pad if the thermal balance in the chambers between the pads, called **Hot Oil Carry Over (HOCO)**, is given by (and see Fig. 7.6).

$$k_o = \frac{\theta_1 - \theta_0}{\theta_{se} - \theta_0}, \tag{a}$$

where k_o is an empirical coefficient obtained from the curve fit equation below

$$k_o = -4 \times 10^{-6} U_1^3 + 4 \times 10^{-4} U_1^2 - 2 \times 10^{-2} U_1 + 0.85, \tag{b}$$

and U_1 is the runner speed.

Also assume that the effective temperature equals the mean of the pad temperature rise

$$\theta_{se} = \frac{\theta_1 + \theta_2}{2} \quad \text{(see Eq. 7.24)} \tag{c}$$

Chapter 8 Questions

8.1. This question is based on Journal bearing experiments by Cole (1957) Given: Lubricant viscosity characteristics

100°C $\eta = 4.2$ cS
50°C $\eta = 34$ cS
20°C $\eta = 440$ cS
$\rho = 850$ kg/m^3

General data: Length 0.0254 m, Diameter 0.0508 m, radial clearance 5×10^{-5} m
Load: 1397 N

Test number	Rev/s of journal (N)	Film supply temperature T_0°C	Film temperature rise $\Delta T = (T_2 - T_0)^\circ$C
1	91	56	21
2	118	61	25.5
3	148	66	28
4	187	72	32

Using Math Cad or by writing a computer program, compare these experimental results with theory using the equivalent viscosity method.

8.2. A uniform rotor is supported at its ends by two identical hydrodynamic journal bearings. Each has the following characteristics:

$L = 50\,\text{mm}$, $D = 100\,\text{mm}$, $c = 0.07\,\text{mm}$, $N = 20\,\text{rev/s}$, assume the effective viscosity of the lubricant is $\eta = 30\,\text{cP}$

(a) For eccentricity ratios $0.4 \leq \varepsilon \leq 0.65$ find an approximate relationship of the form

$$S = a\exp(b\varepsilon).$$

Over the given range of eccentricities:

(b) Show that the operating eccentricity ratio is about 0.6

(c) Obtain an expression for the stiffness of each bearing, $\partial W/\partial e$

(d) Find the stiffness of each bearing

(e) Hence find the undamped natural frequency of the system

Notes:

For (a) Go to Appendix Table 8.2 and for $L/D = 0.5$, plot $1/S$ (on a log scale) against ε (on a linear scale). You should get an approximate straight line over the given range. Then choosing two values of ε (say 0.4 and 0.6) and their corresponding $1/S$ values, obtain the constants a and b.

8.3. Using results from the worked Example 8.8.3, find the temperatures at the beginning and end of the film pressure arc, T_1 and T_2.

The questions below combine some materials covered in Chapters 1, 8 and 12.

8.4. A hydrodynamic journal bearing is expected to supports a maximum load, W, as its journal rotates with an angular velocity ω rad/s. Its bushing is stationary. The bearing diameter-to-length ratio is: $D/L = 3$.

(a) If the attitude angle is measured to be 5°, calculate the eccentricity ratio under the stated condition and comment on its value.

(b) If the designed bearing clearance is 100 μm and the surface roughness of the journal and bushing are 0.5 μm Ra and 0.4 μm Ra respectively, what is the likely regime of lubrication under the stated condition and why?

(c) For this bearing: $\frac{c}{R_j} = \frac{1}{1000}$, where: c is the clearance, R_j is the radius of the journal and L the length of the bearing. The maximum pressure is given by the expression:

$$p_m = \frac{3u\eta_0\varepsilon}{4R_j}\left(\frac{L}{c}\right)^2\frac{\sin\varphi_m}{(1+\varepsilon\cos\varphi_m)^3}$$

where: u is the speed of entraining motion, η_0 the dynamic viscosity of the lubricant at atmospheric pressure, ε the eccentricity ratio and $\varphi_m$the circumferential position for the position of maximum pressure: $\varphi_m = \cos^{-1}\left\{\frac{1-\sqrt{(1+24\varepsilon^2)}}{4\varepsilon}\right\}$.

Show that the maximum pressure is obtained as: $7.59\,\omega\eta_0$ GPa

(d) What would be the new value of the minimum film thickness (under the same maximum pressure), if the bearing bush is replaced by a thin shell of Babbitt of 2 mm thickness, with modulus of elasticity 60 GPa, and Poisson's ratio 0.23? Give your answer in terms of the product $\omega\eta_0$.

8.5. (a) In an engine the crankshaft journal bearings have a radius of 0.05 m, and a width of 0.05 m. The radius-to-nominal clearance ratio for these bearings is 1000. The lubricant used has a dynamic viscosity of 20 cP (Centi-Poise) at the operating contact temperature of 85°C. At an engine speed of 3000 rpm, the load carried by the journal closest to the flywheel position is 15 kN. Assume that the bearing operates under hydrodynamic regime of lubrication. Treat the problem as isothermal.

(i) What simplified journal bearing theory would you use for your analysis, and why?

(ii) Using the chart in figure Q 8.5 (note the vertical axes are given in logarithmic scale), obtain the eccentricity ratio for this bearing. This chart gives the relationship between the Sommerfeld Number and the eccentricity ratio for various L/D ratios.

(iii) What is the attitude load angle under the above stated condition? Comment as to the significance of its value with respect to the state of bearing loading.

(iv) Assuming half-Sommerfeld condition, what is the minimum film thickness under the above stated condition?

(b) (i) If the position of the maximum pressure is given by:

$$\varphi_m = \cos^{-1}\left\{\frac{1-\sqrt{(1+24\varepsilon^2)}}{4\varepsilon}\right\},$$

where φ denotes the circumferential direction around the bearing, and ε is the eccentricity ratio, then:

Find the maximum pressure in the bearing.

(ii) The above stated hydrodynamic bearing is replaced by a thin shell bearing of the same geometric dimensions and nominal clearance, using a Tin-based alloy of modulus of elasticity 60 GPa, Poisson's ratio 0.25 and thickness 2 mm. If the same pressure is assumed to apply what would be the enhanced minimum film thickness on the account of the local deformation of the thin shell?

Chapter 9 Questions

9.1. A circular pad hydrostatic thrust bearing is of outside radius 0.04 m with a central recess of 0.02 m radius. Two different feeding systems are used.

(a) The recess is supplied directly by oil of 25 cP by a constant flow pump producing 10 liters/min. Find the load that can be carried for a uniform film thickness of 0.1 mm and the percentage load increase if the film thickness is reduced by 10%.

(b) Alternatively, the recess is supplied with the same oil via a capillary restrictor such that the film thickness is again 0.1 mm,

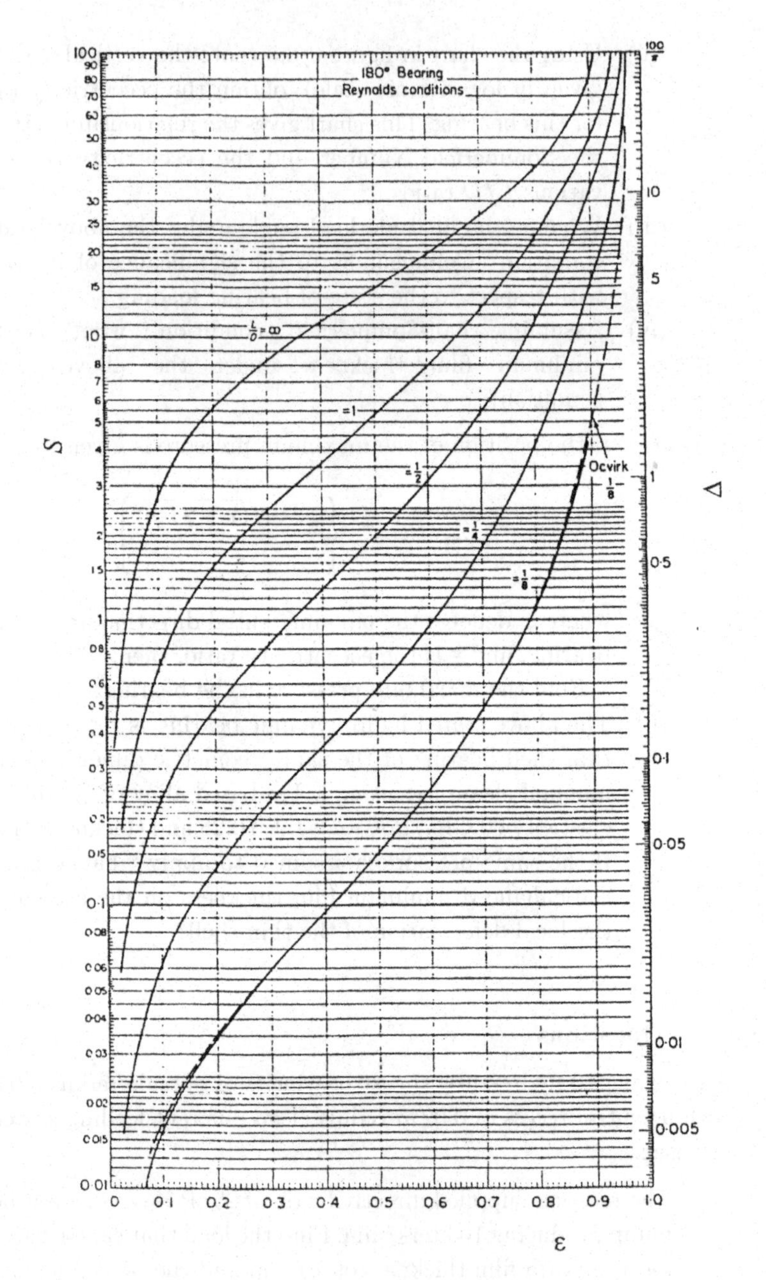
180° Bearing
Reynolds conditions
$\frac{L}{D} = \infty$
= 1
= 1/2
= 1/4
= 1/8
Ocvirk
1/8
S
Δ
ε
100/π
10
5
1
0.5
0.1
0.05
0.01
0.005
0 0.1 0.2 0.3 0.4 0.5 0.6 0.7 0.8 0.9 1.0

Fig. Q8.5.

with a pressure drop across the capillary of $10\,\text{mN/m}^2$ from a pump supply pressure of $15\,\text{mN/m}^2$. Determine the load that can be carried and the percentage increase in this load if the film thickness is reduced by 10%.

9.2. The accompanying figure shows a hydrostatic conical pad bearing which takes axial load only. $l = 75\,\text{mm}$, $R_0 = 25\,\text{mm}$, $R = 50\,\text{mm}$, $h = 0.05\,\text{mm}$. The conical bearing has the same load coefficient, $\bar{A}$,

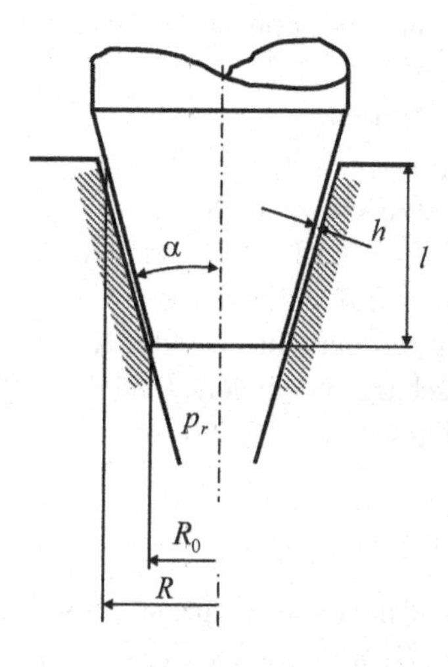

as a flat bearing of identical projected dimensions. Thus, Fig. 9.4 or Eq. (9.5) can be used where $\bar{R} = (R/R_0)$.

The flow leakage from a conical bearing is reduced, compared with a flat bearing of the same R, because of the increased length of the land. Thus, for a conical bearing, $\bar{B}$ in Figure 9.4 or Eq. (9.4) must be multiplied by sin α. The film thickness in a conical bearing is measured normal to its conical surfaces.

If the oil viscosity and density in this bearing, are respectively 53 cSt and $880\,\text{Nm}^{-2}s$ and $h = 0.05\,\text{mm}$

(a) Find the volumetric flow rate, Q, and pumping power absorbed for a supply pressure, p_r, in the recess of $2\,\text{MN/m}^2$.
(b) What is the load capacity of the bearing.

Ans: $\therefore Q = 1.28 \times 10^{-3}\,\text{L/s}$, $H = 256\,\text{W}$ and $\therefore W = 8230\,\text{N}$

9.3. A single orifice aerostatic step bearing must be designed with maximum possible stiffness to accommodate the thrust load from a shaft of equal diameter to the outside diameter of the bearing.

Determine:

The pocket diameter of the bearing
The orifice diameter
The mass flow through the bearing
The power to lift ratio

The following data is supplied:

shaft diameter, and hence outside diameter of the bearing 0.066 m
axial load 1000 N
ambient pressure $10^5\,\mathrm{Nm}^{-2}$
supply pressure $5 \times 10^5\,\mathrm{Nm}^{-2}$
viscosity of air $1.82 \times 10^{-5}\,\mathrm{Nm}^{-2}\mathrm{s}$
gas constant 267 kJ/kg deg K
air supply temperature 293 deg K
orifice discharge coefficient 0.65
film thickness 2.5×10^{-5}m

Chapter 10 Questions

10.1. As part of a scientific experiment, starting from rest, a smooth steel roller, of radius R_1 and length L_1, is made to roll down a smooth steel ramp, which is inclined at 45° and covered by a thin layer of oil. In order to control the roller path down the ramp, two discs of radius R_2 and length L_2, are attached to the roller ends so that each inner face is flush with the ramp sides, as illustrated in Fig. Q10.1.

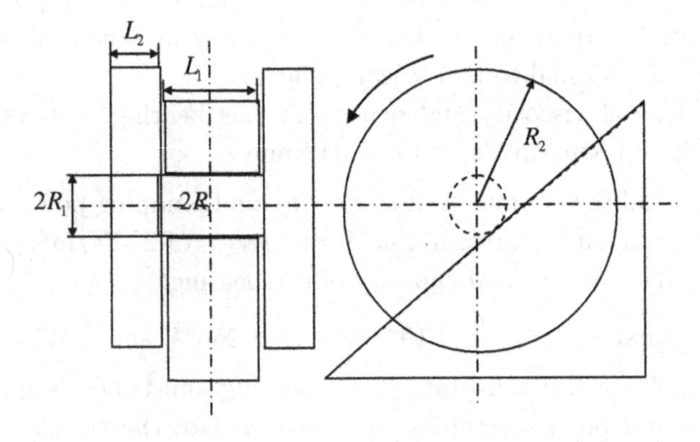

Fig. Q10.1.

Each ramp side incorporates an aerostatic thrust bearing arrangement (not shown), allowing the assembly to be guided there with negligible friction.

The data is as follows:

$R_1 = 0.0127\,\text{m}$, $R_2 = 0.05\,\text{m}$, $L_1 = 0.0254\,\text{m}$, $L_2 = 0.03\text{m}$, Young's Modulus for steel $E^* = 110\,\text{nGN/m}^2$, acceleration due to gravity $g = 9.8\,\text{m/s}^2$, $\rho = 7800\,\text{Kg/m}^3$ and for oil: $\upsilon = 0.3$, $\eta_0 = 4\,\text{cP}$, pressure viscosity coefficient, $\alpha_0 = 2.5 \times 10^{-8}\,\text{m}^2/\text{N}$.

Assuming the roller contact footprint with the ramp is long and the oil film is sufficiently thin to ignore friction at start up and when rolling,

(I) Calculate the speed of the roller and disc assembly center after it has rolled down the ramp (1) 0.01 m, (2) 1 m

(II) Calculate the respective minimum oil film thickness for (1) and (2) above assuming that the roller is (a) rigid (b) elastic

Comment on which of the two assumptions, (a) or (b), is appropriate.

Hints:

(a) You might need your first year mechanics notes to calculate the roller assembly speed

(b) You will need Fig. 10.12 to find approximately which zone is relevant when estimating the appropriate film thickness under the roller

10.2. A four-ball machine is used for finding the properties of a lubricant by submitting the balls to point contact loads under pure sliding conditions. Fig. Q10.2 shows the arrangement. The upper steel ball is vertically loaded and rotated against a nest of three stationary balls confined by a clamping cup containing the oil to be tested. All four balls are the same.

The machine has the following data:

Ball radius, $R = 0.0127\,\text{m}$, Reduced Young's Modulus $E^* = 110\,\text{GPa}$, oil viscosity at the film conjunction entrance temperature of 30°C is $\eta_0 = 0.140\,\text{Pas}^{-1}$, pressure viscosity coefficient, $\alpha = 2.29 \times 10^{-8}\,\text{m}^2\text{N}^{-1}$, rotational speed of top ball, $N = 1000\,\text{rev/min}$, load applied to top ball, $F = 600\,\text{N}$

(a) If the ball radius is R, prove that the radius of the track circle, the contact footprint follows, is $\text{AB} = r = R/\sqrt{3}$.

(b) What is the normal load at each of the ball contacts?
(c) Hence find the EHL conjunction film thickness based on the conjunction entrance temperature.

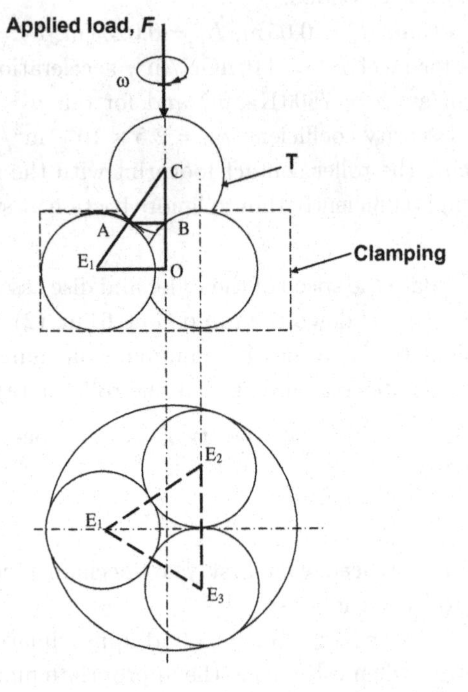

Fig. Q10.2. Four-ball machine. Plan view has top ball removed.

10.3. A disk machine is used to test an oil subjected to a rolling-sliding motion. It comprises two long cylindrical discs, each of radius $R_1 = 0.03\,\text{m}$ and length $L = 0.015\,\text{m}$, loaded circumferentially against each other. The reduced elastic modulus of the disc material is $E^* = 110\,\text{GPa}$ and their controllable combined rolling speed (the sum of their two rotational speeds) is $N = 3000\,\text{rev/min}$. The load per unit length applied to the disc assembly is $F = 2000\,\text{N}$.

The test oil has the following characteristics:

Viscosity at 30°C$\eta_0 = 0.056\,\text{Pas}^{-1}$
Temperature-viscosity coefficient, $\beta_0 = 0.026(°\text{C})^{-1}$
Pressure/viscosity coefficient, $\alpha_0 = 3.3 \times 10^{-8}\,\text{Pa}^{-1}$
Thermal conductivity, $k_t = 0.125\,\text{Jm}^{-1}\text{s}^{-1°}\text{K}^{-1}$
Reference stress, $\tau_0 = 3\,\text{MPa}$

(a) Find the mean pressure based on a dry Hertzian contact
(b) Find the conjunction EHL film thickness based on Grubin's theory
(c) For an average temperature rise of 0.25°C, find the corresponding coefficient of friction, the sliding speed in revs/s necessary to achieve this rise, and the corresponding slide-roll ratio
(d) Repeat (c) for an average temperature rise of 30°C
(e) Repeat (d) for an average temperature rise of 100°C
(f) Comment on the differences in the three coefficients of friction found in (c), (d) and (e).

Chapter 11 Questions

11.1. A radial ball bearing has a load capacity of 20 kN. Estimate its L_{10} life when operating at 500 rev/min under a radial load of 5 kN.

Ans: 2133 hours.

11.2. A machine tool spindle, rotating at 500 rev/min with a radial load of 2 kN at its mid point, is supported by two identical radial ball bearings, each of load capacity 20 kN.

(a) Find the life in hours of one of these bearings under its reaction force
(b) Find the life in hours of the whole assembly.
(c) What is the probability of survival over that life of the whole arrangement?

Ans: 266,000 hours, 168,000 hours and 0.81

Chapter 12 Questions

12.1. Refer to the example for the automobile polynomial cam in Sec. 12.2.1. Use the same cam geometrical data, lubricant rheology and camshaft speed as in the same example. Determine the film thickness at a position very close to the inlet reversal location, corresponding to $\theta/\hat{\theta} = 0.6$. If the composite surface roughness of cam and tappet is 0.56 μm, what is the regime of lubrication in this location?

Hint: Find the extent of lift, s, using Fig. 12.8, and thus the elastic force. Follow the same steps as the example in Sec. 12.2.1.

Chapter 13 Questions

13.1. Refer to Example 13.4.1 and use the same data, except for a salt concentration of 0.005 mols/dm^3 of $CaCl_2$ in water in a gap of 10 nm between two frictionless smooth flat plates. (a) Calculate the Debeye's length, and (b) What would be the electrostatic pressure?

13.2. A roller of radius 2 mm and length 5 mm is coated with DLC and rolls without sliding with a speed of 60 rpm on a flat plate also coated with DLC, with water as the intervening fluid (take the dynamic viscosity to be 0.00025 Pas). The applied load is 2 mN.

The composite roughness of the contacting surfaces is 100 nm. (a) What is the regime of lubrication? (b) If the average asperity tip radius is 1 μm, and we assume 10,000 such asperity-pairs in the contact area what is the adhesive coefficient of friction?
Modulus of elasticity for DLC is 800 GPa and take its shear strength to be 100 MPa

Hint: Use Fig. 13.11 to find the meniscus force for a pair of asperities and thus the total meniscus force. Add this to the applied load, then use the rolling term in Eq. (13.6) to find the film thickness. This will enable you to surmise the regime of lubrication and thus make a judgment for the value of ψ in Eq. (13.41).

Chapter 14 Questions

14.1. Volumetric wear of the acetabular cup can be estimated from Lancaster's wear equation as:

$$V = KWX$$

where V is the volumetric wear in mm^3, W is the contact load in N and X is the sliding distance in m. K is the wear factor, with the unit mm^3/Nm.

The wear factor K is a function of average roughness of the contacting surface, Ra (in μm), as:

$$K = 4(Ra)^{1.2} \times 10^{-5}$$

The average surface roughness of the acetabular cup is 0.05 μm. The contact load varies according to the various stances of the hip replacement recipient. We assume this load to be on average 2.5 times the mass of an average adult of 75 kg. A million steps per annum, with an average 60° rotation per step at the hip are estimated for a typical recipient. The diameter of the Charnley joint is 22.225 mm. What would be the total wear volume in a year?

BOOK OF SOLUTIONS

We have provided solutions to some of the questions set for you previously. You can check your solutions against these.

Solution for 2.2:

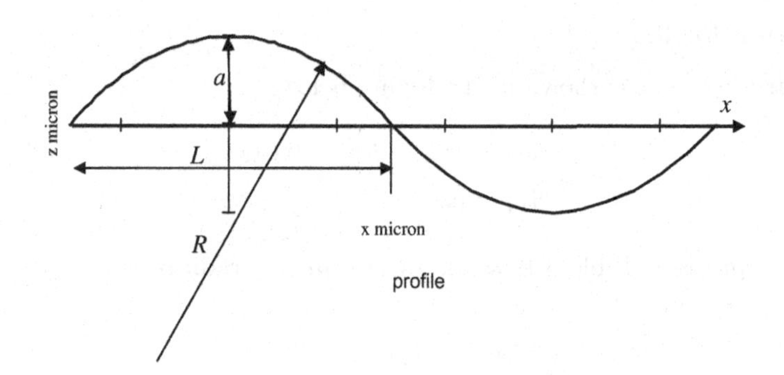

$$z = a - \frac{x^2}{2R} a \ll R \quad \sigma^2 = \frac{1}{L}\int_{-L/2}^{L/2}\left(a - \frac{x^2}{2R}\right)^2 dx$$

$$\sigma^2 = \frac{1}{L}\int_{-L/2}^{L/2}\left[a^2 - \frac{ax^2}{R} + \frac{x^4}{4R^2}\right] dx = \frac{1}{L}\left(a^2x - \frac{ax^3}{3R} + \frac{x^5}{20R^2}\right)_{-L/2}^{L/2}$$

$$\sigma = \left[a^2 - \frac{aL^2}{12R} + \frac{L^4}{320R^2}\right]^{1/2} \qquad \text{But when } x = L/2 \to 0 = a - \frac{L^2}{8R}$$

Substituting for L, $\sigma = \frac{8a}{10} = 0.8 \times 0.03 \times 10^{-6} = 0.024 \times 10^{-6}$ (**Answer**)

Additional Notes:

This solution can be used in the Chapter 3 questions to find the Plasticity Index for this surface

$\Psi = \left(\frac{E}{H}\right)\left(\frac{\sigma}{R}\right)^{1/2}$ Assuming the wavy surface is of steel, contacting a similar wavy surface so that $H = 8\,\text{GPa}$, $E^* = 110\,\text{GPa}$, $\sigma = \sqrt{2} \times 0.024 \times 10^{-6} = 0.04\,\mu\text{m}$ and reduced radius is $R = 75\,\mu\text{m}$:

$\Psi = \left(\frac{110}{8}\right)\left(\frac{\sigma}{R}\right)^{1/2} = \frac{110}{8} \times \left(\frac{0.04}{75}\right)^{1/2} = 0.317$. This is well into the elastic contact range because of the 'super-finish' low value of σ and high radius

Solution for 3.2:

$$p_0 = \frac{3p_m}{2} = \left(\frac{6WE^{*^2}}{\pi^3 R_x R_y}\right)^{1/3}, \quad \text{Hence } W = \frac{\pi^3 R_x R_y}{6E^{*2}} p_0^3$$

From Eq. (3.32) $p_0 = 0.6\,\text{H}$ at yield, $R_x = 69.9\,\text{mm}$, $R_y = 7.2\,\text{mm}$

With these substitutions $W = 2381\,N$ (**Answer**)

Solution for 3.4:

The arrangement is shown in the Figure below

$$R_{x2} = R_{y1} = R_{y2} = 0.01,$$
$$R_{x1} = \infty.$$

Using Appendix Table 3.1, we find the principle radii of curvature in the contact as below

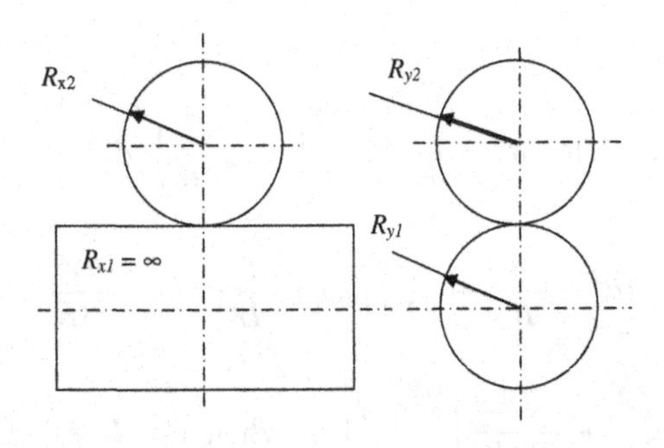

$$R_x = \left(\frac{1}{R_{x1}} + \frac{R}{R_{x2}}\right)^{-1} = \left(\frac{1}{\infty} + \frac{1}{10}\right)^{-1} = 10\,\text{mm}$$

$$R_y = \left(\frac{1}{R_{y1}} + \frac{R}{R_{y2}}\right)^{-1} = \left(\frac{1}{10} + \frac{1}{10}\right)^{-1} = 5\,\text{mm}$$

$$E^* = \left(\frac{1-\nu_1^2}{E_1} + \frac{1-\nu_2^2}{E_2}\right)^{-1}.$$

If a and b are the respectively the major and minor semi axes then

$$\sqrt{ab} = \left(\frac{3W\sqrt{R_x R_y}}{4E*}\right)^{1/3} = \quad \text{and} \quad a/b \approx (R_x R_y)^{2/3}.$$

Solving for a and b : $a = 4.57 \times 10^{-4}\,\text{m}$ and $b = 2.88 \times 10^{-4}\,\text{m}$ (**Answer**)

Solution for 4.3:

$$W = \frac{\pi d^2}{8} H \quad F = \frac{d}{2} hH$$

$$\mu = \frac{F_p}{W} = \frac{8dhH}{2\pi d^2 H} = \frac{4h}{\pi d}$$

now

$$\cot\theta = \frac{2h}{d} \quad \therefore \quad \mu_p = 2\cot\theta/\pi \text{ (Answer)}$$

Comment: The larger the asperity slope, θ, the lower the ploughing friction.

(Try another problem with a metal surface being scratched under 2 different loads (Prob. 4.2))

Solution for 4.4:

Equation (4.30): If Q = the volume collected per meter travelled, $Q = \frac{K_a W}{H}\,\text{m}^2$.
If S = average distance moved by annulus surface.
Then S = average speed of surface × time = $(2\pi \times 5 \times 0.1) \times 20 \times 3600 =$ 22,622 m.

For the steel:

The total volume collected is: $V_s = \frac{8\times10^{-6}}{7800} = 1.02 \times 10^{-9}\,\text{m}^3$.
Then $Q_s = \frac{V_s}{S} = 1.02 \times 10^{-9}/22{,}622 = 4.5 \times 10^{-14}\,\text{m}^2$.
So that $K_s = \frac{H_s Q_s}{W} = \frac{2400\times10^6\times4.5\times10^{-14}}{100} = 1.08 \times 10^{-6}$ (**Answer**)

For the bronze:

Volume collected is $V_b = \frac{250\times10^{-6}}{8400} = 3\times10^{-8}\,\mathrm{m}^3$
Then $Q_b = \frac{V_b}{S} = 3\times10^{-8}/22,622 = 1.33\times10^{-12}\,\mathrm{m}^2$
So that $K_b = \frac{H_b Q_b}{W} = \frac{800\times10^{6}\times1.33\times10^{-12}}{100} = 10.64\times10^{-6}$ (**Answer**)

Solution for 4.5:

In this problem, the ploughing components of friction coefficient are geometrical only, so we can detemine them. The wanted adhesive components depend on the state of the friction surfaces

$$\theta 1 := 70^\circ \quad \cot(\theta_1) = 0.364 \quad \theta_2 := 80^\circ \quad \cot(\theta_2) = 0.176 \quad \tan(\theta_2) = 5.671.$$

$$\mu \mathrm{p1} := 2\frac{1}{\pi - \tan(\theta_1)} \quad \mu \mathrm{p1} = 0.232 \quad \mu \mathrm{p2} : 2\frac{1}{\pi - \tan(\theta_2)} \quad \mu \mathrm{p2} = 0.112.$$

guess $\mu\mathrm{da} := 0.6 \quad \mu\mathrm{la} : 0.1$

Given

$$1.1 = \frac{\mu\mathrm{da} + \mu\mathrm{p1}}{\mu\mathrm{da} + \mu\mathrm{p2}}.$$

$$1.3 = \frac{\mu\mathrm{la} + \mu\mathrm{p1}}{\mu\mathrm{la} + \mu\mathrm{p2}}.$$

$$\mathrm{a} : = \mathrm{Find}(\mu\mathrm{da}, \mu\mathrm{la}), \quad \mathrm{a} = \begin{pmatrix} 1.082 \\ 0.286 \end{pmatrix}.$$

$\mu\mathrm{da} = 1.082$ (**Answer**), $\mu\mathrm{la} = 0.286$ (**Answer**).

Solution for Parts (a) and (b) of question 5.2:

Solution for 5.2:

Fluid 1

(a) find Walther constants for the first fluid

guess $\mathrm{d} := 8 \quad \mathrm{e} := 3$

Given

$$\log(\log(100 + .6)) = \mathrm{d} - \mathrm{e}\log(25 + 273)$$

$$\log(\log(8.5 + .6)) = \mathrm{d} - \mathrm{e}\log(100 + 273)$$

Find (d, e)

$$a := \text{Find}\ (d, e) \quad a \begin{pmatrix} 8.417 \\ 3.28 \end{pmatrix}$$

$d := 8.417 \quad e := 3.28$ **aretheconstantsforthefirstfluid**

To find ν for the first fluid at 35 degr C
guess $\nu := 60$
Given

$$\log(\log(\nu + 0.6)) = 8.417 - 3.28 \log(35 + 273)$$

$b := \text{Find}(\nu)$ $b = 62.065$ ν **of fluid 1 is 62.05 cSt at 35 degr C**

Check $x := \log(\log(\nu + .6)) \quad x = 0.251$
$y := 8.417 - 3.281 \log(308) \quad y = 0.255$

Fluid 2
(b) find Walther constants for the second fluid
guess $d := 8 \quad e := 3$
Given

$$\log(\log(350 + .6)) = d - e \cdot \log(25 + 273)$$
$$\log(\log(17 + .6)) = d - e \cdot \log(100 + 273)$$
$$\text{Find}(d, e)$$

$a := \text{Find}(d, e) \quad a = \begin{pmatrix} 8.281 \\ 3.183 \end{pmatrix}$ $d = 8.281$ **and** $e = 3.183$ **are the constants for the second fluid**

To find ν for the second fluid at 35 degr C
guess $\nu := 190$
Given

$$\log(\log(\nu + .6)) = 8.281 - 3.183 \log(35 + 273)$$

$b := \text{Find}(\nu)$ $b = 194.639$ ν **of fluid 2 is 194.6 cSt at 35 degr C**

We can now blend the two fluids at 35 degrees C at the chosen blend viscosity of of 150 cSt

first fluid

$$\log(\log(\nu + .6)) = 8.417 - 3.28\log(t + 273) \quad t := 20\ldots120$$

second fluid

$$\log(\log(\nu + .6)) = 8.281 - 3.183\log(t + 273) \quad t := 20\ldots120$$

νf and νs define the first and second fluids.

Transpose the Walther equations to get viscosity on the LHS

$$vf(t) := 10^{[10^{(8.417-3.28\cdot\log(t+273))}]} - .6$$
$$vs(t) := 10^{[10^{(8.281-3.183\cdot\log(i+273))}]} - .6$$

Math Cad Plot the visc-temp behavior of both fluids

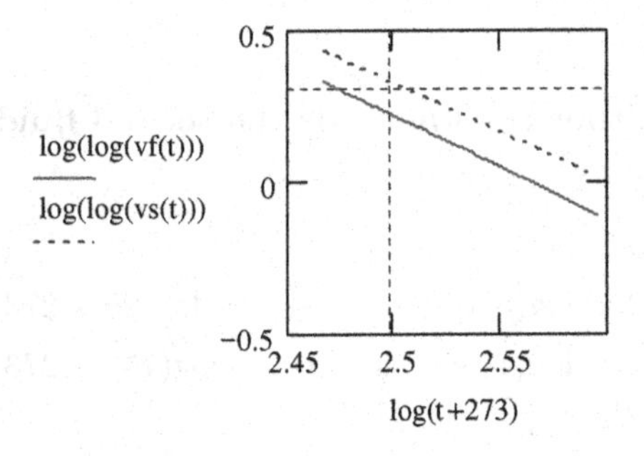

coordinates for the blended oil in terms of the log expressions

let $f(t) := \log(\log(vf(t)))$ $t := 35$ $f(t) = 0.254$let $l(t) := \log(t + 273)$
$l(t) = 2.489$

let $s(t) := \log(\log(vs(t)))$ $t := 35$ $s(t) = 0.36l(t) = \log := (t + 273)$
$l(t) = 2.489$

$v(t) := 115$ $b(t) := \log(\log(v(t)))$ $b(t) = 0.314$

If ν for the blend oil is to be 115cSt at 35 deg C, then bt $= 0.314 =$ (ordinate for blend oil)

The coordinates for the blend oil in terms of the axes expressions are therefore 2.489, 0.314.

To get the solution, the above figure must be printed and the blend proportions measured. These are indicated on the enlarged figure.

Solution to part (c) of 5.2:

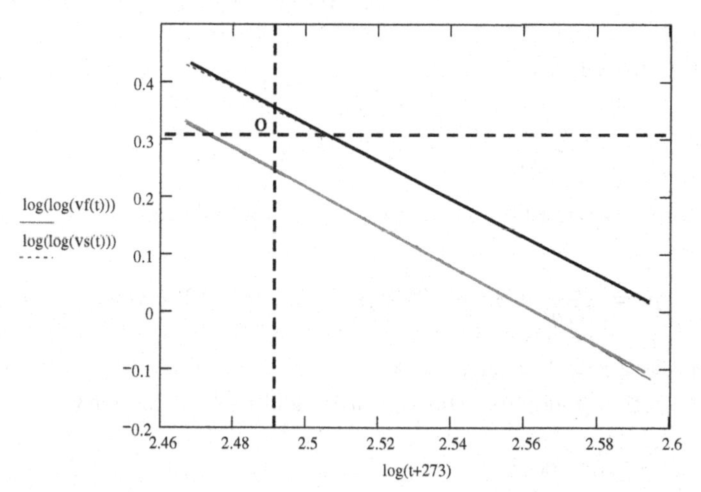

Characteristics of the two components
and the blend coordinates.

$X_b = \frac{OA}{AB} = \frac{1.5}{2.4} = 62.5\%$ of the higher component (**Answer**)

Solution for 6.3:

(a) $\dfrac{dp}{dx} = 6U\eta\dfrac{h - h_c}{h^3} \quad h = h_0 e^{\alpha x} \quad \dfrac{dp}{dx} = 6U_1\eta\left(\dfrac{h_0 e^{\alpha x} - h_0 e^{\alpha x_c}}{h_0^3 e^{3\alpha x_c}}\right)$

$\rightarrow$ Integrate :

$$\frac{h_0^2 p}{6U\eta} = \left[\frac{-1}{2\alpha}e^{-2\alpha x} - e^{\alpha x_c}\frac{1}{3\alpha}e^{-3\alpha x}\right] + C \quad p = 0 \text{ at } x = 0 \text{ and } x = B$$

$$\therefore 0 = -\frac{1}{2\alpha} + \frac{e^{\alpha x_c}}{3\alpha} + C, \qquad (1)$$

$$\therefore 0 = -\frac{1}{2\alpha}e^{-2\alpha B} + \frac{e^{\alpha x_c}}{3\alpha}e^{-3\alpha B} + C. \qquad (2)$$

Subtract and get x_c

$$e^{\alpha x_c} = \frac{h_c}{h_0} = \frac{3}{2} e^{\alpha B} \frac{(1 - e^{-2\alpha B})}{(1 - e^{-3\alpha B})} \rightarrow \text{Let } \beta = \frac{(1 - e^{-2\alpha B})}{(1 - e^{-3\alpha B})}.$$

Thus,

$$h_c = \frac{3}{2} \beta h_0 \text{ (Answer)}$$

(b) Get C from Eq. 1

$$C = \frac{1}{2\alpha} - \frac{e^{\alpha x_c}}{3\alpha} = \frac{1}{2\alpha} - \frac{3}{2}\beta \frac{1}{3\alpha} = \frac{1}{2\alpha}(1 - \beta).$$

Therefore, inserting C into (2) after some manipulation

$$p^* = \frac{h_0^2 \alpha p}{U_1 \eta} = 3[-e^{-2\alpha x} + \beta e^{-3\alpha x} + (1 - \beta)] \text{ (Answer)}.$$

As a check, if $x = 0$, or $B, p^* = 0$.

Integration p along the bearing will yield its load capacity.

Solution for 6.4:

The film shape suggests that the narrow bearing pressure distribution expression is valid because there the film has zero slope at either end.

$$p = 3U_1 \eta \left(\frac{dh}{dx} \right) \frac{1}{h^3} (y^2/4 - L^2 - 4)$$

$$\therefore W = 3U_1 \eta \int_{-1/2}^{L/2} \int_0^B \frac{dh}{dx} \frac{1}{h^3} \left(y^2 - \frac{L^2}{4} \right) dx dy$$

$$\text{or} \quad \frac{W}{3U_1 \eta} = \int_{-L/2}^{L/2} \int_{h_1}^{h_0} \frac{dh}{L^3} \left(y^2 - \frac{L^2}{4} \right) dy = \frac{3}{16} \frac{L^3}{h_0^2}$$

Dividing both sides by B^2 and rearranging, we obtain the solution.

Solution for 7.1:

(a) *Effective viscosity*

We require the effective viscosity, η_e. From Eq. (5.6), the VI of the test oil will give us its viscosity at 40°C.

As VI $= \frac{L-U}{L-H} \times 100$ where here U is the kinematic viscosity of the test oil at 40°C

Then

$$95 = \frac{471.3 - U}{471.3 - 221.1} \times 100. \quad \text{Solving: } U = 233.6\,\text{cSt at } 40°\text{C}$$

Knowing 2 viscosities and temperatures of the test oil we can now find its viscosity at the required 50.5°C. To do this use Walther's equation:

$$\log(\nu + 0.6) = d - e \log T \tag{5.5}$$

Firstly we must find the constants d and e for the test oil. Using $\alpha =$ 233.6 cSt at 40°C and $\nu = 19.5$ cSt at 100°C, insert these values into Eq. (5.5) and solve the resultant simultaneous equations to give $e = 3.42$ and $d = 8.91$. The equation for the test oil is therefore: $\log\log(\nu + 0.6) = 8.91 - 3.42 \log T$. Solving this expression at the effective temperature $T = (50.5 + 273)$°K, we find $\nu = 131$ cSt, giving

$$\eta_e = 0.131 \times 0.87 = 0.113\,\text{Pas}\ (\textbf{Answer}).$$

(b) *Pad dimensions*

From Appendix Table 7.2, plot K against W^* for $L/D = 1$.

The maximum value of $W^* = 0.073$ at $K = 1.2$ and from Appendix Table 7.1

$$W = W^* \left(\frac{\eta_e U_1 B^2 L}{h_0^2}\right).$$

We are given $W = 0.4\,\text{mN}$, $W^* = 0.073$, $B = L$, $h_0 = 50 \times 10^{-6}\,\text{m}$. Therefore

$$B^3 = \frac{0.4 \times 10^6 \times 2500 \times 10^{-12}}{50 \times 0.073 \times \eta_e} = \frac{2.74 \times 10^{-4}}{\eta_e}.$$

Inserting the value of η_e above: $B = L = 0.14$ m (**Answer**).

(c) *Power consumed*

From Eq. (7.28) and knowing that $F^* = \frac{F_0 h_0}{LB\eta U_1}$,

Power consumed $= F_0 U_1 = \frac{F^* LB\eta_e U_1^2}{h_0}$. Turn to Appendix Table 7.2 for $L/D = 1$ to find F^* . We need the power only approximately, so $F^* = 0.728$ at $K = 1$ is sufficiently accurate. Inserting all the values in the power expression. Power consumed = 81 kW (**Answer**).

Q7.2 solution

$$W_p = \frac{6U_1\eta B^2 L}{h_0^2} k_p S = \frac{6U_1\eta k_p}{h_0^2} B^2 L(0.370L/B) = \frac{6U_1\eta k_p}{h_0^2} 0.370BL^2$$

$$W_t = NW_p = N\frac{6U_1\eta k_p}{h_0^2} 0.370BL^2 = \frac{6U_1\eta k_p}{h_0^2} 0.370L(NBL)$$

where (NBL) is the total pad area

$$W_t = \frac{6U_1\eta k_p}{h_0^2} 0.370L\left[\frac{0.85\pi\left(D_o^2 - D_i^2\right)}{4}\right] \quad \text{since from data,}$$

$$NBL = 0.85\left(\frac{\pi}{4}D_o^2 - \frac{\pi}{4}D_i^2\right)$$

Furthermore, $D_o^2 - D_i^2 = (D_o - D_i)(D_o + D_i) = 4(R_o - R_i)(R_o + R_i)$ but $R_m = \frac{R_o - R_i}{2}$ and $R_o - R_i = L$

$\therefore D_o^2 - D_i^2 = 4(2R_m)L = 4D_m L$. With this substitution, W_t becomes

$$W_t = 6\frac{0.370U_1\eta k_p}{h_o^2} L\frac{0.85\pi}{4} 4D_m L \qquad \text{(a)}$$

Answers

Equation (a) shows that the total load depends only on the pad length and the mean diameter and is independent of the number of pads. Substituting the data into (a) and noting that $S_l = 0.370L/B$, we get $W_t = 8.154 \times 10^6 D_m^3$. Furthermore, from the data, $D_m = 0.4\,\text{m}$ so that $W_t = 0.522 \times 10^6\, N$.

Solution for 7.3:

At $L/D = 1$ and $K = 1$, from Table Appendix 7.1 $W^* = 0.0689$ and $T_s^* = 2.96$. From the definition of W^*

$$\frac{\eta_e}{h_0^2} = \frac{1}{W^*}\frac{W}{U_1 LB^2}. \tag{a}$$

All terms on the right hand side of this expression are known. Likewise, from Eq. (7.29)

$$\Delta\theta_s = \theta_s^* K_1 \left(\frac{\eta_e}{h_0^2}\right) U_1 B. \tag{b}$$

Substituting Eq. (a) into Eq. (b)

$$\Delta\theta_s = \frac{\theta_s^*}{W^*} K_1 \frac{W}{LB}. \tag{c}$$

All terms on the RHS of Eq. (c) are known. With these substitutions, $\Delta\theta_s = 45.1°\text{C}$. Therefore, if θ_0 is the supply temperature (= 30°C)

$$\theta_{se} = \theta_o + 45.1 = 75.1°\text{C}\ (\textbf{Answer}).$$

We can now replace the effective value of θ_{s2} in Vogel's expression in example 7.2 with its absolute value of $\theta_{se} = 75.1 + 273 = 348°K$

$$\ln(\eta_e) = -1.845 + \left(\frac{700.81}{348 - 203}\right)$$

$$\therefore \eta_e = 19.8\,\text{cP}(\textbf{Answer})$$

$$\therefore h_0 = \sqrt{\frac{ULB^2 W^* \eta_e}{W}} = \sqrt{\frac{10 \times 0.05^3 \times 0.0689 \times 0.0198}{7000}}$$

$$= 1.56 \times 10^{-5}\,\text{m}\ (\textbf{Answer}).$$

Solution for 7.4:

Solution to tutorial question 4 using Math Cad (inverse of worked ex 2)
θss means Ts star Qss is Qs star Ws is W star θo is supply temperature
η is viscosity $\Delta\theta$s is temperature rise t is the taper
θs1 is effective temperature obtained from from temperature rise, degrees C
Θs2 is effective temperature obtained through Vogel, degrees K. θs2 is in degr C

Input data

$$B := .05 \quad L := .05 \quad h_0 := 2.07.\ 10^{-5} \quad U_1 := 10 \quad W := 7.\ 10^3$$

$$\theta_0 := 30 \quad K_1 := 0.376.\ 10^{-6}$$

$$K := \begin{pmatrix} .5 \\ 1 \\ 1.5 \\ 2 \\ 3 \\ 4 \end{pmatrix} \quad Qs_s := \begin{pmatrix} .122 \\ .2462 \\ .371 \\ .496 \\ .75 \\ 1.01 \end{pmatrix} \quad Ws_s := \begin{pmatrix} 0.0557 \\ .0689 \\ .07 \\ .0671 \\ .0583 \\ .0503 \end{pmatrix} \quad Ts_s := \begin{pmatrix} 6.76 \\ 2.96 \\ 1.767 \\ 1.242 \\ .732 \\ .461 \end{pmatrix}.$$

$$t := 2.07 \times 10^{-5} \cdot \vec{K} \quad \text{etab} := \frac{W}{U_1\ .05\ B^2} \quad \eta := \text{etab}\overrightarrow{\frac{h_0^2}{Ws_s}}$$

$$\Delta\theta_s := (K_1 \cdot U_1 \cdot B)\left(\overrightarrow{Ts_s \cdot \frac{\eta}{h_0^2}}\right)$$

$$\theta_{s1} := \theta_0 + \overrightarrow{\Delta\theta_s} \quad cp := 1000\ \eta \quad \Theta_{s2} := 203 + \frac{700.81}{\ln(\overrightarrow{cp}) + 1.845}$$

$$\theta_{s2} := \Theta_{s2} - 273$$

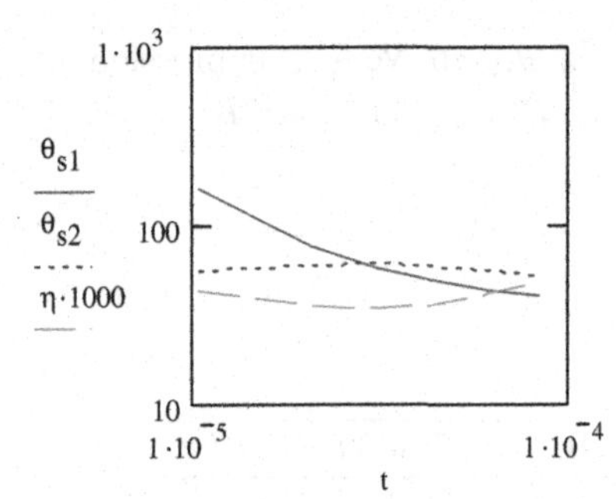

$t_e = 2.76\ 10^{-5}$ $K_e := \frac{t_e}{h_0}$ $K_e = 1.333$

Answers: te=27.6 microns,

$\theta se = 62\,\text{deg}$ **C,** $\eta e = 35\,$**cP**

OK, agrees with inverse question

Solution for 7.5:

(a) $U_1 = 10\,\text{m/s}$.

Hence from Eq. (b), $k_o = 0.7$.

Now from Q7.4,$\theta_{se} = 62°\text{C}$ and $\theta_0 = 30°\text{C}$

Solving Eq. (a) for θ_1

$\theta_1 = k_o(\theta_{se} - \theta_0) + \theta_0 = 52.4°$ C (**Answer**).

Hence, from Eq. (c) $\theta_2 = 71.6°\text{C}$ (**Answer**).

Comment:

It is quite common in thrust bearings to find θ_1 relatively high because of the large HOCO from the trailing edge of the previous pad. Considerable thrust bearing research aims to improve thermal mixing in the chamber in order to reduce θ_1.

(b) Solution using Eqs. (7.25) and (7.26) only

Equating these:

$$Q_s(\theta_{se} - T_0) = Q_s(\theta_{se} - \theta_1) + Q_2(\theta_2 - \theta_1).$$

Also, from Eq. (7.24)

$$\theta_{se} = \frac{\theta_1 + \theta_2}{2}. \quad \text{(c)}$$

Replacing the actual flows by their dimensionless equivalents and eliminating θ_2 from these equations, we get, after some manipulation

$$\theta_1 = \frac{\theta_0 + 2\left(\frac{Q_2^*}{Q_s^*}\right)\theta_{se}}{1 + 2\left(\frac{Q_2^*}{Q_s^*}\right)} \quad \text{(d)}$$

where $Q_2^* = Q_1^* - Q_s^*$ is obtained from Appendix Table 7.2 at the value of K corresponding to the solution of Q7.4 (This will involve some plotting). Then θ_1 can be obtained directly from equation (d) and hence θ_2 from Eq. (c). Using slope factor, $t = 27.6$ micron from Q7.4, the corresponding value of $K = 1.353$. Hence $Q_2^* = 0.63$ and $Q_s^* = 0.35$. Then Eqs. (c) and (d) produce effective temperatures $\theta_{1e} = 55°$ and $\theta_{2e} = 69°$ (**Answer**)

Comment:

The outlet temperature θ_{2e}, is underestimated in this approach because we have assumed that thermal mixing in the chamber has been perfect. Nevertheless, a rough solution of the total temperature rise can be found using this approach.

An additional computer solution for 7.5 (b) for 7.4:

Q7.5 Solution to find θ1 and θ2 for Q 7.4

$$Qs_1 := \begin{pmatrix} 0 \\ 0.5 \\ 0.68 \\ 0.847 \\ 1.01 \\ 1.165 \\ 1.47 \\ 1.769 \end{pmatrix} \quad Qs_s := \begin{pmatrix} 0 \\ 0.122 \\ 0.246 \\ 0.371 \\ 0.496 \\ 0.75 \\ 1.01 \end{pmatrix} \quad Qs_2 := Qs_1 - Qs_s \; Qs_2 = \begin{pmatrix} 0.5 \\ 0.558 \\ 0.601 \\ 0.639 \\ 0.669 \\ 0.72 \\ 0.759 \end{pmatrix}$$

$$K := \begin{pmatrix} 0 \\ 0.5 \\ 1 \\ 1.5 \\ 2 \\ 3 \\ 4 \end{pmatrix} \qquad t := 2.07 \times 10^{-5} \cdots \vec{K}.$$

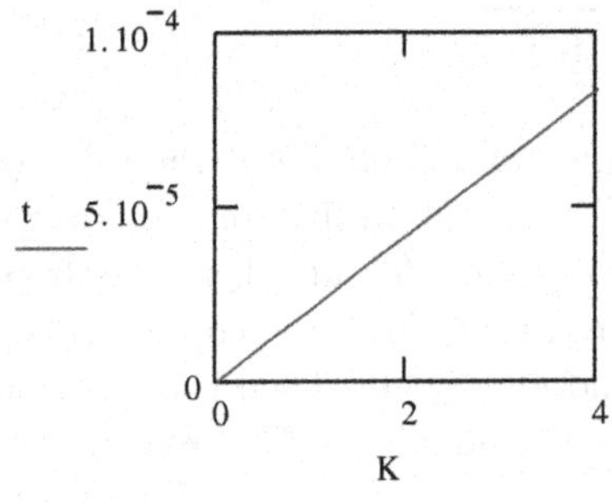

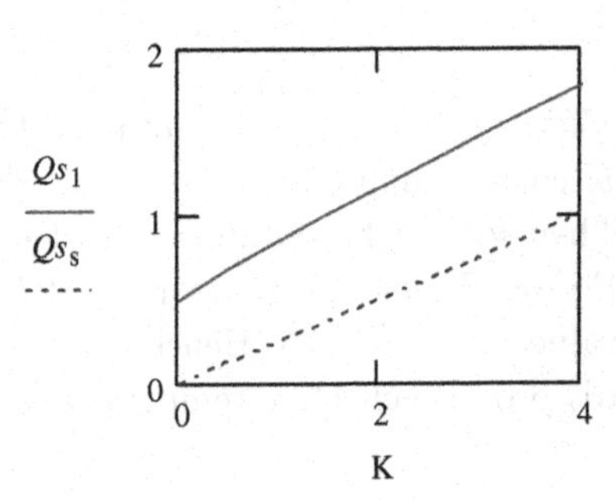

$\theta_0 := 30 \;\; Qs_{1e} := 97 \;\; Qs_{se} := 0.35 \;\; \theta_e := 62 \;\; Qs_{2e} := Qs_{1e} - Qs_{se} \;\; Qs_{2e} = .$

$$\theta_{1e} := \frac{\theta_0 + 2.\left(\frac{Qs_{2e}}{Qs_{se}}\right) \cdot \theta_e}{1 + 2 \cdot \left(\frac{Qs_{2e}}{Qs_{se}}\right)} \theta_{1e} = \quad Q_{2e} := 2 \cdot \theta_e - \theta_{1e} \; \theta_{2e} = .$$

$$t_e := 2.76 \times 10^{-5}$$

$$\theta_1 = 55°\text{C}, \quad \theta_2 = 69°\text{C}, \quad t = 2.76 \times 10^{-5}\,\text{m}$$

Solution for 8.1:

The worked example 8.8.3 in Chapter 8 used Math Cad for a solution. However, in this case some additional equations are needed to find the total temperature rise. In this case the film outlet temperature T_2 is required. To do this we will need the internal power flow Eq. (7.26), along the film. This is equation applies to both thrust and journal bearings, it being also possible to deduce it directly from Fig. 8.8.

$$H = \rho\sigma[Q_s(T_e - T_1) + Q_2(T_2 - T_1)]. \tag{7.26}$$

First we have to replace T_e by an experimental expression that estimates where along the film T_e can be placed to represent the effective temperature of the side flow. It is Eq. (7.24) for thrust bearings, which also can be used journal bearings. If the total temperature rise *along the film* is given by $\Delta T = T_2 - T_1$

$$T_e = T_1 + k_2\Delta T. \tag{7.24}$$

Journal bearings, experiments have shown that $k_2 = 0.8$.

Substituting Eq. (7.24) into Eq. (7.26) and noting that $Q_2 = Q_1 - Q_s$, after some manipulation we get

$$H = \rho\sigma[Q_1 - Q_s(1 - k_2)].$$

Replacing the flows by their dimensionless equivalents, we can now write Eq. (8.23) as

$$\Delta T = \left(\frac{2K_1W}{RL}\right)\frac{\mu^*}{Q_1^* - Q_s^*(1 - k_2)}. \tag{a}$$

Here, the flow terms in the denominator have replaced Q_s^* in Eq. (8.23). Using the data supplied above, proceed with the solution exactly as in the worked example 8.8.3. In the CAD solution:

tf1 $\equiv T_1$, tf2 $\equiv T_2$, k2 $\equiv k_2$, p0 $\equiv p_0$, Ss $\equiv 1/S$, eps $\equiv \varepsilon$,

Qs $\equiv Q_s^*$, Mus $\equiv \mu^*$, Qts $\equiv Q_1^*$, eta $\equiv \eta$

(1) input data

L : = 0.0254   D := .0508   R := 0.0254   K := $0.484\ 10^{-6}$

W : = 1397   N := 91

k : = 1 ti := 56   c := 5×10^{-5} d := 4.10^{-3} $\Omega := 2\pi$ N   NB: Ss = **1/S,**

so W is least at Ss = **4.327**Ω = 571.77 $k_2 = 0.8$

NB ti is the bearing supply temperature.

(2) enter known coefficient vectors

$$\text{eps} := \begin{pmatrix} 0.4 \\ 0.45 \\ 0.5 \\ 0.55 \\ 0.6 \\ 0.65 \\ 0.7 \\ 0.75 \\ 0.8 \\ 0.85 \\ 0.9 \\ 0.95 \end{pmatrix} \quad \text{Ss} := \begin{pmatrix} 0.7855 \\ 0.6337 \\ 0.5093 \\ 0.4059 \\ 0.3192 \\ 0.2463 \\ 0.185 \\ 0.1338 \\ 0.0916 \\ 0.0576 \\ 0.0312 \\ 0.0119 \end{pmatrix} \quad \text{Qs} := \begin{pmatrix} 0.7507 \\ 0.8446 \\ 0.9385 \\ 1.032 \\ 1.126 \\ 1.2202 \\ 1.3146 \\ 1.4093 \\ 1.5041 \\ 1.5995 \\ 1.6955 \\ 1.7916 \end{pmatrix} \quad \text{Mus} := \begin{pmatrix} 17.0961 \\ 14.1996 \\ 11.8145 \\ 9.8131 \\ 8.1022 \\ 6.6262 \\ 5.3425 \\ 4.2197 \\ 3.2334 \\ 2.3521 \\ 1.5191 \\ 0.8916 \end{pmatrix}$$

(3) calculations

$$\text{h0} : = \text{c} \cdot \left(1 - \overrightarrow{\text{eps}}\right) \quad \Delta\text{T1} := \left(2.K.\frac{\text{W}}{\text{LR}}\right) \cdot \overrightarrow{\left(\frac{\text{Ms}}{\text{Qs}}\right)}$$

$$\text{eta} : = \frac{\text{W}}{(\text{N.L.D.R.}^2)} \overrightarrow{(\text{c}^2\text{Ss})} \quad \text{cp} := 1000\,\text{eta}$$

(3) calculate c vector and temp rise

$$\text{tel} := \text{ti} + \overrightarrow{\Delta\text{T1}}$$

(4) oil visc/temp properties

Given

$$\ln(7.65) = \ln(a) + \frac{b}{373 - c1} \quad \ln(38.25) = \ln(a) + \frac{b}{323 - c1}$$

$$\ln(170) = \ln(a) + \frac{b}{293 - c1}$$

$$\text{Find}(a, b, cl) \rightarrow \begin{pmatrix} 2.5769747971287902 22^{-2} \\ 129.6358395910853027 \\ 146.12894339861691747 \end{pmatrix}$$

$$a := 0.025769 \quad b := 1292 \quad c1 := 146.13$$

$$T = \frac{1292}{\frac{cp}{.0257}} + 146.13 \quad Te2 := c1 + \frac{b}{\ln\left(\frac{\vec{cp}}{a}\right)} \quad te2 := \overrightarrow{Te2} - 273$$

(5) solution graph

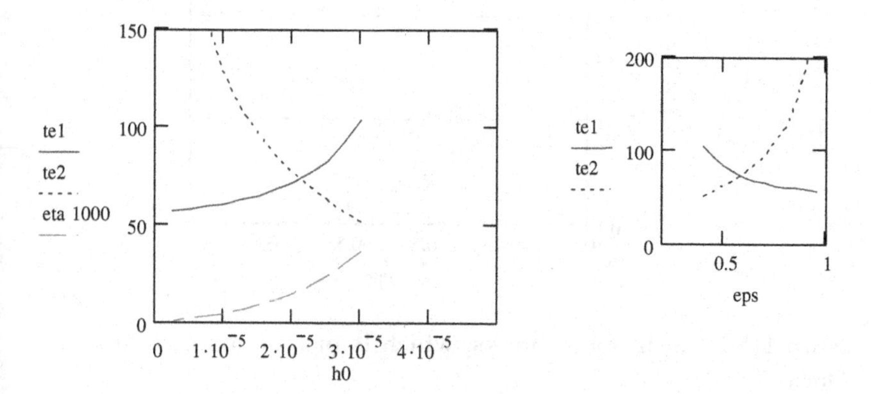

$$te := 75 \quad \Delta Te := te - ti \quad \Delta Te = 19 \quad h0 := 2.11 \cdot 10^{-5} \quad eps := 0.588$$

$$cp := 3.163 \quad eta := 0.0316.$$

(This is temperature rise from supply hole to the mean temperature and the equivalent viscosity).

(6) Total temperature rise

$$\mathrm{ti} := 56 \quad \mathrm{k2} := 0.8 \quad \mathrm{Qb} := 0.552 \quad \mathrm{te} := 75 \quad \mathrm{t1} := \frac{\mathrm{te} + \left(\frac{\mathrm{k2}\cdot\mathrm{Qb}}{1-\mathrm{Qb}}\right)\mathrm{ti}}{1 + \left(\frac{\mathrm{k2}\cdot\mathrm{Qb}}{1-\mathrm{Qb}}\right)}$$

$\mathrm{t1} = 65.568$

$$\mathrm{t2} := \frac{\mathrm{te} - \mathrm{t1}\cdot(1-\mathrm{k2})}{\mathrm{k2}} \quad \mathrm{t2} = 77.358 \quad \Delta\mathrm{t} := \mathrm{t2} - \mathrm{ti} \quad \Delta\mathrm{t} = 21.358 \ \textbf{(Answer)}$$

cf 21degr C in experiment

Question 8.2 solution details

Stiffness of a journal brg

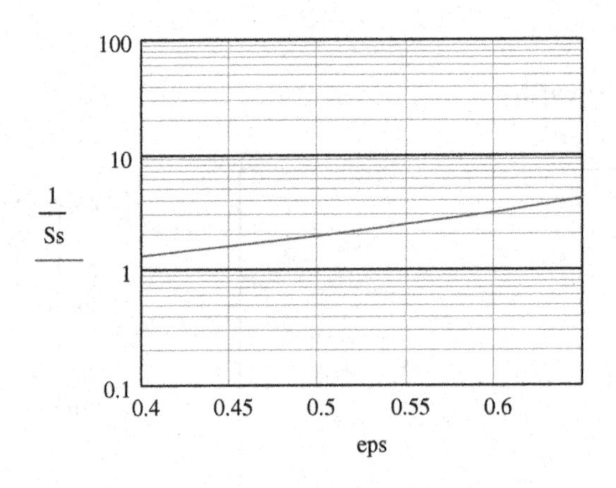

Note 1/Ss = S, in most books, which is proportional to W
Data

$\mathrm{L} := 0.05 \quad \mathrm{D} := 0.1 \quad \mathrm{N} := 20 \quad \eta := 0.03 \quad \mathrm{W} := 4905$

$\mathrm{a} := 0.276 \quad \mathrm{b} := 4.13 \quad \mathrm{R} := 0.05 \quad \mathrm{C} := .00007 \quad \varepsilon := 0.6 \quad \mathrm{e} := 2.718$

For (a), plot result for L/D = 0.5 from Table 2 between eps = 0.4 to 0.65, as above.
Assume

$\mathrm{S} := \mathrm{a}(\mathrm{e})^{\mathrm{b}-\varepsilon}$

From Figure, at $\varepsilon = 0.4\mathrm{S} = 1.44$
and at $\varepsilon = 0.6, \mathrm{S} = 3.29$

Therefore $1.44 = a\exp(0.4b)$
and $3.29 = a\exp(3,29b)$

Solving: a=0.276, b=4.13

Thus **(a) S=0.276 exp(4.13ε) Answer**

(b) From the definition of $w := \left(\frac{R}{c}\right)^2 0.276\exp(4.13\varepsilon)L.D.N.\eta$ Therefore

$$\text{Solving } \varepsilon := \frac{1}{b}\ln\left[\frac{W(c)^2}{(R)^2 L.D.N.\eta.a}\right] \quad \varepsilon = 0.6 \quad \textbf{Answer}$$

Check value of c

$$c := \left[\frac{(L.D.N.\eta.R^2.a.e^{2.478})}{W}\right]^{\frac{1}{2}} \quad c = 6.999 \times 10^{-5}\,\text{m}\ \textbf{OK}$$

(c) bearing stiffness expression

$\varepsilon := 0.6$ ks = dW/de

$$ks := \left(\frac{L \cdot D \cdot N \cdot \eta \cdot R^2}{c^3} a \cdot b \cdot e^{b \cdot \varepsilon}\right) \quad \textbf{Answer}$$

(d) stiffness of each bearing or complete assembly
Solving ks:
ks = 2.971×10^8 **N/m Answer. Alternatively:**

ks := W.$\frac{b}{c}$ ks = 2.971×10^8 **N/m**

(e) Natural frequency

g := 9.81 b := 4.13

Cameron gives: $n := \frac{1}{2\pi} \cdot \sqrt{\frac{g.b}{c}}$ n = 121.09 **vibn/s Answer**

Solution for 8.3:

Procedure:

Equating Eqs. (7.25) and (7.26)

$$Q_s(T_e - T_0) = Q_s(T_e - T_1) + Q_2(T_2 - T_1). \quad \text{(b)}$$

And from Eq. (7.24)

$$T_2 = \frac{T_e - T_1(1 - k_2)}{k_2}. \tag{c}$$

Substituting for T_2 into Eq. (b), after some adjustment we get

$$T_1 = \frac{T_e + \left(\frac{k_2 Q_s}{Q_2}\right) T_0}{\left(1 + \frac{k_2 Q_s}{Q_2}\right)}. \tag{d}$$

Remember that:

$$Q_2 = Q_1 - Q_s. \tag{e}$$

We need to find Q_s and Q_2 at the operating eccentricity $\varepsilon = 0.763$, obtained from the solution to the worked example 8.3. The flows at this eccentricity ratio can then be found by plotting ε against them using Math Cad or similar. Knowing also k_2, T_0 and T_e from the worked example 8.3, Eqs. (d) and (e) enable us to find T_1. Finally, Eq. (c) can then be used directly to find T_2.

Solution is in 360 deg brg 'effective flows'

Q8.3 360 degr brg example Q 8.3
Calculation of T1 and T2

$$\text{eps} := \begin{pmatrix} 0.1 \\ 0.2 \\ 0.3 \\ 0.4 \\ 0.5 \\ 0.6 \\ 0.7 \\ 0.8 \\ 0.9 \end{pmatrix} \quad \text{Ss} := \begin{pmatrix} 4.327 \\ 2.03 \\ 1.22 \\ 0.785 \\ 0.506 \\ 0.319 \\ 0.185 \\ 0.092 \\ 0.0312 \end{pmatrix} \quad \text{Qs} := \begin{pmatrix} 0.188 \\ 0.376 \\ 0.561 \\ 0.751 \\ 0.938 \\ 1.126 \\ 1.314 \\ 1.504 \\ 1.695 \end{pmatrix}$$

$$\text{Mus} := \begin{pmatrix} 69.7 \\ 32.3 \\ 19.3 \\ 12.9 \\ 8.88 \\ 6 \\ 4.1 \\ 2.42 \\ 1.22 \end{pmatrix} \qquad \text{Qts} := \begin{pmatrix} 1.0921 \\ 1.1837 \\ 1.275 \\ 1.365 \\ 1.455 \\ 1.545 \\ 1.634 \\ 1.723 \\ 1.811 \end{pmatrix}$$

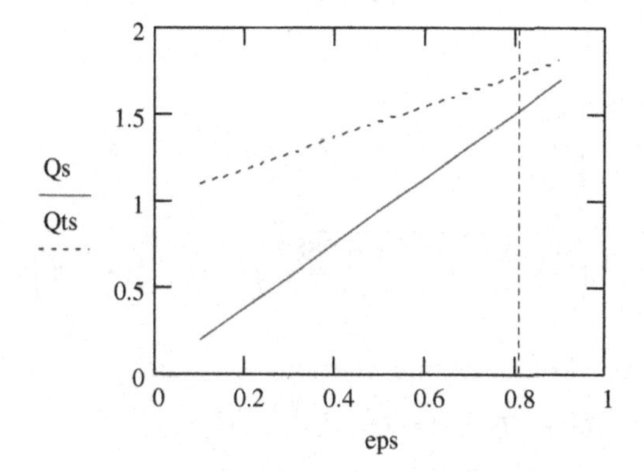

From 360 degr brg solution, eps := 0.763
Hence:

$$\text{Qts} := 1.708 \quad \text{QS} := 1.45 \quad \text{ce} := 88$$

$$\text{k2} := 0.8 \quad \text{Q2} := \text{Qts} - \text{Qs} \qquad \text{Qst} := \frac{\text{k2} \cdot \text{Qs}}{\text{Q2}} \quad \text{T0} := 40 \quad \text{Te} := 91.13$$

$$\text{T1} := \frac{\text{Te} + \text{Qst} \cdot \text{T0}}{\text{Qst}} \quad \text{T1} = 60.269 \ (\textbf{Answer})$$

$$\text{T2} := \frac{\text{Te} - \text{T1} \cdot (1 - \text{k2})}{\text{k2}} \quad \text{T2} = 98.845 \ (\textbf{Answer})$$

Solution for 8.4:

(a) The attitude angle is given as:
$\tan\psi = \frac{\pi\sqrt{1-\varepsilon^2}}{4\varepsilon}$

$$\text{Thus: } \varepsilon = \sqrt{\frac{\pi^2}{\pi^2 + 16\tan^2\psi}} = \sqrt{\frac{9.869}{9.869 + 0.1225}} = 0.988.$$

The value is very close to unity, indicating highly loaded bearing, where very close condition to metal-to-metal contact is expected.

(b)

$$h = c(1 + \varepsilon \cos\phi)$$
$$\text{min film, when: } \cos\phi = -1$$
$$\therefore h = c(1 - 0.988) = 0.012c = 0.012 \times 100 = 1.2\,\mu\text{m}.$$

Composite Ra: $\sigma = \sqrt{\sigma_1^2 + \sigma_2^2} = \sqrt{0.4^2 + 0.5^2} = 0.64\,\mu\text{m}$
Lambda ratio, $\lambda = \frac{h}{\sigma} = \frac{1.2}{0.64} = 1.875$, thus a mixed regime of lubrication is expected, since: $1 < \lambda < 3$ (see Chapter 1).

(c) First find:

$$\varphi_m = \cos^{-1}\left\{\frac{1 - \sqrt{(1 + 24\varepsilon^2)}}{4\varepsilon}\right\}$$

$$= \cos^{-1}\left(\frac{1 - \sqrt{1 + 24 \times 0.988^2}}{4 \times 0.988}\right) = \cos^{-1}(-0.997) = 175.5^\circ$$

Now find the speed of entraining motion: $u = \frac{1}{2}\omega R_j$.
Note that: $\frac{c}{R_j} = \frac{1}{1000}, \frac{2R_j}{L} = 3, \therefore L = \frac{2R_j}{3}, \therefore \frac{L}{c} = \frac{2000}{3}$.
Thus:

$$p_m = \frac{3u\eta_0\varepsilon}{4R_j}\left(\frac{L}{c}\right)^2 \frac{\sin\varphi_m}{(1 + \varepsilon\cos\varphi_m)^3}$$

$$= \frac{3(\omega R_j)\eta_0(0.988)}{4R_j}\left(\frac{2000}{3}\right)^2 \frac{0.0785}{(1 - 0.988 \times 0.997)^3}$$

$$= 7.59\,\omega\eta_0 \times 10^9 = 7.59\,\omega\eta_0 \text{ GPa}.$$

(d) The shell will deform, and the film thickness will be obtained as:

$$h = c(1 + \varepsilon\cos\varphi) + \delta = h_i + \delta,$$

where h_i is the film thickness as before, thus: $h = 1.2 + \delta\,\mu\text{m}$

We need to find δ from column method, given as (see Chapter 12):

$$\delta = \frac{(1 - 2\upsilon)(1 + \upsilon)d}{E(1 - \upsilon)}p = \frac{(1 - 0.46)(1 + 0.23) \times 2 \times 10^{-3}}{60 \times 10^9(1 - 0.23)}7.59\,\omega\eta_0 \times 10^9$$

$$\approx 220\,\omega\eta_0\,\mu\text{m}$$

Thus: $h = 1.2 + 220\omega\eta_0\,\mu\text{m}$

Solution for 8.5:

(a)

(i) Diameter to width (or length) ratio is: $\frac{2\times0.05}{0.05} = 2$, thus we will use the short-width bearing approximation.

(ii) We can use either forms of the Sommerfeld number. For example, we can use:

$$S = \frac{(W/L)c^2}{N\eta R_j^2}, \quad N = \text{speed of revolution of the journal.}$$

$\frac{c}{R} = \frac{1}{1000}$, Also: $\eta = 20cp = 0.02$ Pa.s thus: $S = 5000$, which on the log-scale of the chart gives us the value $\log S \approx 3.7$.

Now using the chart, we use the curve for $L/D = 1/2$, and find that: $\varepsilon \approx 0.65$

(iii) The attitude angle is given by:

$$\psi = \tan^{-1}\frac{\pi\sqrt{1-\varepsilon^2}}{4\varepsilon} = \tan^{-1}\frac{\pi\sqrt{1-0.4225}}{2.6} = \tan^{-1}(0.918) = 44.4°$$

As this angle reduces the bearing becomes further loaded. The calculated value corresponds to a moderately loaded bearing.

(iv) Assuming half-Sommerfeld condition the extent of film is between 0–180 degrees, and the minimum film is where $\cos\varphi = -1$ in: $h = c(1+\varepsilon\cos\varphi) = c(1-\varepsilon)$. Note that: $c = \frac{R}{1000} = \frac{0.05}{1000} = 0.00005 = 50\ \mu\text{m}$

Thus: $h = 50(1-0.65) = 17.5\ \mu\text{m}$

(b)

(i)

$$\varphi_m = \cos^{-1}\left\{\frac{1-\sqrt{(1+24\varepsilon^2)}}{4\varepsilon}\right\}$$

$$= \cos^{-1}\left\{\frac{1-\sqrt{1+24(0.65)^2}}{2.6}\right\} \approx 154°$$

Now we can find the maximum pressure as:

$$p_m = \frac{3u\eta_0\varepsilon}{4R_j}\left(\frac{L}{c}\right)^2 \frac{\sin\varphi_m}{(1+\varepsilon\cos\varphi_m)^3}$$

$$= \frac{3\times 15.71\times 0.02\times 0.65}{0.2}\times 10^6 \frac{\sin 154}{(1+0.65\cos 154)^3} = 18.7\text{ MPa}$$

Where: $u = 2\pi\frac{N}{60}R = 2\pi\frac{3000}{60}0.05 = 15.71\,\text{m/s}$ is the speed of entraining motion.

(ii) The shell deflection is obtained as (see Chapter 12):

$$\delta = \frac{(1-2\nu)(1+\nu)d}{E(1-\nu)}p_m = \frac{0.5\times 1.25\times 2\times 10^{-3}}{60\times 10^9\times 0.75}\times 18.71\times 10^6 \approx 0.5\,\mu\text{m}$$

Of course note that this is a very approximate solution as a numerical solution is required for simultaneous solution of pressure, film and deflection as in Chapter 12. Now the minimum film thickness is: $h = c(1-\varepsilon)+\delta = 17.5+0.5 = 18\,\mu\text{m}$

Solution for 9.2:

(a) $\tan\alpha = \frac{R-R_0}{l} = \frac{25}{75} = 0.333$. Therefore $\alpha = 18.41°$.
From Eq. (9.2)

$$Q = \frac{p_r\pi h^3}{6\eta}\left[\frac{1}{\ln(R/R_0)}\right]\sin\alpha = \frac{2\pi\times 10^6\times 0.05^3\times 10^{-9}}{6\times 0.447}\left(\frac{1}{\ln 2}\right)\sin 18.41$$

$$\therefore Q = 1.28\times 10^{-3}L/s \ (\textbf{Answer}).$$

From Eq. (9.8) $H = Qp_r = 1.28\times 10^{-6}\times 2\times 10^6 = 256\,\text{W}$ (**Answer**).

(b) From Eq. (9.5)

$$\therefore W = \frac{p_r\pi}{2}\left[\frac{R^2-R_0^2}{\ln(R/R_0)}\right] = \frac{2\times 10^6\pi}{2}\left[\frac{50^2-25^2}{\ln\left(\frac{50}{25}\right)}\right]\times 10^{-6}$$

$\therefore W = 8230\,\text{N}$ (**Answer**).

Solution for 9.3:

Q9.3

Input data
Given: outside diameter, find pocket diameter for max stiffness

$$pa := 10^5 \quad pab := 0.2 \quad b := .066 \quad W := 1000 \quad R := 267 \quad T := 293$$

$$Cd := .65 \quad Kg := 0.69 \quad \eta := 1.82 \cdot 10^{-5} \quad h := 2.5 \cdot 10^{-5} \quad ps := 5 \cdot pa$$

$$ab := 0.5 \quad A := \pi b^2$$

$$Fp := \frac{pab}{\frac{1}{2}} \quad Fp = 0.186$$

$$(1 - pab)^2 (1 + pab)$$

Iteration loop

$bb := 2$ (**Guess**)

Given

$$W = Kg \frac{Pa \cdot (1 - pab)}{pab} \pi \cdot b^2 \frac{(1 - ab^2)}{2ln(bb)} \quad fbb := Find(bb) \quad fbb = 4.122$$

$$bb := fbb \quad bb = 4.122 \quad a := \frac{b}{bb} \quad a = 0.016 \quad m \text{ (\textbf{Answer})}$$

To get orifice dia, *d*

$$G := \frac{Kg}{(1 - Kg)^{\frac{1}{2}}} \quad G = 1.239 \quad d := \frac{G \cdot pa \cdot 8 \cdot h^3}{F_p \cdot 24 \eta \sqrt{2 \cdot R \cdot T} \cdot Cd \cdot ln(bb)}$$

$$d = 7.229 \times 10^{-4} \; m \text{ (\textbf{Answer})}$$

Mass flow

$$m := 0.69(Ps^2 - pa^2) \frac{\pi \cdot h^3}{12 \cdot \eta \cdot R \cdot T \cdot in\left(\frac{1}{pab}\right)}$$

$$m = 2.956 \times 10^{-4} \quad \textbf{Kg/s (Answer)}$$

$$Bb := \frac{\pi}{6ln(bb)} \quad Ab := \frac{1}{2}\left(\frac{1 - ab^2}{in(bb)}\right) \quad Bb = 0.37 \quad Ab = 0.265$$

Power to lift ratio

$$\textbf{Eqn(9.3)} \quad PL := \frac{W \cdot h^3}{A^2 \eta}\left(\frac{Bb}{Ab^2}\right) \quad PL = 0.024 \quad m/s \text{ (\textbf{Answer})}$$

Solution for 10.1:

QUESTION 1 CHPT 10: Solution

let

$$R1 := 0.0127 \quad L1 := 0.0254 \quad Es := 110 \cdot 10^9 \quad \upsilon := 0.3 \quad \rho := 7800$$

$$g := 9.8 \quad \eta 0 := 0.004 \quad \alpha := 2.5 \cdot 10^{-8} \quad \Delta s := 0.001$$

$$\Delta h := \Delta s \cdot \sin(45) \quad \Delta h := \Delta s \cdot \sin(45) \quad L2 := 0.03 \quad R2 := 0.05$$

$$M1 := \pi \cdot R1^2 \cdot L1 \cdot \rho \quad M1 = 0.1 \quad M2 := \pi \cdot R2^2 \cdot L2 \cdot \rho \quad M2 = 1.838$$

$$W := (M1 + 1 \cdot M2) \cdot g \quad W = 37.005$$

$$Vc := \left[\frac{2(M1 + 2M2) \cdot g \cdot \Delta h}{\frac{3}{2} \cdot M1 + M2 \cdot \left[2 + \frac{1}{2}\left(\frac{R2}{R1}\right)^2\right]}\right]^{\frac{1}{2}} \quad Vc = 0.187$$

$$h0 := \frac{2 \cdot \eta 0 \cdot Vc \cdot R \cdot L}{W}$$

$h0 = 2.242 \times 10^{-8}\, m^3K$ **Rigid-isoviscous solution**

EHL solution

$$US := \frac{\eta 0 \cdot Vc}{Es \cdot R1} \quad Us = 5.345 \times 10^{-13}$$

$$Ws := \frac{W \cdot \sin(45)}{Es \cdot R1 \cdot L1} \quad Ws = 8.874 \times 10^{-7}$$

$$Gs := Es \cdot \alpha \quad Gs = 2.75 \times 10^3$$

$$hms := 2.58 \cdot Us^{.7} \cdot Gs^{.54} \cdot Ws^{-0.13} \quad hms = 2.919 \times 10^{-6}$$

$hm := hms \cdot R1 \quad hm = 3.707 \times 10^{-8}$ **ANSWER CHART**

$$As := \left(\frac{W^2}{\eta 0 \cdot Vc \cdot Es \cdot L1^2 \cdot R1}\right)^{\frac{1}{2}} \quad Bs := \left(\frac{\alpha^2 \cdot W^3}{\eta 0 \cdot Vc \cdot L1^3 \cdot R1^2}\right)^{\frac{1}{2}}$$

$$As = 1.426 \quad BS = 4.006$$

$Cs := 6 \quad hm := \dfrac{L1 \cdot Cs \cdot (\eta 0 \cdot Vc \cdot R1)}{W} \quad hm = 3.906 \times 10^{-8}$ **Chart Answer**

Solution for 10.2:

Question 10.2 solution

$$\theta 0 := 30 + 273 \quad \beta 0 := 0.026 \quad \text{R1} := 0.0127 \quad \eta 0 := 0.540$$

$$\alpha 0 := 29 \cdot 10^{-8} \quad \text{Es} := 110 \cdot 10^{9}$$

$$\text{r} := \frac{\text{R1}}{\sqrt{3}} \quad \text{r} = 7.332 \times 10^{-3} \quad \textbf{(Answer)} \quad \text{F} := 600$$

$$\text{W} := \frac{\text{F}}{3 \cdot \sqrt{\frac{2}{3}}} \quad \text{W} = 244.949 \quad \textbf{(Answer)}$$

Enter rolling speed from data N := 1000

Solve for film thickness

$$\text{R} := \frac{\text{R1}}{2} \quad \text{U1} := \overrightarrow{\frac{2\pi \cdot \text{N} \cdot \text{r}}{60}} \quad \text{U} := \frac{\overrightarrow{\text{U1}}}{2}$$

$$\text{Ws} := \frac{\text{W}}{\text{Es} \cdot \text{R}^2} \quad \text{Us} := \frac{\eta 0 \cdot \vec{\text{U}}}{\text{Es} \cdot \text{R}} \quad \text{Gs} := \alpha 0 \cdot \text{Es}$$

$$\text{h0s} := 2.77 \cdot \overrightarrow{(\text{Us}^{.68} \cdot \text{Gs}^{.49} \cdot \text{Ws}^{-.073})} \quad \text{h0} := \overrightarrow{\text{h0s}} \cdot \text{R}$$

$$\text{h0} = 5.546 \times 10^{-7} \quad \textbf{(Answer)}$$

Solution for 10.3:

Question 10.3 solution

(a) Given Data for 2 disc machine

$$\theta 0 := 30 \quad \Theta 0 := \theta 0 + 273 \quad \beta 0 := 0.026 \quad \text{R1} := 0.03 \ \text{R} := \frac{\text{R1}}{2}$$

$$\eta 0 := 0.0456 \quad \alpha 0 := 3.3 \cdot 10^{-8} \quad \text{Ge} := 1.5 \cdot 10^{8}$$

$$\tau 0 := 3 \cdot 10^{6} \quad \text{Es} := 110 \cdot 10^{9} \quad \text{kt} := 0.125 \quad \text{L} := 0.015 \quad \text{F} := 2000$$

$$\text{P} := \frac{\text{F}}{\text{L}} \quad \text{P} = 1.3333 \times 10^{5}$$

N := 3000 **(constant rolling speed)** $\text{U} := \frac{\text{R1}}{2}\left(\overrightarrow{\frac{2 \cdot \pi \cdot (\text{N})}{60}}\right) \quad \text{U} = 4.7124$

(b) conjunction film thickness (Based on rolling speed) Solve for film thickness

$$\mathrm{Ws} := \frac{\mathrm{P}}{\mathrm{Es} \cdot \mathrm{R}} \quad \mathrm{Us} := \frac{\eta 0 \cdot \mathrm{U}}{\mathrm{Es} \cdot \mathrm{R}} \quad \mathrm{Gs} := \alpha 0 \cdot \mathrm{Es} \quad \mathrm{Gs} = 3.63 \times 10^{3}$$

$$\mathrm{h0s} := 2.076 \cdot \left[\left(\frac{\alpha 0 \cdot \eta 0 \cdot \mathrm{U}}{\mathrm{R}} \right)^{.727} \cdot \left(\frac{\mathrm{Es} \cdot \mathrm{R}}{\mathrm{P}} \right)^{.091} \right]$$

$\mathrm{h0} := \overrightarrow{\mathrm{h0s}} \cdot \mathrm{R} \quad \mathrm{h0} = 1.8501 \times 10^{-6}$ **Answer**

All above this line are isothermal

(c) Coefficient of friction and average film temperature

$$\mathrm{pm} := \frac{\pi}{4} \cdot \left(\frac{\mathrm{P} \cdot \mathrm{Es}}{\pi \cdot \mathrm{R}} \right)^{\frac{1}{2}} \quad \mathrm{pm} = 4.3816 \times 10^{8} \quad \eta \mathrm{s} := \eta 0 \cdot \exp(\alpha 0 \cdot \mathrm{pm})$$

$$\eta \mathrm{s} = 8.6813 \times 10^{4} \quad \mathrm{A} := \overrightarrow{\left[\frac{1}{\tau 0 \cdot \mathrm{h0}} \left(\frac{2 \cdot \mathrm{kt} \cdot \eta \mathrm{s}}{\beta 0} \right)^{\frac{1}{2}} \right]} \quad \mathrm{A} = 164.6109$$

(c) Choose temperature rise for numerical procedure

Let $\Delta\theta := 0.25 \quad \mathrm{X} := (\exp(\beta 0 \cdot \Delta\theta) - 1)^{\frac{1}{2}} \quad \mathrm{Xs} := \overrightarrow{\frac{\mathrm{asinh}(\mathrm{X})}{\sqrt{1+\mathrm{X}^2}}} \quad \mathrm{X} = 0.0808 \quad \mathrm{Xs} = 0.0804$
$\mathrm{z} := 4 \cdot (\mathrm{A}^2 \cdot \mathrm{Xs}^2) \quad \mathrm{z} = 700.7097$ **z is RHS of equation in numerical procedure below**
ΔN will now be deternined from chosen value of $\Delta\theta$
Procedure solves for factor Yb from:

$$\mathrm{Yb} = \frac{4\mathrm{A}^2 \cdot \mathrm{X} \cdot \mathrm{Xs}}{\Delta \mathrm{Us}}$$

We seek Yb, the unknown variable in equation below (see notes)

As a guess, let $\mathrm{Yb} := 1$

Given
$(\mathrm{Yb} \cdot \sinh(\mathrm{Yb})) = \mathrm{z} \quad \mathrm{q(z)} := \mathrm{Find(Yb)}$
$\mathrm{q(z)} = 9.2364$ **q(z) was found from this numerical procedure. It is the correct value of Yb**

(c) Coefficient of friction

$\tau b := q(z) \cdot \tau 0 \quad \mu := \left(\frac{\tau b}{pm}\right) \quad \mu = 0.0379$ **This is coefficient of friction, answer r to (c) We can now find the required sliding speed from q(z).**

(c) sliding speed

$$\Delta Us := \frac{4 \cdot A^2 \cdot X \cdot Xs}{q(z)} \quad \Delta Us = 127.1623 \quad \Delta U \frac{\Delta Us \cdot (\tau 0 \cdot h0)}{\eta s}$$

$$\Delta U = 8.13 \times 10^{-3}$$

$$\Delta N := \frac{60 \cdot \Delta U}{2 \cdot \pi \cdot R1} \quad \Delta N = 2.5879 \quad \text{rev/min (\textbf{Answer})}$$

Slide-roll ratio $\delta := \frac{\Delta U}{U} \quad \delta = 1.7252 \times 10^{-3}$ (**Answer**)

Answers to (d) and (e))

$$\textbf{At 30 deg C rise } \mu = 0.0628, \Delta N = 1.67 \text{ rev/min}, \delta = 0.11$$

$$\textbf{At 100 deg C rise } \mu = 0.0607, \Delta N = 472 \text{ rev/min}, \delta = 0.315$$

Condition (d) has a higher friction coefficient than (c) because of the shear stress initially rises with speed. Then for condition (e), there has been a fall in μ because of the excessive temperature rise.

Solution for 11.1:

From Eq. (13.7):$L_{10} = (Q_c/Q)^3$ in millions of revolutions. As $Q_c = 20\,\text{kN}$

$$\therefore L_{10} = \left(\frac{20}{5}\right)^3 \times 10^6 = 64 \times 10^6 \quad \text{revolutions}$$

The bearing life is therefore $\frac{64\times10^6}{500\times60} = 2133$ hours (**Answer**)

Solution for 11.2:

(a) For one bearing $L_{10} = (Q_c/Q)^3 = \left(\frac{20}{1}\right)^3 = 8 \times 10^9$ revolutions,
The life of one bearing is $8 \times 10^9/500 \times 60 = 266,000$ hours (**Answer**)

(b) As there are two identical bearings supporting the spindle, from Eq. (13.8), L_{10} of the whole assembly in revolutions is

$$L_{10}(ass) = [2L_{10}^{3/2}]^{-2/3} = 2^{-2/3} \times 8 = 5.04 \times 10^9 \quad \text{revolutions}$$

The life of the whole assembly is therefore $5.04 \times 10^9/500 \times 60 = 168000$ hours (**Answer**).

(c) The probability of survival of the whole arrangement is $0.9 \times 0.9 = 0.81$ (**Answer**).

Solution for 13.2:

Speed of entraining motion: $U = \frac{1}{2}\left(2\pi \frac{N}{60} R\right) = \pi(2 \times 10^{-3}) = 6.28 \times 10^{-3}$m/s.

Refer to Fig. 13.11 (which is already calculated for asperity pair interactions with tip radii of 1 μm) to find that for DLC: contact angle 60° and Meniscus force $= 0.5 \times 10^{-4}$ mN per asperity pair (or per micro-meniscus), thus for $n = 10,000$ asperity pairs: $F_m = 0.5$ mN. Thus, the total contact force is: $F_n = W + F_m = 2.5$ mN

Now use Eq. (13.6) to establish any hydrodynamic film thickness:

$$h = \frac{2bU\eta_0 R}{W} = \frac{2 \times 5 \times 10^{-3} \times 6.28 \times 10^{-3} \times 0.00025 \times 2 \times 10^{-3}}{2.5 \times 10^{-3}}$$

$$= 12.5 \times 10^{-9}\text{m} = 12.5\,\text{nm}.$$

Therefore (from Chapter 1): $\lambda = \frac{h}{\sigma} = \frac{12.5\,\text{nm}}{100\,\text{nm}} = 0.125 < 1$, which means that the prevailing regime of lubrication is boundary. We can make a reasonable judgement that $\psi \approx 1$ for boundary interactions, thus Eq. (13.41) simplifies to:

$$F_a = 3.2 \frac{\tau_s F_n}{E^*} \sqrt{\frac{r}{\sigma_p}} = 3.2 \frac{100 \times 10^6 \times 2.5 \times 10^{-3}}{800 \times 10^9} \sqrt{\frac{1 \times 10^{-6}}{100 \times 10^{-9}}}$$

$$= 3.16 \times 10^{-6}\ \text{N}$$

Thus, the coefficient of adhesive friction is:

$$\mu_{ad} = \frac{F_a}{F_n} = \frac{3.16 \times 10^{-6}}{2.5 \times 10^{-3}} \approx 0.0013.$$

Note that with such smooth surfaces as DLC the coefficient of friction is still quite low, even under boundary lubrication condition. This is the reason why DLC is used as a wear resistant coating.

Solution for 14.1:

Sliding distance per step is, therefore: 1/5 of the perimeter of the femoral head, or: $\frac{\pi D}{5} = 13.96\,\text{mm} = 0.01396\,\text{m}$:

Total sliding distance per annum: $X = 0.01396 \times 10^6\,\text{m}$
The contact load per hip joint is: $W = 2.5 \times 75 \times 9.81 = 1839.375\,\text{N}$.

The wear factor is obtained as:

$$K = 4(Ra)^{1.2} \times 10^{-5} = 4(0.05)^{1.2} \times 10^{-5} = 1.09856 \times 10^{-6}\text{mm}^3/\text{Nm}$$

Thus, wear per annum is:

$$V = KWX = 1.09856 \times 10^{-6} \times 1839.375 \times 0.01396 \times 10^6 = 28.2\,\text{mm}^3$$

Note that with such a [illegible] DLC the coefficient of friction is still quite low even under [illegible] condition. This is the reason why DLC [illegible] a good [illegible] material.

Solution for 14.[illegible]

Sliding distance [illegible] of the femoral head, [illegible] $\times 10^{-3}$ mm = 0.0[illegible] m

Total sliding distance per annum [illegible] $= 0.0[illegible] \times 10^{6}$ m

The normal load per [illegible] $= [illegible] \times 75 \times 9.81 = 1839.[illegible]$ N

[illegible]

[illegible]

[illegible]

Index